Radiation Risk Estimates in Normal and Emergency Situations

NATO Security through Science Series

This Series presents the results of scientific meetings supported under the NATO Programme for Security through Science (STS).

Meetings supported by the NATO STS Programme are in security-related priority areas of Defence Against Terrorism or Countering Other Threats to Security. The types of meeting supported are generally "Advanced Study Institutes" and "Advanced Research Workshops". The NATO STS Series collects together the results of these meetings. The meetings are co-organized by scientists from NATO countries and scientists from NATO's "Partner" or "Mediterranean Dialogue" countries. The observations and recommendations made at the meetings, as well as the contents of the volumes in the Series, reflect those of participants and contributors only; they should not necessarily be regarded as reflecting NATO views or policy.

Advanced Study Institutes (ASI) are high-level tutorial courses to convey the latest developments in a subject to an advanced-level audience

Advanced Research Workshops (ARW) are expert meetings where an intense but informal exchange of views at the frontiers of a subject aims at identifying directions for future action

Following a transformation of the programme in 2004 the Series has been re-named and re-organised. Recent volumes on topics not related to security, which result from meetings supported under the programme earlier, may be found in the NATO Science Series.

The Series is published by IOS Press, Amsterdam, and Springer, Dordrecht, in conjunction with the NATO Public Diplomacy Division.

Sub-Series

A. Chemistry and Biology	Springer
B. Physics and Biophysics	Springer
C. Environmental Security	Springer
D. Information and Communication Security	IOS Press
E. Human and Societal Dynamics	IOS Press

http://www.nato.int/science
http://www.springer.com
http://www.iospress.nl

Series B: Physics and Biophysics – Vol. 9

Radiation Risk Estimates in Normal and Emergency Situations

edited by

Arrigo A. Cigna
Past-President Union Internationale de Radioecologie,
Cocconato (Asti), Italy

and

Marco Durante
University Federico II, Napoli, Italy

Springer

Published in cooperation with NATO Public Diplomacy Division

Proceedings of the NATO Advanced Research Workshop on
Impact of Radiation Risk Estimates in Normal and Emergency Situations
Yerevan, Armenia
8–11 September 2005

A C.I.P. Catalogue record for this book is available from the Library of Congress.

ISBN-10 1-4020-4955-2 (PB)
ISBN-13 978-1-4020-4955-2 (PB)
ISBN-10 1-4020-4954-4 (HB)
ISBN-13 978-1-4020-4954-5 (HB)
ISBN-10 1-4020-4956-0 (e-book)
ISBN-13 978-1-4020-4956-9 (e-book)

Published by Springer,
P.O. Box 17, 3300 AA Dordrecht, The Netherlands.

www.springer.com

Printed on acid-free paper

Dedicated to the memory of
Vladimir Andreevich Shevchenko
and
Vladimir Ivanovich Korogodin

Table of Contents

PREFACE

Nikolaj Timofeeff-Ressovsky left a significant mark in the history of science in hte XX century. Educated in zoology, he had the opportunity to become assistant of Prof. N.K. Koltosov in Moscow in the early twenties. Those were exciting years for science – and Russia was leading mainstream research in the field. Unfortunately, in those same years a young Trofim D. Lysenko got his degree as agronomist, and soon he wrecked the science of genetics in the Soviet Union. Yet Timofeeff-Ressovsky, emigrated to Germany, became one of the founders of radiobiology and radioecology.

An International Conference, the second one of the series, was organised in Yerevan, Armenia, dedicated to the 70th anniversary of the publication of the so called "green pamphlet", published by N.W. Timofeeff-Ressovsky, K. Zimmer and M. Delbruck in 1935, whose title was "On the origin of gene mutations and gene structure". The importance of this paper for genetics and radiobiology was enormous – not only for radiobiology, but for genetics in general. The "target" of the radiation action was indeed the "gene", that for the first time found a clear, operational definition.

The organisers of this Conference proposed to convene scientists from all over the world to discuss the impact of radiation risk estimates in normal and emergency situations. This was the starting point of the NATO Advanced Research Workshop held in Yerevan. While the Organising Committee of the Timofeeff-Ressovsky Conference should be commended for proposing a parallel workshop on a related topic, the NATO Science Committee is to be acknowledged for accepting the proposal and providing support. On behalf of the participants to the NATO ARW we express our gratitude to the NATO Science Committee and to everyone contributed to its success.

The parallel development of the Conference and the Workshop resulted in a true synergism, both events taking advantage from the mixed environment of molecular biologists, genetists, radiobiologists, radioecologists, biophysicists etc. gathered in Yerevan. We were delighted to include in these proceedings a few excellent papers dealing directly with Timofeeff-Ressovsky's life, and the "green pamphlet". The life and scientific achievements of Timofeeff-Ressovsky are often not well known in western countries, hence we deemed advisable to include those papers in this book.

Unfortunately, our workshop was saddened by the untimely loss of two great Russian friends and colleagues: Vladimir Andreevich Shevchenko passed away in July, and Vladimir Ivanovich Korogodin in October. These distinguished scientists and gentlemen had played for years a decisive role in reviving genetics in Russia, in the track of Koltsov, Vavilov, and Timofeeff-Ressovsky. Their last papers are included in these proceedings, which are dedicated to their memory.

February 2006 The Editors

VLADIMIR ANDREEVICH SHEVCHENKO

(17 July 1936 - 29 July 2005)

Vladimir Andreevich Shevchenko was born in the village of Dulovo, Kalininskaya oblast, in 1936. After secondary school, he graduated from Timiryazev Moscow Agricultural Academy in 1960 and immediately started his scientific work, first as a senior laboratory assistant and subsequently as a junior scientist in the Laboratory of Radiation Genetics of the Institute of Biophysics, headed by academician N.P. Dubinin. From 1966 till his last breath he worked in the Vavilov Institute of General Genetics in Moscow. He got his Ph.D. degree in 1966, and in 1974 he became Doctor of Biological Sciences. In 1985, Dr. Shevchenko was awarded the title of Professor in Genetics. Since 1978 he was the head of the Laboratory of Radiation Genetics of the Institute of General Genetics.

Professor V.A. Shevchenko was a leading expert in the field of radiation genetics, radiobiology, and radioecology. He dedicated his entire life to research into radiation genetics of flora and fauna in the territory of East Urals Radioactive Trail, Semipalatinsk Test Site, Chernobyl, and other radiation-polluted areas. His research contributed greatly to the modern radiation genetics. He was instrumental in the description of the radio-adaptation phenomenon; he analysed microevolution processes taking place over a number of generations of radiation-exposed populations, and studied the dose-response relationship of genetic effects after exposure to various radionuclides.

Over many years V.A. Shevchenko was studying genetic effects in humans exposed to radiation.

Right after the Chernobyl accident Prof. Shevchenko organised cytogenetic examination of clean-up workers and residents of the contaminated area. Analysis of chromosome aberrations in peripheral blood lymphocytes from more than 3000 persons was performed in order to assess the radiation exposure. The data obtained helped in the assessment of carcinogenic risks in exposed persons, and to estimate long-term hereditary effects. For his work in Chernobyl, V.A. Shevchenko was honoured with the Order of Red Labour Standard.

Over the last 15 years Professor Shevchenko was leading the project on cytogenetic examinations of residents of Altai region, exposed to nuclear bomb tests at Semipalatinsk, and victims of radionuclide contamination of the Techa River. The results of these studies enabled dose reconstruction after 40-45 years from exposure.

In 1995, V.A. Shevchenko, as an expert, took part in the examination of the victims of Three Mile Island accident, occurred in 1979, assessing absorbed radiation doses.

For a long time he was studying the effects of space flight environment on various biological objects. He selected high-yield strains of *Chlorella* algae for air regeneration systems to be used for long-term space missions and acquired a certificate of authorship. Over recent years, along with his associates, Prof. Shevchenko was participating in cytogenetic studies of space crews from MIR station and International Space Station, developing biological dosimetry and assessing genetic consequences of long-term space missions.

Together with his colleagues from the Topchiev Institute of Nonchemical Synthesis RAS, Professor V.A. Shevchenko took part in developing and testing various efficient antimutagens (3 inventions patented).

Vladimir Andreevich Shevchenko is also known as an expert in water ecosystems. In 1989 he was co-author of the book "Ecology of sea phototrophic phikoplankton", as a result of the expedition to the Atlantic Ocean on the research ship called "Professor Vernadsky".

Professor Shevchenko was a brilliant organiser and a great populariser of science. He is author of 420 scientific publications, 7 monographs, and 8 patents in the field of radiation genetics and radioecology. He was organiser and inspirer of numerous scientific conferences and symposiums, which were enthusiastically attended by scientists and experts from all over the world to discuss urgent problems of radiation genetics and radiobiology.

Beyond the scientific research activity, V.A. Shevchenko was a master in his field. He created a scientific school in the field of radiation genetics, and was scientific advisor of 26 Ph.D. theses.

V.A. Shevchenko had a reputation as an outstanding highly skilled expert, able to raise important questions and to solve them, to inspire his associates, and to lead them in accomplishing the tasks set forth.

The achievements of Professor V.A. Shevchenko are highly appreciated by the native and international biological communities. In recognition of his eminent progress in radiation genetics, he was awarded the N.V. Timoffeef-Ressovsky Medal in 2003.

V.A. Shevchenko was corresponding member of the Russian Academy of Natural Sciences, honoured scientist of the Russian Federation, president of Russian Radiobiological Community, member of Russian Nuclear Community, chairman of Genetic section of the Moscow Society of Nature Investigators, member of the International Union of Radioecology (Belgium), presidium member of Eco-Ethics International Union (Germany), and for many years a member of the editorial board of the journal "Radiobiology and Radioecology".

Vladimir Andreevich had a bright mind and many-sided personality. Not only was he a brilliant scientist, but also an artist and connoisseur of minerals. One of his merits was his respectful and friendly attitude towards his colleagues, friends, and the people surrounding his life. He was always ready to provide assistance and believed that benevolence and mutual understanding were the corner stone for every scientific team. He will be sorely missed by the scientific community, his friends, and relatives.

Prof. R.M. Alexakhin,
Russian Academy of Agricultural Sciences

Prof. A.F. Tsyb,
Russian Academy of Medical Sciences

Prof. S.G. Inge-Vechtomov,
Russian Academy of Sciences

VLADIMIR IVANOVICH KOROGODIN
(4 January 1929 – 31 October 2005)

V.I. Korogodin was born in 1929 in Donetsk. In 1947 he entered the Faculty of Physics at the Moscow State University. In 1948 he moved to the Faculty of Soil Biology, and in 1952 he graduated from the Department of Genetics. He worked for one year in the Far North as a zootechnician deer breeder, and in 1953 returned to Moscow, where he worked as a laboratory assistant in the Department of Biophysics of the MSU Faculty of Soil Biology.

In his research on the effects of radiation on yeast, V.I. Korogodin discovered (Diploma on discovery N 115, priority of March 1957) and then extensively investigated post-irradiation recovery from lethal damage induced by ionizing radiation. This discovery contradicted classical target theory and the view that cells are unable to recover from potentially lethal damage. Later, numerous reports from many laboratories confirmed the observation of post-irradiation recovery in bacteria, plants, and mammalian cells.

From 1962, V.I. Korogodin established and headed the Laboratory of Cell and Tissue Radiobiology at the Institute of Medical Radiology of the Academy of Medical Sciences; the laboratory was part of the Department of Radiobiology and Genetics headed by N.W. Timoffeef-Ressovsky. Together with Yu.G. Kapultsevich, he concluded a mathematical analysis of post-irradiation cell recovery and continued his studies concerning factors affecting post-irradiation recovery and their dependence on cell ploidy. In the late 1960s, together with K.M. Blisnik and Yu.G. Kapultsevich, he discovered the "cascade mutagenesis" effect, a heritable mode of colony-morphology instability. In 1966 he defended his doctoral thesis, and in 1970 he was appointed Professor of Radiobiology.

V.I. Korogodin headed the Laboratory of Genetics and Yeast Selection at the Institute of Genetics and Selection of Industrial Microorganisms in Moscow from 1972. He carried out extensive studies demonstrating that post-irradiation repair is associated with the presence of homologous chromosomes, a process now called homologous recombination repair and described in many kinds of organisms.

At the end of 1977, V.I. Korogodin moved to Dubna to work at the Joint Institute for Nuclear Research, where he organised a Department of Biological Research. Together with E.A. Krasavin, S. Kozubek (Czech Republic), and a group of young graduates from the Moscow Institute of Physical Engineering (A.V. Glazunov, P.N. Lobachevsky, and K.G. Amirtaev), he proved that the biological effects of radiations with different quality critically depend upon the ability of cells to repair radiation damage.

Together with V.L. Korogodina, he performed research on the dependence of yeast mutability on gene expression, using a set of genes controlling adenine synthesis and varying the amounts of adenine in the media. They observed that the mutation frequency in active *ade2* genes is about 100-fold higher than in repressed *ade2* genes. The frequency of suppressor mutations whose activity does not depend on the amount of adenine in the medium also changed, but less dramatically.

Since 1990, radiation-induced adaptive mutagenesis was studied in plant seeds in both laboratory and indigenous populations. Irradiated seeds could be divided into two subpopulations, in one of which non-targeted effects were observed. Such untargeted damage enhances mutant accumulation and selection for adaptive resistant variants. A correlation was observed between the amount of untargeted effects and the dose, as well as the antioxidant status of the seeds, which influences the cell sensitivity to low-dose irradiation.

The first paper on radioecology was written by V.I. Korogodin (together with A.L. Agreh) in 1957 and published in 1960. It discussed the question of the amount of radionuclide contamination, which can be released into a reservoir without disturbing its work as a reservoirdeactivator. Together with Yu.A. Kutlakhmedov and G.G. Polikarpov, he showed that the amount of radionuclides, which an ecosystem can tolerate without disturbing its functioning, is a measure of the radiocapacity of the ecosystem.

Studies of the impact of glucose load on malignant Erlich ascite carcinoma cells were conducted in 1986 together with N.L. Shmakova and other researchers from the All-Union Oncological Scientific Centre, Academy of Medical Sciences. These studies addressed the question of why glucose load sometimes increases the radiosensitivity of tumour cells. The experiments showed that a glucose load under hypoxia caused massive death of tumour cells: the action of the glucose load and irradiation is additive. Under conditions of hypoxia, cell respiration changes to glycolysis, which leads to lactic acid accumulation, self-oxidation, and cell death. This phenomenon is characteristic of malignant (and not benign!) tumours and is demonstrable both in vitro and in vivo, and in humans.

One more focus in V.I. Korogodin's work at that time was the role of information in biological processes. In his books, he suggested that information could be regarded as a "guide to action" which determines a system's ability to conduct purposeful actions. He showed that all information systems possess two mutual and independent properties: autocatalysis and heterocatalysis. Autocatalysis determines system self-reproduction and the information that encodes the system, while heterocatalysis defines the environmental changes caused by a system.

V.I. Korogodin educated several generations of scientists. All those close to him were constantly busy solving scientific tasks. He paid much attention to the training of the younger generation in science. As a result, in his last years he was an informal leader of the Timoffeeff conferences on "Modern Problems of Genetics, Radiobiology, Radioecology and Evolution", supporting the People's Prize commemorating N.W. Timoffeef-Ressovsky – The Teacher.

V.I. Korogodin was a Full Member of the Russian Academy of Natural Sciences and a member of numerous Russian and international scientific societies. For many years he served in the editorial board of the journal *Radiobiology* (called *Radiation Biology & Radioecology* since 1993). He was among the first laureates of the N.W. Timoffeev-Ressovsky Medal.

Prof. R.M. Alexakhin,
Russian Academy of Agricultural Sciences

Prof. A.F. Tsyb,
Russian Academy of Medical Sciences

Prof. S.G. Inge-Vechtomov,
Russian Academy of Sciences

Nikolay Wladimirovitch Timoféeff-Ressovsky:

the man and the scientist

(1900-1981)

TIMOFÉEF-RESSOVSKY: A SHORT ACCOUNT OF HIS LIFE

VASSILY BABKOFF
Institute of the History of Science & Technology,
1/5 Staropansky, Moscow K-12,
E-mail: heba@ihst.ru

Abstract. This paper provides a short account of the life and work of Nikolay Vladimirovitch Timoféeff-Ressovsky (1900-1981), with special reference to the 1935 monograph on the nature of the gene mutation structure by Timoféeff-Ressovsky, Zimmer, and Delbrück, its origin, its role in the development of molecular biology, and the part played in Timoféeff's own life.

Keywords: Personalia: Timoféeff-Ressovsky, molecular biology, biophysics, Russian science.

1. Introduction

The title of the Conference held in Yerevan in September 2005, "Genetics, radiobiology, radioecology, and evolution", indicates the broad scope of scientific interests and contributions of Timoféeff-Ressovsky, which are discussed at length in the book by Babkoff and Sakanian (2002). Here I will focus on the events concerning the 1935 monograph known as TZD, whose 70th anniversary is marked by the Conference.

It was X-ray crystallography that made, in the early 1950s, a decisive contribution to the deciphering of the structure of DNA. But for the first time X-ray crystallographers (Astbury and Bernal) and geneticists (Timoféeff-Ressovsky and others) jointly discussed the problem of the chromosome and the gene in Klampenborg in April 1938, on the first of a series of meetings initiated by Timoféeff-Ressovsky and supported by the Rockefeller Foundation.

The history of the discovery of DNA has been told many times by its creators in their memories and discussed by historians in their writings. "With one exception... the historical aspects... have been adequately treated in the book "Phage and the Origin of Molecular Biology". The exception is due to the fact that the contribution to this book by N.W. Timoféeff-Ressovsky ...could not be included in the book", noted Max Delbrück (1970). "Perhaps molecular biology owes more to the geneticist

A.A. Cigna and M. Durante (eds.),
Radiation Risk Estimates in Normal and Emergency Situations, 3–14.

who began the work – N.W. Timoféeff-Ressovsky – than has so far been admitted", – maintained R. Olby (1974).

While first visiting Niels Bohr in the early 1930s, Timoféeff-Ressovsky admired the scientific style of Bohrs Kreis (which was close to that of his Russian circles). In August, 1932 Bohr delivered address "Licht und Leben", and stressed that physical approaches were desirable for, and needed by biology (Bohr, 1933). While European physicists were not ready to share Bohr's views, Timoféeff was well prepared to discuss them, because of his earlier cooperation with his dear teacher N. K. Koltzoff founder and Director of the Institute for Experimental Biology in Moscow. Koltzoff's efforts to introduce physical and chemical methods into biology, and thus to give more solid basis to life sciences, made a great impact on Timoféeff in his green years. In 1920, Koltzoff suggested Timoféeff exposing Drosophila flies to X-ray with the goal of producing artificial mutations. (The early results were uncertain, since he had no strains with known genetics history in his disposal, and could not estimate exact doses.) In 1925, Oscar Vogt came to Moscow to study V.I. Lenin's brain. He had interest to establish genetic studies in his Berlin Brain Research Institute, and asked Koltzoff to recommend a young Russian geneticist. Timoféeff-Ressovsky was a natural choice, both because of his talent and ability to work by himself in his 25 years, and because Koltzoff's wish to protect his student from troubles with political police. Over a year before the appearance of H.J. Muller's 1927 classic paper on genetic mutations, Timoféeff was systematically irradiating Drosophila flies in Siemens labs. His experiments with Drosophila led Timoféeff to the conclusion that ionizing radiation can cause remote damage to röntgenologists working in early nuclear medicine. Timoféeff was the first who, in the beginning of 1930s, prescribed röntgenologists to use protective lead aprons. Timoféeff was very active in the dissemination of his findings, and it is because of his radiation experiments with Drosophila flies that the method of temporary sterilization of women by radiation, practiced then by many women's doctors, was banned in Germany.

Following the 1932 Genetics Congress in Ithaca and a couple of months in Cold Spring Harbor Laboratory, back in Buch, Timoféeff had a guest, Hermann Joseph Muller, his friend since his visit to one of Koltzoff's zoological stations near Moscow in 1922. Timoféeff found for Muller a good German expert in dosimetry; they discussed a broad variety of questions; they were planning a Circum-Scandinavian journey with lectures. But they were late to start for the journey: Nazis seized power on Jan. 30, 1933. It would be wiser for Timoféeff not to go now, wrote Muller to O.L. Mohr on March 13, 1933, though Muller felt himself safe in Germany. Ironically, he himself was arrested by SA-men on March 15 (by mistake, released in the morning). Another Timoféeff's guest during these months was N.I. Vavilov, who was coming back home via Buch-Berlin from his last foreign travel. He suggested that Muller should head a laboratory of genetics in his Institute, and soon Muller took the Circum-Scandinavian root to Leningrad.

In 1934 the Scientific Council, die Notgemeinschaft der Deutschen Wissenschaft, set a commission on the problem of hereditary damage by radiation, headed by A. Kühn. Thus the Council responded to Timoféeff's proposal made in the fall of 1933 for supporting his studies in radiation genetics, which he later expanded to include Alfred Kühn, Hans Stubbe and other radiation geneticists.

Back in Buch after visiting Bohr, Timoféeff started private discussions on the nature of the gene mutation and the structure of the gene, with Karl Günter Zimmer (1911-1988), expert in radiation dosimetry, and Max Delbrück (1906-1981), a young theoretical physicist. Timoféeff's principle of "convariant reduplication" (inspired by Koltzoff's general idea of autocatalysis in the reproduction of "hereditary molecules") became the organizing center; his radiation genetic experiments gave the material for discussions. Their cooperation resulted in 1935 paper "On the nature of the gene mutation and the structure of the gene".

The article, written first in Russian (it was a rule to Timoféeff, who had no troubles with German, English, French, Italian, but wished to come back one day to Russia with scientific works), consisted of four chapters. Chapter One (TR) expounded results of the qualitative and quantitative biophysical analysis of the mutational process, and theoretical considerations arguing the conclusion that the genes were macromolecules and mutations were intermolecular changes; on this basis the problems of the theory of the gene mutations and the structure of the gene were formulated. Chapter Two (Z) applied the hit principle to Timoféeff's results; Zimmer's formula for the number of mutations supposed a one-hit process; under the action of X- or γ-rays, the "hit event" meant one single ionization or excitation event. Chapter Three (D) described the model of gene mutation from the standpoint of atomic physics needed to estimate limits of the stability of the gene (regarded as "Atomverband", association of atoms, rather than a molecule) and to find correlates in the known atomic properties which would correspond to this stability. Jointly written Chapter Four argues that the view that mutation of the gene is an individual elementary process in the sense of quantum theory can be used to explain both spontaneous and induced mutation process.

Timoféeff delivered this paper to the Göttingen Scientific Society (presented by Alfred Kühn) on its meeting on April 12, 1935. It was published in its "Nachrichten" on June 29, 1935. In spite of low circulation of the "Bulletin" the work became known to a certain circle of biologists and physicists. They referred to the paper either as "TZD" (after the authors' names), or "die Grüne Pamphlet" (because of the cover color), or, being astonished by interdisciplinary cooperative work, the "Drei-Männen-Werke" (while one man was enough to make the whole work with his woman). This paper had immediate and long-lasting consequences.

Soon after the paper was published, and in part because of it, Timoféeff-Ressovsky received from the Rockefeller Foundation an invitation to run a laboratory in the Carnegie Institution. He negotiated for a while and finally rejected the offer in the Fall of 1936. As a recognition to his contribution to modern sciences (and with the goal of preventing him from leaving the country), the Kaiser-Wilhelm-Gesellschaft appointed him Director of the autonomous Department of Experimental Genetics, directly subordinated to KWG. In May, 1937 Timoféeff-Ressovsky's Soviet passport was canceled by the Soviet Consulate in Berlin, and for five years he was playing hide-and-seek with the authorities of German science and education, successfully trying to avoid acceptance of the German citizenship as the pay for the very welcomed appointment.

On April 22, 1938, the Italian Society of Experimental Biology elected Timoféeff-Ressovsky its Honorary Member. In May, 1939 he was awarded the Lazzaro Spallanzani Medal by the University of Pavia. In May, 1938 KWG Senate elected Timoféeff-Ressovsky its Scientific Fellow – he cherished this honor, and, I witness, he was always keeping the KWG badge on his coat lapel. On April, 1940, he was elected Member of the German Academy of Naturalists Leopoldina (Timoféeff produced a broad variety of scientific results, other than TZD, and his election to Leopoldina was due, in part, to other accomplishments).

Max Delbrück won, in 1937, the Rockefeller Fellowship to work with T.H. Morgan in Caltech, and brought Timoféeff's principle of "convariant reduplication" to the New World. Timoféeff has opened the new field on the basis of physical approaches (radiation genetics experiments) and higher organisms (insects). Delbrück participated in the transforming of the field to use chemical approaches and microorganisms.

2. Memorandum

The nearest consequence of TZD was the start of European discussions designed by Timoféeff-Ressovsky on the model of the circle for informal discussions with S.S. Chetverikov's students in the early 1920's. Brilliant scientist and a charming personality, Timoféeff-Ressovsky was recognized in the 1930s as the 5th best geneticist in the world (after T.H. Morgan and three Morganists), and the 1st one in Europe. He became the moving spirit of the new discipline in mid-1930s. In TZD he had formulated the central problem of "convariant reduplication" which permitted to consolidate efforts of experts in genetics and X-ray crystallography, cytology and electron microscopy, cell physiology and embryology, chemistry and biochemistry, theoretical and experimental physics, with the goal of laying the bases of what he provisionally called "biophysics".

His memorandum to the European Bureau of the Rockefeller Foundation, signed in Paris by Boris Ephrussi, H.J. Muller, and himself on October 24, 1937, stated (as cited in Babkoff and Sakanian, pp. 464-466):

> "Geneticists in general agree that the most fundamental problem of genetics is that of "convariant reduplication", that is the induction by each gene of the formation adjacent to itself of a daughter gene having an internal pattern identical with its own, even though that pattern itself is capable of being altered by mutation, without loss of this property of reduplication. They regard this problem as of great importance not only for genetics but for biology in general, in as much as they see in this property the most unique characteristic of living matter, which in turn has given rise to the possibility of origination of many other of the distinctive properties of protoplasm. In the gene, as in the simplest living things, this primary property should exist in a form more or less separable from these secondary complications."
>
> "...In all described cases the material of the experiment remains biological, but the approach should be made with the aid of the methods and knowledge of modern physics and chemistry..."
>
> "Specific problems which should be taken up by such a group in the near future are as follows:
>
> 1. The mechanism of mutations of genes, both spontaneous and induces; this involves also the problem of the energy levels necessary for different types of gene changes, which retain or destroy the property of convariant reduplication.
> 2. The mechanism of gene rearrangement and its relation to that of gene mutation.
> 3. The problem of the nature of conjugation between genes and chromosomes.
> 4. The structures of chromosomes and of organic fibers.
> 5. The problem of the method of reaction of the gene with the protoplasm ("gene function") and of the influence of neighboring genes upon these reactions ("position effect").
> 6. All the above problems are interrelated and bear more or less directly upon the central problem of convariant reduplication. But it might also be well to take up this problem in more direct form at a special meeting..."

The project to make cooperation of young physicists and biologists more permanent and efficient, by holding a regular series of conferences, got support from the European Bureau of the Rockefeller Foundation and of Warren Weaver, director for the natural sciences in the RF. Timoféeff's project to hold four meetings during two years, 1938-39, found financial support ($ 2400) from the Rockefeller Foundation.

"Members already decided upon:	Provisional list from which members will be eventually chosen:
Pierre AUGER	Astbury
Hans BAUER	Bernal
DARLINGTON	Bawden
DELBRÜCK	A.Fischer
EPHRUSSI	A.Kühn
MÜLLER (if in Europe)	Langmuir (when awailable)
RAPKINE	Needham
TIMOFÉEFF-RESSOVSKY	Stubbe
WADDINGTON (if in Europe)	Wrinch
ZIMMER	Serologist ??..."

3. European Meetings

The first meeting took place in Klampenborg, near Copenhagen, in the beginning of April, 1938, with the introductory reports presented by H. Bauer (on "Chromosome structure") and by C.D. Darlington (on "Chromosome division and conjunction").

The physical-chemical nature of forces determining the motion of chromosomes during meioses and mitoses and their conjugation were discussed. Astbury was absorbed in the idea of the role of nucleic acid in the replication of the genes, and discussed a possibility of external matrix molecule. Darlington discussed double-thread structures in terms of "coils" and "spirals", as described with the use of optical microscope (Klampenborg Conference, 1939; Darlington, 1969; Babkoff & Sakanian, pt. 1, ch. 6).

On the second meeting, in Spa (Belgium), in the end of October, 1938, the major reports were done by N.W. Timoféeff-Ressovsky on "Gene mutations", and by H.J. Muller on "Chromosome and the position effect of the gene" (Babkoff & Sakanian, pt. 1, ch. 6).

While drafting 1938 Annual Report, Warren Weaver entitled the section, dedicated to the Klampenborg and Spa conferences, "MOLECULAR BIOLOGY", and used this term for the first time (Weaver, 1970).

Taking into account that many of the participants of these meetings (and other pioneers in the field, e.g. Avery or Landsteiner) were supported by the Rockefeller Foundation, either individually or institutionally, we may speak of molecular biology as a Rockefellerian science, initiated and headed by Timoféeff-Ressovsky.

The 3rd conference scheduled in Melrose (near Edinburgh), August 31 – September 1, 1939, with a very rich scientific program, was associated with the ill-fated 7th International Congress of Genetics: scheduled for 1937 in Moscow, delayed to 1938, rescheduled for late August in Edinburgh, it was finally canceled by the approaching World War II. The war made the 3rd conference impossible (Babkoff & Sakanian, pt 1, ch. 6).

4. "What is Life ?"

In the beginning of 1940s, when Erwin Schrödinger was Professor in Dublin, he met another German theoretical physicist, E.E. Evald, who gave him a copy of the 1935 paper by Timoféeff-Ressovsky, Zimmer, and Delbrück. For some time Schrödinger paid attention to this circle of questions, but TZD impressed him so deeply that he dedicated to its analysis a series of lectures in the Trinity College in February, 1943, and published them as a book entitled "What is Life?" (Schrödinger, 1944). The author of this short fascinating book was one of fathers-founders of the "new" quantum mechanics, Nobel Prize winner, whose wave function was being studied in all universities. No wonder that the book got broad attention and interest from young physicists, chemists, and biologists in the post-war years, and attracted them to the new field. Translated into Russian and published in the beginning of 1947, the book was being discussed on special sessions in the Moscow University and in the Physical Institute of the USSR Academy of Science, and initiated a series of discussions aimed to analyzing the chemical nature of the gene. Public discussions came to an end after August, 1948 Session of the Lenin Agricultural Academy (VASKHNIL), when Trofim Denisovich Lysenko (Joravsky, 1970), who stressed that his report was completely supported by the Central Committee of the Party, accused Schrödinger of ideological sins. Meanwhile, physicists involved in the Soviet Uranium project, learned from the book that biology was a sound science and, after Stalin's death, used their influence to support genetics in Russia.

5. " Biophysik"

When the war was going to an end, the whole Brain Research Institute was evacuated to Göttingen in the Western Occupation Zone Remained in Buch the Department of Genetics and Biophysics only. Director Dr. Spatz demanded that Timoféeff should also join them. He stayed in Buch, and worked on the problem of dismantling of neutron generator. Timoféeff was receiving offers to go to America, England, Italy and France, that he rejected, and stayed in the future Soviet Occupation Zone.

In war-time Germany, Timoféeff summed up his vision of the new science of biophysics (as he used to designate what later came to be known as molecular biology) in "Biophysik", vol. 1, co-authored by Zimmer. Written in 1943, the book was published in Leipzig in 1947 under permission of the Soviet Military Administration in Germany (Timoféeff-Ressovsky and Zimmer, 1947).

In April, 1945 the Red Army entered Buch, north-east of Berlin, which had never been bombed by the Allies, probably because of Timoféeff's friend Paul Rosbaud, one of the leaders of the Julius Springer publishing house and the editor of "Die Naturwissenschaften", who was believed to be in contact with the British Intelligence and was known as "the Griffin". Timoféeff was made Burgomaster of Buch by the Soviet Military Administration. The staff of his Department was 12 persons, but he managed to organize regular three meals a day for more that 100 persons and their relatives.

There were many visitors to Buch from the Soviet Union in May–August, 1945. The physiologist Leon Orbeli invited Timoféeff to join his Institute; there were other suggestions, but Timoféeff accepted the first one, made by A.P. Zavenjagin, vice-director of the Soviet Uranium Project, and was waiting for the evacuation of his Department. But in September 1945, a *donos* (false political accusation) made by N.I. Nuzhdin, geneticist turned Lysenkoite, who visited Berlin-Buch, resulted in the arrest of Timoféeff, who spent several months in the Karlag concentration camp (death camp, as a matter of fact).

After the war, "Biophysik" circulated in the United States in a number of copies. There was a project to translate "Biophysik" into English, and leading geneticists supported it. There was also one geneticist, admirer of Stalin, who appeared to be able to block the road for translating the book.

Frederic Joliot-Curie, who praised Timoféeff high both because of his scientific accomplishments and his part in the European Resistance, sent a letter to Stalin concerning Timoféeff-Ressovsky. While visiting Moscow physicists he spoke to the head of Soviet Uranium Project L.P. Beria about him (Beria: "Well, is he really so gifted?" – Joliot: "He is a genius!"). Timoféeff-Ressovsky was found in a camp, taken to a hospital in Moscow, and placed to the Sungul *sharashka* in the South Urals (originally planned for his studies), where he continued his genetics and biophysical studies and started studies in radioecology.

In August, 1948 his genetics was outlawed in the USSR but in Sungul, he continued the genetics studies. Adriano Buzzati-Traverso and Luigi Luca

Cavalli-Sforza dedicated their 1948 book, entitled "La teoria dell'urto e le unità biologiche elementari", "*A Nikolay Vladimirovich Timoféeff-Ressovsky, amico e maestro, con l'augurio che egli possa continuare la sua opera*" (to NVTR, friend and teacher, with the wishes that he will be able to continue his work).

Timoféeff was recognized as founder or co-founder of several most modern scientific problems and disciplines of the 20th century. There remained, besides genetics and among others, his radiobiology and biophysics: the physicists were making the Bomb and they knew Schrödinger's book, in 1947 Russian translation. So what should be his social standing?

The answer was given by most popular weekly *Ogonjok* in March, 1949: the article entitled "Fly-lovers are Man-haters" was accompanied by cartoons, ordered and edited by Stalin himself, which presented his negative attitude towards Timoféeff (and the Rockefeller philanthropy). The cartoons, which came too late to mark the August, 1948 victory over genetics, and were not in accord with the announced goal of the article (to condemn H.J. Muller's and Henry Dale's resignation from the USSR Academy of Sciences) were: ugly scientist in a bow-tie, hanging a test-glass (with poisonous bacteria) standing in the shadow of a gallows; ugly fat financial boss in a morning-coat leading a small army of puppet-scientists under the "banner of pure science" with the dollar sign; scientist in a white doctor's smock and black academy hat with a siring and anthropometric instrument, and test-glasses in his pocket, in between a Ku-Klux-Klan man with a rope for hanging and a club, and a gangster-looking policeman with a Thompson machine-gun; thick book entitled "Genetics" (in English) with two snakes, one in the form of the dollar sign, the other one, in the form of fly-cross, swastika; finally, someone in a morning-coat, with a gun and a fly-cross marked book in his pocket, is looking though the microscope (Fig. 1).

The cartoons series appeared at the moment when the personal prohibition of genetics studies reached Timoféeff-Ressovsky in his laboratory in Sungul, and coincided with the prohibition of "Biophysik".

In 1950, while prisoner at a Gulag secret laboratory in the South Urals, Timoféeff was nominated for the Nobel prize, for his research in biophysics/molecular biology, by Boris Nikolajevitch Rajevsky (Director of the Max-Planck-Institut für Biophysik).

6. Type of Behaviour

"Radiation, especially with X-rays, is the principal means of producing such new kinds of, or rather monsters, of animals and plants, and the wizard in this business was a Russian, Dr. Timoféeff-Ressovsky, who has found an asylum at the Kaiser Wilhelm Institute for Brain Research. An industrial concern has presented him with the enormous machines with which he radiates the minute sex cells of tiny little Drosophila flies. Timoféeff is a fanatic and enthusiast. I was really spellbound while he gave me a three-hour lecture on his work, incessantly gesticulating as he walked up and down the floor. The German staff of the Institute looked at this strange and temperamental Russian with amusement and sincere admiration. They even granted him a freedom of speech and opinion they would deny any other human being", reminded an American journalist about his visit to Germany during the first war winter, 1939/1940 (Ellinger, 1942).

Figure 1. I.V. Stalin's elucidation of Timoféeff's official standig. Cartoon by. Bor. Efimov in *Ogonjok*, March 1949.

Arne Müntzing, the well-known Swedish geneticist, was in Germany at a conference in 1936 or 1937. "The meeting was interrupted. A speech by Hitler was being broadcast. Everyone was supposed to stand and listen in silence. Everyone rose, and amid the universal silence Nikolay Wladimirovitch's voice thundered out: "Wann wird denn dieser Wahnsinn endlich aufhören?" (When will this madness finally cease?)" (Berg, 1988).

It was the type of behavior Timoféeff-Ressovsky demonstrated in Nazi Germany, in the Soviet Union, always and everywhere. After the Prague Spring, Soviet authorities found Timoféeff-Ressovsky to be the major source of ideological threat, and in May 1969 he was forced to retire. He was appointed the smallest possible sum of money per month since "the years in Germany were not included into record of scientific service". In December that year Max Delbrück received his Nobel Prize in Stockholm and rushed to Russia (Fig. 2), to visit his old friend and teacher, and to tell the Soviet authorities that there was a job waiting for Timoféeff in the States if he was not needed at home. In January 1970 Timoféeff accepted a post of Senior Researcher-Consultant at the Institute for Medico-Biological Problems directed by Oleg Gazenko.

Figure 2. "Hooray for Delbrück" Max Delbrück the Nobelist back home from Stockholm and Moskow, California, 1970 (N.W. Timoféev-Ressovsky's personal collection).

7. Rehabilitation

While working on our Timoféeff' book and reading protocols of interrogations kept in his personal file at the FSB (former KGB) Archives, I was intrigued by its mild character and loyalty. For instance: Investigator: "You are charged in *nevozvrashchenii* (not having come back) and in espionage." – TR: "I am guilty in not having come back. I am not guilty in espionage." – Investigator: "H-mm. You are charged in *nevozvrashchenii.*" – The protocols of interrogations and other related materials from the FSB Central Archives first appeared in Part III of our Timoféeff book (Babkoff & Saknian, 2002). – Yes, Joliot-Curie was right: after Hiroshima, Timoféeff-Ressovsky and his art were badly needed by the Soviet Uranium Project.

Timoféeff's personal file at the KGB/FSB Archives consists of 11 volumes, vol. 1 for 1945/1946, vols. 2-11 for 1987-1992 (Elena Sakanian applied for his rehabilitation on Aug. 8, 1987, and the rehabilitation was given on June 28, 1992). I was not permitted to study vol. 11, which contains KGB's faked resolution on the 1945/1946 case. (The KGB backed publication of the protocols in a book form (Rokitjanskij, Goncharov, Nekhotin, 2003) does not include the resolution). On the basis of this resolution the secret hearings of Timoféeff-Ressovsky's case resulted in the refuse of rehabilitation. Elena Sakanian undertook an investigation of her own, produced her four-hour "Cinema-Trilogy about the Zubr" (Sakanian,

1988-1991), (Zubr, *Bison bonasus*, Timoféeff-Ressovsky, after D. Granin's glasnost novel Zubr, 1987), exposed the KGB falsification, and gained posthumous rehabilitation for Timoféeff-Ressovsky. The official rehabilitation was the cherished desire of his life. Are you interested in the slanderous accusations, I was not permitted to read? – Have a look at 1949 cartoons, and you'll see all of them.

Acknowledgements

This paper, based mostly on materials of our 2002 Timoféeff book, was supported by the Russian Fund for Fundamental Sciences, Grant no. 04-06-80174.

References

Babkoff V. & Elena Sakanian, 2002, *Nikolay Wladimirovitch Timoféeff-Ressovsky*, 1900-1981, Moscow, PIM: 1–672.
Berg, R., 1988, *Acquired Traits*, Viking .
Bohr N., 1933a, Licht und Leben, *Naturwiss*., **21**: S. 245–250.
Bohr N., 1933b, Light and Life, *Nature*, **131**: 421–423; 457–459.
Buzzati-Traverso A. & L.L. Cavalli-Sforza, 1948, *Teoria dell'urto ed unità biologiche elementari*, Milano, Longanesi & C.
Delbrück M., 1970, A physicist's renewed look at biology: twenty years later, *Science,* **168:** 1312–1315.
Ellinger T., 1942, On the breeding of Arians, and other genetics problems of war-time Germany, *The Journal of Heredity*, **33** (4): 141–144.
Joravsky D., 1970, *The Lysenko Affair*. Hatvard Univ. Press: 1–459.
Klampenborg Conference, 1939, *Report of the Klampenborg Conference*, April 2nd to 5th, 1938, The Rockefeller Archive Center.
Olby R., 1974, *The Path to the Double Helix*, Macmillan: 1–510.
Rokitjanskij J.G., V.A. Goncharov, and V.V. Nekhotin, 2003, *Rassekrechennyj Zubr. Sledstvennoje delo N.V. Timofejeva-Resovskogo*. Moscow. (in Russian).
Sakanian Elena, 1988, *Behind the Zubr*. Moskow, TzNF (90 min. film).
Sakanian Elena, 1990, *Chasing the Zubr*. Moskow, TzNF (60 min. film).
Sakanian Elena, 1991, *Heroes and Traitors*. Moskow, TzNF (90 min. film).
Schrödinger E., 1944, *What is Life? The Physical Aspect of the Living Cell,* Cambridge; (Russian translation, Moscow, 1947).
Timoféeff-Ressovsky N.W., and K.G. Zimmer, 1947, *Biophysik*, Bd. I. Das Trefferprinzip in der Biologie, Leipzig, Hirzel, 317 SS.
Timoféeff-Ressovsky N.W., K.G. Zimmer, and M. Delbrück 1935, Über die Natur der Genmutation und der Genstruktur, *Nachr. Ges. Wiss. Göttingen,* Math.-phys. Kl., Fg. IV Biologie, N.F., **1**(13): 189–245.
Waddington C.D., 1969, Some European contributions to the prehistory of molecular biology, *Nature*, **221** (5128): 318–321.
Weaver W., 1970, Molecular biology: the origin of the term, *Science*, **170**: 581–582.

FROM THE MUTATION THEORY TO THE THEORY OF THE MUTATION PROCESS

SERGEY G. INGE-VECHTOMOV
Department of Genetics and Breeding, St-Petersburg State University, University embankment, 199034 St-Petersburg, Russia
E-mail: inge@SI2444.spb.edu

"The general theory of inheritance seems to me the same way impossible and unnecessary as the general theory of variability".
K.A. Timiryazev, 1890

"...science is the art of doubt, not of certainty"
F. Ashcroft, 2005

Abstract. The main contributions to the biology made by N.V. Timofeev-Ressovsky and his co-authors (in "three gentlemen paper", 1935) were: radiobiological approach to the mutation process, materialization of the gene as a macromolecule and foundation of molecular biology. These directions appeared as development of the template principle, offered previously by N.K. Koltzov, the teacher of Timofeev. Finally Timofeev formulated his principle of con-variant reduplication, which united two main biological features - inheritance and variations as a single one. Study of primary lesions and repair was added to the theory of mutations from this point of view. Nevertheless, we have no satisfactory definition of mutation so far and even contemporary classification of types of variations is contradictory now. The situation is explainable by the fact that the classification was introduced rather from the phenomenological approach than from the mechanisms underlying the variation phenomena. It is proposed to divide variations for two groups: those connected with replication of genetic material and those connected with expression of genetic information. This classification should be introduced without a-priory division of variations on inherent and non-inherent because the same mechanisms may be involved both in inherent and non-inherent variations, depending upon taxonomic position and stage of ontogenetic development of the organism.

Keywords: mutation, variation, template principle, primary lesions, repair, variations classification

A.A. Cigna and M. Durante (eds.),
Radiation Risk Estimates in Normal and Emergency Situations, 15–26.

1. Introduction

The tremendous contribution of N.V. Timofeev-Ressovsky to different fields of biology is very well known. It is partially reflected in the title of our meeting. Now it is 70 years of the "Green Pamphlet" (Timfeev-Ressovsky, Zimmer & Delbruck, 1935) issued by three scientists: a biologist, a physicist and a mathematician, who strongly influenced future development of Biology. Among other consequences of this paper was its influence on E. Schrödinger, who wrote his "What is life? The Physical Aspect of the Living Cell" (Schödinger, 1945). The paper of the three gentlemen should be considered as the first ideological step in development of molecular biology and of molecular genetics in particular.

We have to remember that the main questions discussed in the paper were: the mechanism of mutagenesis and the nature of the gene. These two problems are tightly linked to each other. Moreover their interpretation reflects the very status of genetics during all periods of its history. Both the problems were resolved in the paper in a very stimulating manner for that time. The gene had been identified as a macromolecule and mutation was described as a change in the structure of that molecule.

I am not going to discuss the contemporary problem of the gene here and will consider it only in some aspects, namely in the aspect of mutational variability and in the aspect of general theory of variability. Our modern knowledge of genetic processes and of their molecular mechanisms shows that in understanding of mutations and of the other types of variability (including modifications) we are still far from having a perfect understanding of what they are.

2. Template Principle

Interest in the nature of the gene and its variation originated in the works of Timofeev-Ressovsky from the template principle in biology. This principle had been formulated by his teacher N.K. Koltsov (Fig. 1) in 1928 in his paper in which he offered a hypothesis of template reproduction of the chromosome (Koltson, 1936).

From the modern position we shall say that Koltsov's definition of the gene as a protein molecule appeared to be wrong, but the template principle proved to be absolutely right. Finally it was extended by Timofeev as the "principle of convariant reduplication" of genetic material. This principle united the two processes as a single one: reproduction and variation of the gene. It supposed that the gene reproduction (reduplication) is accompanied by variation of its structure (mutations) and these mutant variants are capable of reproduction and so on. Since that times the template principle

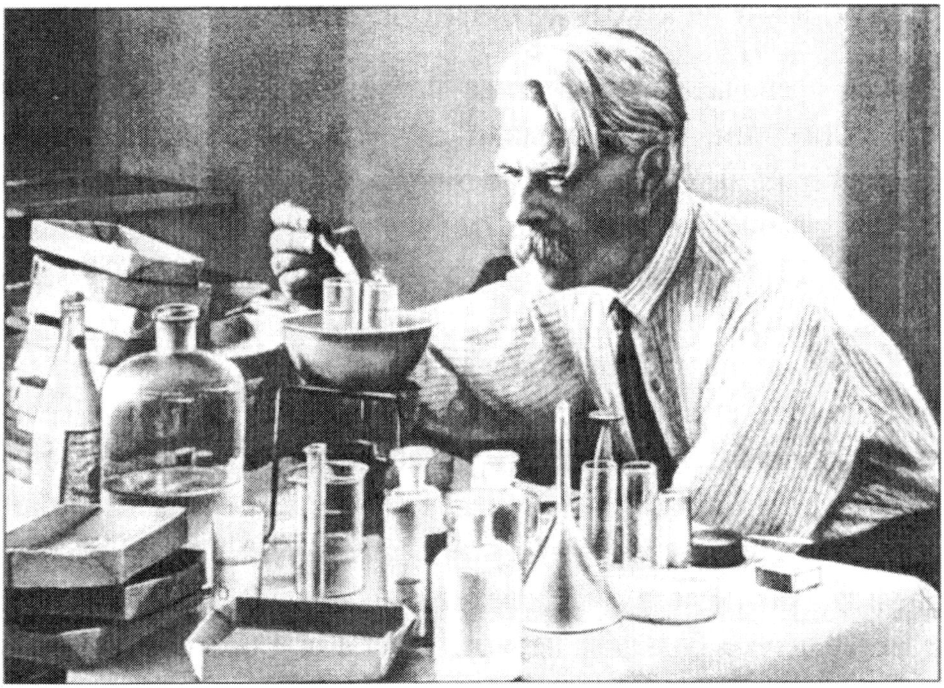

Figure 1. N.K. Koltsov (1872-1940) (from Soyfer, 2001).

became the central one in development of molecular biology (Inge-Vechtomv, 2003).

Highly promising in the study of inherent variations was the discovery of radiation-induced mutagenesis in *Drosophila* by H. Müller (1927). Timofeev (in Germany) (Fig. 2) and another coworker of Koltsov – A. Serebrovsky (in Russia) appreciated the meaning of this discovery and utilized the method of induced mutagenesis in their works. Serebrovsky and his coworkers (with H. Muller in that number) finally demonstrated the fine structure of the gene in Drosophila in the 1930's (Müller & Prokofyeva, 1934). Timofeev came to what we now call the "Green Pamphlet", working with the same object and utilizing the same method as Muller – registration of recessive lethal mutations in X-chromosome of *D. melanogaster*.

Figure 2. N.V. Timofeev-Ressovsky (1900-1981) Berlin-Buch, 1940 (Timofeev-Ressovsky, 2000).

Finally the template principle had been proved by demonstration of genetic functions of DNA, by discovery of DNA structure and the mechanism of its semi-conservative replication. So the substrate of "convariant reduplication" was identified. Later we could see development of the template principle in Francis Crick's Central Dogma of molecular biology (Crick, 1958). There are many contradictory theses in the contemporary interpretation of the Central Dogma, but it is completely valid to consider it as a symbol of template principle in biology. Even the recent discovery of so-called protein inheritance (Prusiner, 1998) may be included in this scheme, This is the prion mechanism of inheritance in lower eukaryotes (namely in fungi). In this mechanism of protein inheritance we again deal with template processes. The difference is that there is a protein template, which does not code the sequence of a daughter molecule, as in nucleic acids, but defines the conformation of the sister protein. Here we deal with inheritance on the level of protein conformation (Inge-Vechtomov, 2003). Today therefore, we can see a triumph of the template principle founded by Koltsov (1936) and extended by Timofeev-Ressovsky.

3. Mutations

An interesting episode accompanied the development of the other achievement, published in the “Green Pamphlet”. It is interesting from natural scientific and from historical perspective. It concerns the very mechanism of mutations. Timofeev, Zimmer and Delbruck first considered gene mutation as a monomolecular reaction, which changed gene structure in accordance with the “treffer teorie” or the target theory, developed by the authors. In accordance with this theory they got a one-hit dependency of mutation frequency from the dose of irradiation. It was a nice generalization for that time. Later on it became evident, that the great majority of recessive lethal mutations in X chromosome of *Drosophila*, induced by radiation were small chromosome rearrangements, predominantly small deletions. They needed two hits, two breakage points to appear. Taking this into consideration one-hit dependency appeared to be a puzzle from the point of “traffer teorie”.

This contradiction was picked up by M. Lobashev, who tried to approach the mechanism of mutation process from the other side. He considered mutation as a result of non-adequate or non-identical repair of genetic material. Even as a result of the repair of the cell as a whole (Lobashev, 1947). It is necessary to say that Lobashev considered protein as genetic material the same way as Koltrsov did, and thought that the protein was a substrate of the repair process. Nevertheless he was the first who put together two terms: mutation and repair. He did it in 1946 in his theses of dissertation for the degree of Doctor of Biological Sciences (Lobashev, 1946). Now we know that all three main template processes: replication, transcription and translation include mechanisms of correction or repair (Inge-Vechtomov, 2003). Later it was proven that very often mutations start from primary lesions in DNA and that mutations are fixed in the process of repair as inherent traits of genetic material. Remember “mistakes of three R” (Replication, Recombination, Repair) offered by Jack Von Borstel in late sixties of the last century (von Borstel, 1969).

Timofeev-Ressovsky and Lobashev did not know each other until the sixties, when Nikolai Vladimirovich was allowed to visit big cities as Moscow, Leningrad, Kiev etc. after his sentence to prisoner camp and liberation from it. They met first in the apartment of Daniil Granin, the author of a popular novel “A Bison” about Timofeev-Ressovsky’s life. Granin told me the story of meeting of these two scientists. It was a really dramatic situation. These two men were completely different. One of them – Timofeev-Ressovsky had noble roots in his origin, he was a well-educated person, and he was still in Berlin-Buh in 1945.

Lobashev was of completely proletarian origin. He was a very soviet person. Hee also was a hero, of another novel, written by V. Kaverin – “Two captains”, a very popular book in Soviet Union. The part of Lobashev’s life –before the University is shown in this book. The dramatic episode of the meeting of the two classics of mutagenesis was presented in our previous paper (Inge-Vechtomov, 2004). Since that meeting they became friends and every year Lobashev invited Nikolai Vladimirovich to Department of Genetics in Leningrad University (Fig. 3) and we (the students of that period) had unforgettable opportunities to hear his brilliant lectures on population genetics, evolution, mutations and radiobiology. Many of us, students of Lobashev in Genetics Dept of Leningrad University consider Timofeev as our teacher as well.

Figure 3. N.V. Timofeev-Ressovsky (left) and M.E. Libashev (right) in Department of Genetics and Breeding of Leningrad State University. Late 1960-ies. (Archive of the Department).

So, let us get back from history to natural science. These two scientists, very different in their approaches, studied mutation process from different sides and described very important features of it. In spite of time passed and huge amount of information obtained since their time, we are still in a contradictory situation in understanding the very nature of mutations.

Now we at least understand that mutation is not an abrupt event, but it is a multi-step process. We even think about mutagenic pathway, which starts with a step of formation of a primary lesion in genetic material (DNA). The primary lesion is a substrate for several repair systems, which watches for the native DNA structure. Than in the process of repair there may be “mistakes” and the primary lesions are processed to the stable mutations.

Experimental evidence for the existence of the primary lesions in genetic material was shown in the study of photo-reactivation of DNA, damaged by UV light by A. Kelner and R. Dulbecco (1949) and little more than one decade later it was shown that photo-reactivation is an enzymatic process and its substrates are pyrimidine dimers (Friedberg, 1995).

Now we know that only a few of the primary lesions are processed to true mutations. Also the fraction of the primary lesions, which are processed into mutations, is possibly different for different mutagens. Only a few experimental systems allow us to score this fraction directly. The first attempt to show it was by M. Reznick and R. Holliday (1971), utilizing the genetic system of nitrate reductase in *Ustilago maidis*. They showed that after UV irradiation without photo-reactivation a fraction of inactive enzyme, which was encoded by the damaged gene, appeared in the cells (Reznick & Holliday, 1971).

The next decade after that we utilized another system in *Saccharomyces* yeast to calculate the ratio of the primary lesions and of the real mutations after UV irradiation in a single locus. It was a system of mating types in *S. cerevisiae*. There are two mating types in haploid yeast: α and β. It was shown that in "illegitimate" crosses α Xβ± hybrids appear predominantly through phenotypic expression of primary lesions within the locus MAT. These primary lesions express themselves as a transient a-mating type. And after mating these lesions are repaired in more than 99% and the original mating type is restored (Inge-Vechtomov & Repnevskaya, 1989) (Fig. 4). So it was evident that only less than 1% of primary lesions after UV mutagenesis are processed in real mutations. Now we are studying this ratio for different mutagens.

Another intriguing problem of mutagenesis is distribution of mutations and susceptibility to mutagenesis within a cell population. The standard view is that mutability is randomly distributed among cells in genetically homogeneous populations. From this point the probability of multiple mutants should be calculated by multiplication of the probabilities of every mutation event. But it is not the case in some instances.

Sometimes we encounter the phenomenon of so called, multiple mutability, which we studied in the 19's, also in yeast. It looks like there may be some cells in genetically homogeneous cell population, which are more likely to undergo the mutation process (Arifyeva & Inge-Vechtomov, 1977) Unfortunately, we could say nothing about the mechanism of this phenomenon. Some kind of similar phenomenon is described and discussed in paper of J.Drake in this issue (2006).

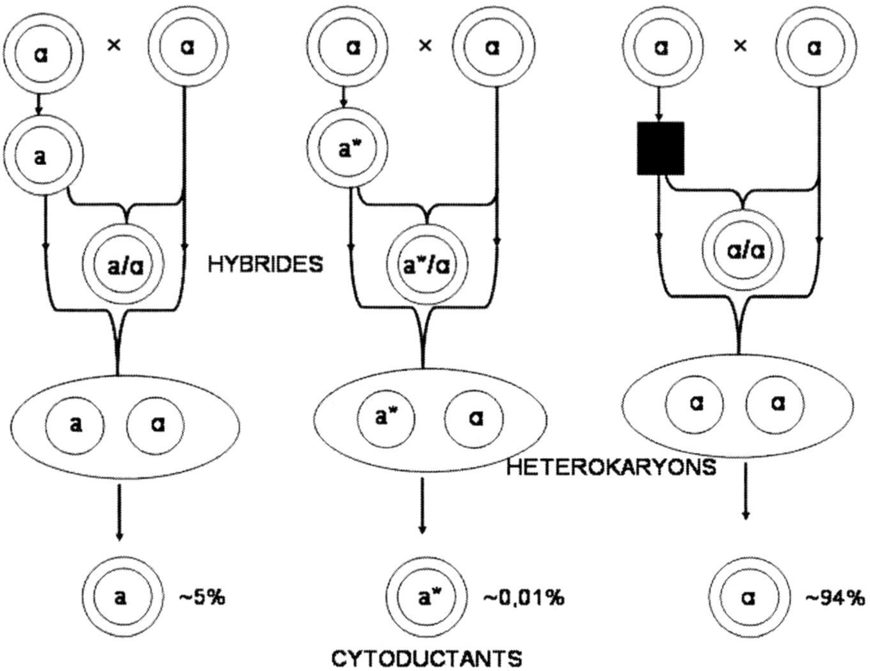

Figure 4. Phenotypic expression of the primary lesions of genetic material. Hybridisation and cytoduction among yeast (*Saccharomyces cerevisiae*) of the same α-mating type [8]. See commentary in the text.

4. Some More Problems and Contradictions

Besides that we still encounter a lot of contradictions in the classification of not only the mutation types, but even in the general classification of the types of variations. The widely accepted classification of variability is presented on Fig. 5. It is not satisfactory now, though it is what we teach our students. Let us take as an example so-called ontogenetic variability. It includes mechanism of mutations and mechanism of recombination. If we remember differentiation of immunoglobulins we encounter recombinational rearrangements of genetic material and site directed elevated mutagenesis. At the same time we have many examples of modifications – regulation of gene expression in ontogenesis and such events as genome imprinting and other effects, which we hide now under the term of epigenetic variation. Epigenetics now is nothing more than one more word, which does not clarify the situation, but makes it more complicated.

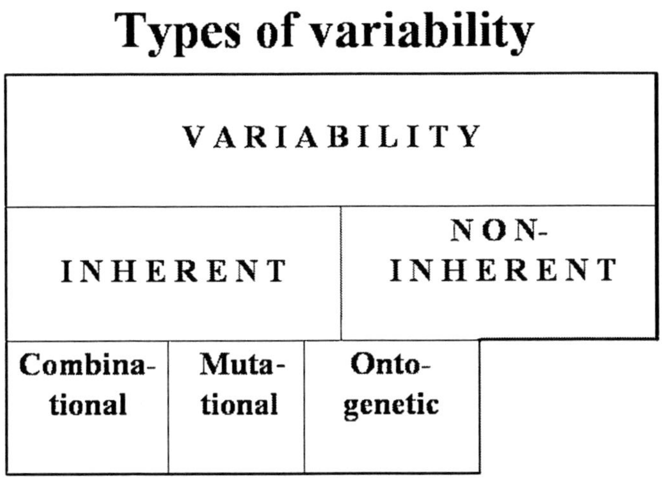

Figure 5. Contemporary classification of variability.

Even if we return to the problem of mutation the situation is not simpler. Chromosome mutations or chromosome aberrations usually appear as a result of either illegitimate or ectopic recombination either among non-homologous chromosomes or between different regions of the same chromosome. We can use a popular combination of words – “transposon mutagenesis”. In reality “transposon mutagenesis” is an example of combinational variability because transposon insertion is a result of recombination. So a great part of what we used to call mutation in reality is a result of recombination – of the other type of variability.

We may mention also so-called genome mutations, for example polyploidisation. It is a change of cell content (change of copy number) of genetic material. It happens regularly in ontogenetic differentiation of some tissues. These, so-called mutations (genomic mutations) are connected with disturbance of micro-tubules in the cytoskeletal apparatus of the cell, but not of DNA structure. I am not going to discuss modification of DNA bases (e.g. methylation) which we do not consider as mutations, again trying to hide the problem using the term “epigenetics”.

So, what is mutation? There is no satisfactory definition so far.

These contradictions are understandable because our accepted classification of variability and of mutations in particular is based historically upon phenomenology, but not on mechanisms, which became evident only later on. Now we understand that the same mechanism may be involved in different phenotypic events in different organisms or at different stages of its development, and a vice versa, since different mechanisms may cause the same phenotypic effect. The most intriguing example is connected again with phenomenon of prions. It presents an example of typical modification of protein molecule on the level of its secondary and tertiary structure, but not of its primary structure. It is a typical modification (non-inherent change) in mammals (Prusiner, 1998). At the same time it is an inherent variation in lower eukaryotes (Wicner *et al.,* 1995) So, to think either of the general theory of mutations or even of the general theory of variations we must start to understand their mechanisms rather than rely on pure phenomenology.

Probably by asking "What is mutation?" we are putting a wrong question. It is well known from the history of science that it is necessary to put a right question to get a proper answer. The history of mutagenesis study gives us a nice example of this. Remember that H. DeVries (1901) and even S. Korzhinsky (1899) beforehand proposed their mutation (or heterogenesis - S.K.) theory as explanation of evolution, considering mutations as an elementary events of evolution. Now as a consequence of Timofeev-Ressovsky's works we understand that the elementary event in evolution is a change of allele frequency in a population (Tymofeev-Ressovsky *et al.,* 1969) DeVries defined mutation as phenotypic variations, but only when we started to discuss mutations as a change in genetic material, was the first real step toward understanding of mutation variability made. The same way it was incorrect to ask, how organs are inherited from generation to generation. Only when Mendel asked: how are elementary characters inherited, was it possible to come to the general theory of inheritance (contrary to Timiryasev's opinion. See the first epigraph to this paper).

5. Prospects

The same way if we want to understand the very nature of mutations (or of whatever it is) we should return to the general theory of variation. It looks reasonable to me:

(1) to classify different types of variations in connection with mechanisms of template processes. All of them posses a characterstic of ambiguity level. It means that variations are already included in the very mechanisms of template processes.

(2) I would suggest for this purpose to consider only two types of variations: those connected with reproduction (replication) of genetic material and those connected with expression of genetic information (transcription, translation and some other events in processing their products) (Fig. 6).

(3) Whether this or that type of variation would be inherent or non-inherent depends on taxonomic position, the stage of development of an organism and on a specific process in which this or that type of variation would be involved in.

It is possible to suggest that only this way we would be able to understand the real nature of both mutation process and of inherent and non-inherent variations.

REPLICATION OF GENETIC MATERIAL ----→ **VARIABILITY**

EXPRESSION OF GENETIC INFORMATION ----→ **VARIABILITY**

Figure 6. Alternative classification of variability, proposed in this paper.

Acknowledgements

Some results presented in this paper were obtained in the work supported by the Program of Russian Academy of Sciences "Dynamics of genofond of plants, animals and men".

References

Arefyeva A.Ya, Inge-Vechtomov S.G., 1977, Multiple mutants of *Saccharomyces cerevisiae*. IV. Mutability of yeast cultures at different growth stages *Genetika* **XIII**: 1237–1245 (in Russian).

von Borstel R.C., 1969, On the origin of spontaneous mutations. *J. Genet* (suppl) **44**: 102–105.

Crick F.H.C., 1958, On protein synthesis. *Symp Soc Exptl Biol* **12**: 138–163.

Drake J.W., 2006, Mutation and DNA repair: From the green pamphlet to 2005. (This issue).

Friedberg E.C., 1995, Out of the shadows and into the light: the emergence of DNA repair. *TIBS-20*: 38116.

Inge-Vechtomov S.G., 2003, The template principle in biology (past, modern, future?) *Ecological Genetics* **0**: 6–15 (in Russian).

Inge-Vechtomov S.G., 2004, Story about how Filipchenco got along with Morgan and sent to him Dobzhansky, how Koltsov directed Timofeev-Ressovsky to Germany, who advised Muller to go to Vaviliv in Leningrad and what came out of it. *Ecological Genetics* **II**: 5–11 (in Russian).

Inge-Vechtomov S.G. & Repnevskaya M.V., 1989, Phenotypic expression of primary lesions of genetic material in *Sacchoromyces* yeast. *Genome* **31**: 497–502.

Koltsov N.K., 1936, Inherent molecules (in Russian). In: Organization of the cell. *Biomedgiz*, Moscow.

Korzhinsky S.I., 1899, Heterogenesis and evolution: To the theory of origin of species. *Zapiski Imperatorskoy Academii Nauk* **9:** 1–94 (in Russian).

Lobashev M.E., 1946, *On the nature of action of external conditions on the dynamics of mutation process*. Dissertation Thesis for the degree of Doctor of Biol Sci. Leningrad (in Russian).

Lobashev M.E., 1947, Physiological (paranecrotic) hypothesis of mutation process *Vestnik Leningradskogo Universiteta* **8**: 10–29 (in Russian).

Müller H.J., 1927, Artificial transmutation of the gene. *Science* **66**: 84–87.

Müller H.J. & Prokofyeva A., 1934, Continuity and discontinuity of the hereditary material. *Doklady Acad.. Sci USSR* **4:** 8–12 (in Russian).

Prusiner S.B., 1998, Prions. *Proc Natl Acad Sci USA* **95**: 13363–13383.

Reznick M.A. & Holliday R., 1971, Genetic repair and synthesis of nitrate reductase in *Ustilago maydis* after irradiation.. *Mol Gen Genet* **111**: 171–178.

Schrodinger E., 1945, What is Life? The Physical Aspect of the Living Cell. Cambridge, The Macmillan Co., New York, cited in Atomizdat, 1972, Moscow. (in Russian).

Soyfer V.N., 2001, *Communist regime and science* Publ House "CheRo", Moscow (in Russian).

Timofeev-Ressovsky N.V., 2000, *Stories told by himself, with letters, photos and documents*. Soglasiye, Moscow (in Russian).

Timofeeff-Ressovsky N.W., Zimmer K.G. & Delbruck M., 1935, Über die Natur der Genmutation und der Genstruktur. *Nachr Gess Wiss Gottingen*, 6, N.F.. Bd 1, **13**: 189–245.

Timofeev-Ressovsky N.V., Vorontsov N.N. & Yablokov A.V., 1969, *A brief sketch of evolution theory*. Nauka, Moscow (in Russian).

de Vries H., 1901, *Die Mutationstheorie: Versuche und Beobachtungen uber die Enstechung von Arten im Pflanzenreich.* Leipzig: Veit,. Bd. 1.

Wickner R.B., Masison D.C. & Edskes H.K., 1995, [*PSI*] and [*URE3*] as yeast prions. *Yeast* **11**: 1671–1685.

N.V. TIMOFEEFF-RESSOVSKY'S VIEWS AS THE BASIS OF RADIOECOLOGICAL STUDIES

INNA MOLCHANOVA*, ELENA KARAVAEVA, ALEXANDER TRAPEZNIKOV
Institute of Plant and Animal Ecology
Russian Academy of Sciences, Ural Division, Russia
e-mail: molchanova_i_v@mail.ru

*To whom correspondence should be addressed.

Abstract. Theoretical fundamental and methodological approach to the investigations of the radionuclide behaviour in biogeocenosis was made by the outstanding researcher N.V. Timofeeff-Ressovsky. He performed and supervised the first investigations of the radionuclide behaviour into the simple links as soil-solution, soil-plant, water-hydrobionts. Later his students and adherents started a new trend in radioecology, studying the radionuclide migration and biological effects on natural ecosystems of land and internal reservoirs. N.V. Timofeeff-Ressovsky's views and ideas are corresponded with theory of the sick plates of the Planet which is being elaborated as a part and parcel of the contemporary ecological doctrine of Russia.

Keywords: radionuclides, biogeocenosis, radioecology, ecosystem

In our country history of radioecology is closely connected with general history of natural science. Ideas of V.I. Vernadsky about living substance and biosphere of the Earth made a valuable contribution to development of this science. V.N. Sukachev brilliantly developed the concept of organization of biosphere. As for evaluation of the results on geochemical activity of living organisms, it became a constituent part of the doctrine on biochemistry and geochemistry of landscapes. Genetic soil science developed by V.V. Dokuchaev formed the basis for these scientific concepts. V.V. Dokuchaev paid special attention to natural and historical origin of soil. Among all factors contributing to soil origin he especially noted time factor, which plays an important role in creation of unique body of nature.

A.A. Cigna and M. Durante (eds.),
Radiation Risk Estimates in Normal and Emergency Situations, 27–34.

N.V. Timofeeff-Ressovsky synthesized and summed up the achievements of the Dokuchaev school. He was the first who marked similar approach of the above naturalists in understanding nature and problems of general science. It is these natural and historical, and biosphere approaches that formed in mid - 1950s the basis for a new scientific discipline, i.e.: radiation biogeocenology later called radioecology. It was dictated by constantly increasing amount of natural and artificial radionuclides in biosphere of the Earth, which were released during large-scale nuclear weapon tests, and developing nuclear industry and power engineering. Since then the world scientific community had to recognize a new global ecological factor, i.e.: man-caused radionuclides and the related ionizing radiation. Evolution in understanding the radiation factor involved a number of sciences, a viz.: ecology, agrochemistry, geochemistry, hydrobiology, and nuclear physics. "Such method of formation of new scientific disciplines as noted by Aleksahin (2000), could lead to heterosis: a qualitatively new discipline appears due to fruitful interaction between different scientific disciplines". As a rule, to develop a new scientific discipline requires activity of "creative personalities", whose ideas form the basis for future study. Actually, two outstanding personalities, N.V. Timofeeff-Ressovsky and V.M. Klechkovsky, were founders of radioecology. Symbolically, both of them, being contemporaries of the century, in 40-50s founded new scientific schools at the Urals, which is the birthplace of home nuclear project. To treat both of them with due respect, today we dwell upon N.V. Timofeeff-Ressovsky.

After World War II Nikolai Timofeeff-Ressovsky, being a prisoner, who passed Butyrka and Karlag, run a Department in one of the "closed" cities in South Urals. In 1955 together with a group of collaborators he was deported to Institute of Biology (presently Institute of Plant and Animal Ecology, Ural Division of Russian Academy Science) located in Sverdlovsk city (now Yekaterinburg). Standing at the beginning of creation of nuclear project he noted a very serious problem of a comprehensive study of the matters related to effect of fast development of nuclear industry on biosphere. Being a follower of his promoted predecessors and with account for vast experience accumulated destructive effect of industrial wastes on environment, he stressed (1962) that "any serious problem related to influence of a human being and his activity on environment should be based on Vernadsky doctrine on biosphere as well as on the biogeocenological ideas of Sukachev".

At the first stage a large number of experiments were arrived out under the leadership of N.V. Timofeeff-Ressovsky, who worked excitedly and with great optimism. Radionuclides were treated as "radioactive tracers" to study the behaviour of chemical elements in different components of biosphere, while ionizing radiation as a convenient and easy in dosage factor to affect organisms and their communities. Accessible investigations in accordance with scientific methodology of the author were performed with a wide range of radionuclides on such simplified systems as solution-soil-solution, water– bottom sediments, and water- hydrobionts. Investigation of these simple systems was started from the study of cause-and-effect relation. There is few number of cause-and-effect relations in a simple system. Besides, they are well-defined and more easy to study. Investigation results allowed Timofeeff-Ressovsky et.al. (1966) to classify radionuclides according to their behaviour in a primary ecological links and to find out common factors and mechanisms responsible for their capacity to migrate in them.

On the basis of biogeocenological studies of water reservoirs Timofeeff-Ressovsky (1962) made the following conclusion: "In water reservoirs, as distinct from soil with important role of soil formation and relation between soils and drainage, element concentration in bottom sediment is of special importance. It depends on flow conditions, climate, biological productivity and composition of biocenoses in water reservoir". This thesis was supported by a number of studies, in which general distribution of radionuclides over main reservoir components (i.e.: water, bottom sediments, and biomass) and accumulation of various radioactive elements by freshwater animals and plants as well as decrease in radiation of water passing through low-running reservoirs was discussed in detail. It is to be noted that E. Timofeeffa-Ressovskaya (1963) was the key person in these studies. It is worth mentioning that these studies were of acute importance because of the practical matters related to purification of water from radioactive contaminations. Generally the first-stage phenomenological studies in hydrobiology outlined the program in radioecology of freshwater reservoirs further successfully developed by the followers and students of Timofeeff-Ressovsky.

At the first stage studies special attention was paid to ability of living organisms to accumulate radionuclides. To qualitatively estimate and compare accumulative ability of organisms a formal criterion- accumulation coefficient- was used. This coefficient shows the ratio between concentration of the given radionuclide in organism and environment under equilibrium condition. It was shown that the accumulation limits are very wide for different types of radionuclides and elements. Those with the highest accumulative factors were named specific accumulators. To determine these accumulators the following criterion was proposed: more than a 4 σ deviation from the average value for the corresponding variational series. To define "concentrators" of radionuclides special attention was paid to organisms-pioneers that form residual soil and initial soil, i.e. bacteria, fungi and lichens. Later these organisms, "specific accumulators" of this or that radionuclide, were used as bioindicators of radioactive contamination of natural ecosystems. Later owing to these studies a specific scientific branch appeared, a viz.: lichen- and brio-indication of environment, one of the tasks of which was to trace the status of natural ecosystems in highly man-caused zones. The advantage of such a monitoring is as follows: high concentration ability and sensitivity of living organisms to pollutions allow to exclude labor-intensive analysis of the ecosystem components with low concentration of pollutants.

On top of this, in 50-60s of the previous century a numerous experiments on comparative radiosensitivity for more than 100 types of plants were performed by Timofeeff-Ressovsky and Poryadkova (1956) and Timofeeff-Ressovsky (1957). The obtained data supported stimulation effect of radiation on growth and evolution of plants. First attempt was made to explain this phenomenon. Experiments on artificial communities of terrestrial plants and fresh water periphiton showed that a comparatively low radiation doses result in negligible stimulation of the communities, without definite transformation. Alternatively, high radiation doses result in deep damage of composition and structure of the communities.

Timofeeff-Ressovsky (1962) analyzed huge analytical material obtained and presented it as thesis for his Doctor's degree. This thesis contains quintessence of the author's views on the novel field of knowledge, its general basis, and enumeration of problems, which are still of acute importance. They are as follows:

– migration mechanisms, distribution and biological effect of radioactive elements in various biogeocenoses (ecosystems);

– scientific prognosis on the consequences of radioactive pollutions of biosphere;

– ecological regulation of these pollutions in nature components and means of weakening their harmful effect.

In Autumn 1957, 74 PBq were polluted into atmosphere as a result of the accident at "Mayak" plant, later called Kyshtym accident. As a result of the Kyshtym accident a trace of radioactive contamination was left over the territory of 23,000 sq. km. This territory was called the East-Ural Radioactive Trace (EURT), which became a unique testing area for radio- ecological studies. With this aim in 1958 upon supervision by V.M. Klechkovsky was arranged Experimental scientific-research laboratory at "Mayak" plant. He was a scientific leader and coordinator of all studies done at this laboratory under strict secrecy. As a result, in 1974 V.M. Klechkovsky together with a group of scientists involved in radioecological studies was awarded the State Prize.

Unfortunately, Timofeev-Ressovsky was not allowed to guide studies of the accident consequences in 1957. He was pushed aside and deeply disappointed. But he continued the so-called "open" studies on this theme with all his enthusiasm and energy at the biological station "Miassovo" and in the Institute of Biology that belonged to the Ural Division of the Academy of Sciences USSR (Sverdlovsk). Even then he foresaw the possibility of global contamination of environment and in this connection he stressed the following acute problems:

– to trace the behaviour of radionuclides in various biogeocenoses, a viz.: routes and scale of migration from contamination centers;
– to study effect of ionizing radiation on living organisms;
– to develop measures to fight harmful radioactive pollutions.

At the same time Timofeeff-Ressovsky (1962) noted that «the problem of biosphere protection as well as all the problems connected with development of conservancy measures belong to a problem of biogeocenology. And to solve this problem one should introduce the present-day ideas on biosphere of the Earth and biogeochemical processes, especially to technicians."

The huge biogeochemical circulation of substances including radionuclides is caused by interaction of living organisms and inert components of the medium as well as by interaction between living communities and more or less independent objects of the nature. According to the concept of life organizational levels Timofeeff-Ressovsky and Tyuryukanov (1966), suggested that biogeocenoses, i.e. elementary links of current biosphere, belong to such objects. Biogeocenotic level of life being determined, Timofeeff-Ressovsky together with his students made qualitative estimation of radionuclides distribution over components of natural biogeocenoses. Even then he marked the role of forest litter, aboveground mass and root system of plants as well as physical and chemical peculiarities of radionuclides in their behaviour in biogeocenoses. Timofeeff-Ressovsky was keen on quantitative studies that finally led him to mathematical simulation of radiation-induced ecological processes. Thus, the problem to mitigate effect of contaminated biogeocenoses on a human being was set and successfully solved by changing its structure. Later the models obtained were used for agricultural radioecology as well as to characterize global processes in atmosphere induced by nuclear explosion.

Large-scale tests of nuclear weapon performed in 60s of the previous century resulted in global radioactive contamination of terrestrial and water ecosystem components. To study natural systems along with reduction-analytical principle, systems approach was widely used. Systems approach is based on the idea that all objects, though being independent, interact between each other. This interaction may occur, for instance, via geochemical flow of substances. Proceeding from this point of view Timofeev-Ressovsky treated migration of substances in a biogeocenoses chain as a series of conjugate processes of dissipation and concentration of substances in organisms, soil, water, bottom, and atmosphere. Later this became the ground to develop methods and principles for radioecological and geochemical studies of landscapes. They consist in the following: one determines certain objects of landscape (e.g. watershed, slopes, river flood plains, swamped low lands) located at contiguous by flow off areas to study routes and migration rates of chemical elements and radionuclides. With this method first data about contamination rate of soil - vegetative cover within the former USSR were obtained by Tyuryukanova (1968); Pavlotskay (1974), Aleksahin and Naryshkin (1977). On top of this, the distribution zone and zones of secondary accumulation of radionuclides in natural conditions were found; and dependence of radionuclide migration rate and their occurrence in a human body via digestive chains on local ecological conditions was evaluated by Rassel (1971); Tyuryukanova (1974); Moiseev and Ramzaev (1975). Technique to study freshwater ecosystems was developed and distribution of radionuclides over basic components was calculated by Trapeznikov and Trapeznikova (1979); Kulikov and Chebotina (1988); Chebotina et al. (1992).

The accumulated data on peculiar behavior of radionuclides depending on habitat of living organisms resulted in differentiation within radioecology and appearance of a number of independent scientific branches, such as: agricultural, forest, freshwater and marine radioecology, each branch with its separate specific problems. In 70s on the basis of Timofeeff-Ressovsky studies and vast accumulated experimental data a concept of continental radioecology was formulated with the aim to study terrestrial ecosystems and internal water reservoirs. At the same time it was underlined by Kulikov and Molchanova (1975) that natural environment cannot be considered as passive diluent of radioactive contaminants. In some links of ecosystem radionuclide concentration may be fairly high. Thus, hydrobiontes and lowest-order plants accumulate such amounts of radionuclides that are by orders higher than in environment.

Actually, the so-called “closed” results on contaminated territories in the Urals received publicity only after Chernobyl accident. “State Program of Russian Federation on Radiation Rehabilitation of the Ural Region” has been developed, in which one of the authors of this report (A. Trapeznikov) took part. Within this Program large-scale radioecological investigations have been started in Sverdlovsk, Chelyabinsk, and Kurgan regions. Researchers from Department of Continental Radioecology, the successor of Timifeeff-Ressovsky laboratory, also took an active part in these studies. Thus, with mathematical models concentration and integral reserves of radionuclides in terrestrial ecosystems and freshwater reservoirs, being under radioactive contamination, were verified and estimated. Of special importance are investigation results on transcontinental transfer of radionuclides in flood plain - river systems of Ob - Irtysh’ basin executed by Trapeznikov et al. (1993; 2005a; 2005b).

Global character of Chernobyl accident put forward the problem at which Timofeeff-Resovsky had concentrated his scientific attention for the whole of his life, and that is the problem called “Biosphere and Humanity”. Timofeeff-Ressovsky made a great contribution to world science by formulating and investigating this problem and hence, took a well-deserved place among outstanding Russian naturalistic scientists. Nowadays mankind has to account for the parameters responsible for stability of biosphere under increasing man-caused conditions. Thus, Ecological Doctrine of Russia aimed at support of biosphere stability includes as noted Chereshnev (2002), “a conception about the necessity not only to know all the hot points of the Planet, but to elaborate international transregional collaboration strategy for protection”.

N.V. Timofeev-Ressovsky was creating radiation biogeocenology as the science about regularities of radionuclide behaviour in biosphere. Hence, the problems of radioecology fall under the formulated global problem “Biosphere and Humanity”. And its priority (i.e. biospheric imperative of human behavior) is the main scientific testament left for us by the outstanding scientist.

Acknowledgement

This study was supported by Russian Fund of Fundamental Investigations, Contract No 04-05-64172, by Ministry of Industry Sverdlovsk Region, Contract No P-73, by International Scientific Technical Center, Contract No 2841.

References

Alexakhin, R.M., 2000, At the source of home radioecology (On scientific activity and heritage of Timofeeff-Ressovsky and Klechkovsky), *Biosphere and Mankind*, Obninsk:11–20.

Alexakhin, R.M., and Naryshkin, M.A., 1977, *Migration of Radionuclides in Forest Biogeocenosis*, Nauka, Moscow: 1–141.

Chebotina, M.Ya., Trapeznikov, A.V., Trapeznikova, V.N., and Kulikov, N.V., 1992, *Radioecological Studies of Beloyarsk Reservoir*, Ural Division of Academy Science, USSR, Sverdlovsk: 1–80.

Chereshnev, V.A., 2002, Ecology doctrine of Russia: from plans to pioneer projects, *Izvestiya of State Ural University*. **23:** 6–22.

Kulikov, N.V., and Molchanova, I.V., 1981, *Continental Radioecology*, Plenium Press and Nauka, Moscow: 1–174.

Kulikov, N.V., and Chebotina, M.Ya., 1988, *Radioecology of Freshwater Ecosystems*, Ural Division of Academy Science, USSR, Sverdlovsk: 1–129.

Moiseev, A.A., and Ramzaev, P.V., 1975, *Cesium-137 in Biosphere*, Atomizdat, Moscow: 1–182.

Pavlotskaya, F.I., 1974, *Migration of Radioactive Substances in the Soils as a Result of Global Fallouts*, Atomizdat, Moscow: 1–215.

Rassel, R., ed., 1971, *Raioactivity and Nutrition of a Man*, Atomizdat, Moscow: 1–373.

Timofeeff-Ressovsky, N.V., 1957, Application radiations and emitters in experimental biogeocenology, *Botanichesky Zhurnal*. **42:**161–194.

Timofeeff-Ressovsky, N.V., 1962, *Some Problems of Radiation Biogeocenology*, Thesis for Dr.'s Degree, Sverdlovsk: 1–53.

Timofeeff-Ressovsky, N.V., and Poryadkova, N.A. 1956, On a plant radiostimulation, *Botanichesky Zhurnal*. **41:**1620–1623.

Timofeeff-Ressovsky, N.V., Titlyanova, A.A., Timofeeva, N.A., et al. 1966, Behaviour of radioactive isotopes in the soil-solution system, *Radioactivity of Soils and Methods of Its Determination*, Nauka, Moscow: 46–80.

Timofeeff-Ressovsky, N.V., and Tyuryukanov, A.N., 1966, On elementary biochorological branches of biosphere, *Byull., MOIP*. **72:**123–132.

Timofeeffa-Ressovskaya, E.A., 1963, *Distribution of Radioisotopes between the Principal Components of Freshwater Reservoirs*, Ural Division of Academy Science, USSR, Sverdlovsk: 1–78.

Trapeznikov, A.V., and Trapeznikova, V.N., 1979, Accumulation of ^{60}Co by freshwater plants under natural conditions, *Ecologia*, **2:**104–106.

Trapeznikov, A.V., Pozolotina, V.N., Chebotina, M.Ya., Chukanov V.N., Trapeznikova V.N., Kulikov N.V., Nielsen S.P. and Aarkrog A., 1993, Radioactive contamination of the Techa River, the Urals, *Health Phys*. **65**:.481–488.

Trapeznikov, A.V., Nikolkin, V.N., Trapeznikova, V.N. et al., 2005a, Radioecological study of the Ob' River and its flood plain sediments *Medical and Ecological Aspects of Ionizing Radiation*, Seversk-Tomsk: 207–208.

Trapeznikov, A.V., Trapeznikova, V.N., Korzhavin, A.V. et al., 2005b, Comparative estimate of annuals radionuclide flow to the Ob and the Irtysh rivers, *Medical and Ecological Aspects of Ionizing Radiation*, Seversk-Tomsk: 209–210.

Tyuryukanova, E.B., 1968, *To Investigation Method of ^{90}Sr Behaviour in Soils of Various Geochemical Landscapes*, Atomizdat, Moscow: 1–25.

Tyuryukanova, E.B., 1974, *Radiogeochemistry of the Soils Forests of Russian Valley*, Nauka, Moscow: 1–156.

LIVING MATTER AND BIOSPHEROLOGY

GUENZE V. GUEGAMIAN
1, Place du Petit Pont,
F- 94140 Alfortville, France
E-mail: ggnalfort@aol.com

Abstract. Without question biospherology is primarily a Russian discipline, as its appearance is closely related to Russian scientists such as V.V. Dokuchaev, V.I. Vernandsky, and N.V. Timoffeev-Ressovsky. The role of V.I. Vernadsky, the student of the founder of genetic pedology, V.V. Dokuchayev, in its creation is especially great, for he discovered the phenomenon of living matter which forms the biosphere. With his theory of living matter he was also the first in the history of science to place life in its proper position in the general picture of the universe and discovered fundamental laws which control geochemical acitivities of living matter in the biosphere. In the first half of the previous century V.I. Vernadsky founded a vitally important science for the future of mankind that we have named biospherology.

N.W. Timoffeef-Ressovsky, even as a young scientist, correctly assessed the importance of V. Vernadsky's ideas on biospherology. Timoffeev-Ressovsky was the first scientist to re-establish Vernandski's forgotten discipline of the biosphere. He left us a short but very persuasive program with his report called "The Biosphere and mankind." It is now possible to include this program in any national or international program for "Sustainable Development."

Keywords: Biosphere, Living Matter

1. Preface

As the title of my paper implies I will speak about “living matter” and the science that is called “biospherology”.

In December 1977, I sent an article titled “On V.I. Vernadsky Biospherology” from France to the editorial office of Zhurnal Obshchei Biologii of the Academy of Sciences of the USSR. In this article I was the first to write about the fact that the doctrine of living matter and biosphere created by V.I. Vernadsky in the twenties and thirties had become an independent science about three decades after the author’s death that can be called *biospherology*.

A.A. Cigna and M. Durante (eds.),
Radiation Risk Estimates in Normal and Emergency Situations, 35–46.

After almost three years of written controversy with my opponent, now deceased, corresponding member of the Academy of Sciences of the USSR N.B. Vassoevitch, the article was published in the above mentioned journal in the mid - 1980s.*

Today I deal with the origin of the doctrine about living matter and biospherology as real science and will speak about the importance of V.V. Dokuchaev's, V.I. Vernadsky's and N.V. Timofeeff-Resovskij's ideas in the development of this science.

2. The Doctrine of Living Matter and Origin of Biospherology

At the end of 19th century a man began his scientific research in Russia, generalised the observations of a number of eminent scientists of Europe, and offered the mankind a teaching that can be treated today as a science of the third millennium.

I refer to Vladimir Ivanovich Vernadsky (1867-1945) and his teaching about living matter and the biosphere – about biospherology (Guegamian, 1980, 1981).

It is not common in the history of any science for both the teaching and ideas developed by one scholar to become eventually an independent science. V.I. Vernadsky's teaching about living matter and the biosphere is one of such lucky exceptions.

In 1931, V.I. Vernadsky wrote in his diary: "My concepts rein is yet to come". He was right. In fact it took decades for the global scientific community to understand and appreciate the importance of Vernadsky's work. In my view, the following stages can be outlined during the development of biospherology in Russia:

1) Development of V.V. Dokuchaev's doctrine of the soil as a special natural body, during whose genesis the role of living organisms is dominant;
2) Living matter as planetary matter, discovered by V.I. Vernadsky and,
3) General teaching on the biosphere developed by V.I. Vernadsky.

V.V. Dokuchaev's teaching of soil as a specific natural body was undoubtedly one of those great works that determines the development of scientific thoughts for many generations.

Was there anyone who could be the precursor of another fundamental teaching – the teaching of the biosphere developed by his disciple V.I. Vernadsky?

---

* About this polemic see: A.N. Tjurjukanov, V.M. Feodorov "N.V. Timofeev-Resovskij: Thoughts on Biosphere", M. 1996, p. 263–267.

It was obvious to B.B. Polynov, one of the leading Russian soil scientists and follower of. Dokuchaev's ideas, (Polinov, 1956) that "the great ideas on which Vernadsky's biogeochemistry is based were the direct and powerful development of those by Dokuchaev." One can hardly find arguments against the statement of B.B. Polynov.

"Dokuchaev determined soil as the compound body that was the result of interaction between climate, nature of rock subjacent, organised world, life existed in it and on it, the relief", wrote Vernadsky in 1904 (Vernadsky, 1992).

When he wrote these words he had not realised fully the importance of living matter in the evolution of the earth's crust and in the formation of the earth's soil cover. He subsequently dedicated the rest of his life to this concept. He could not find answers to questions he had at that time. For example, he wrote: "It is not clear, where it is necessary to outline the soil bottom boundary, the difference between soils and friable products of some chemical erosion on the Earth are not clear either " (Vernadsky, 1992).

In search of the answer Vernadsky then turned to mineralogy, his speciality. He believed that "in order to have knowledge about soils it is necessary to study mineralogy carefully and in detail, i.e. to study properties, genesis, minerals changes and changes of its components" (Vernadsky, 1992). At that time however, he understood the differences in the following and believed that "solid products have special physical state in soil... *They are permanently dressed and processed by organised living matter; they are indissolubly mixed with products of dying down behaviour and vital functions"* (Vernadsky, 1992; emphasis added by the author).

The previous phrase can be considered the beginning of Vernadsky's great ideas in biogeochemistry and hence, it must be admitted that Vernadsky had realised the importance of living matter, biogeochemistry and his teaching of the biosphere through the Pedology, with the help of his teacher V.V. Dokuchaev's concepts of soil as a special natural body, the formation of which is determined by climate, the texture of rocks, relief and, of course, by "the organised world, the life existing in it and on it".

Certainly, his famous teacher influenced young Vernadsky but in my opinion, it would be erroneous to say of that Vernadsky had paid special attention to the vital factor in soil formation process because the help or instruction of V.V. Dokuchaev. Most likely, he achieved it by himself.

The interesting point is that V.V. Dokuchaev saw that V.I. Vernadsky was an extraordinary student. Otherwise he would not have directed the following words to his 24-years old student: "Dear Vladimir Ivanovich, I kindly ask you just to outline your point of view concerning saline soil on a sheet of letter-paper for me" (quotation Sitnik K.M., Apanovitch E.M. and Stoyko C.M., 1988).

There is no doubt then that in his very first works Vernadsky had already noticed the role of a biological factor on the soil formation process but it took him a quarter of a century to develop and formulate his concept about the living matter.

3. "Living Matter"

The concept of living mater is doubtlessly the cornerstone of Vernadsky's biospherology. It is consequently strange but in the literature about the scientist's work which is quite extensive, this concept has not been discussed enough, except some works (Aksionov, 1997; Guegamian, 1980, 1981; Kaznacheyev, 1989; Mirzoyan, 1994; Lapo, 1977, 1987; Moiseev, 1994; Nazarov, 1988; Sitnik K.M. and al., 1988).

Nevertheless, it is my strong conviction that the "living matter" concept developed by basis of the genius Vernadsky is fundamental for science and mankind and will continue to be the study of many scientific researchers to come.

V.I. Vernadsky began his methodical work on living matter in 1916 in Ukraine and continued working on it for nearly 30 years, i.e., for the rest of his life. He himself understood the importance of his ideas and in 1918 noted in his diary: "I'm working well on the living matter. Many new ideas in understanding of nature… I'm trying to criticise myself and my research: I acknowledge that the nature has never been considered this way" (Vernadsky, 1997).

Eight years later, in 1924, he wrote from Paris to B.L. Lychkov and mate these clarifications: " I think that my ideas about the living matter make the understanding of nature fresh and significant and logical presentation of these ideas cannot yet be considered a science but a "teaching" in the scientific scheme that has not been completely discussed yet. Anyway, the teaching of the living matter is a specific comprehension form of both life and the environment. The results are significant" (Kirzhaev, 1979).

I would like to remind you that these words belong to a venerable scholar who possessed encyclopaedic knowledge and definitely knew the material.

The questions arise:

1- Why did Vernadsky believe that the teaching of living matter was a ***"specific form of environment comprehension***?" and,

2 - What did he mean by saying that ***"the results are significant***"?

Before answering these questions, it should be mentioned that Vernadsky was the first to consider life and living organisms as planetary and cosmic phenomenon. Furthermore, V.I. Vernadsky left the complete scientific characteristics of the living matter. He defined the functions, essential properties and differences of living matter from the “inert” (Vernadsky’s term) natural bodies of the biosphere.

J.B. Lamark had the theory in the same sense (Guegamian, 1981) but it was only the beginning of such an approach and Vernadsky could definitely and completely define the place of life in Cosmos. He was the first to understand that all the living organisms on the Earth should be considered as an integrated system, a formation of all organisms, which is very important for the existence of the earth's crust. Vernadskiy showed that the life plays completely unique role in space, that, it has its "own" space, named for the planet the Earth the biosphere. According to Vernadsky’s words, it is a cosmic phenomenon in the form of a “natural body” – living matter that can be expressed in numbers by its chemical composition, weight and energy. But I would like to mention, however, that Vernadsky’s notion of living matter underwent some changes. Sometimes it can be difficult to find accurate or specific definitions to understand his statements on this notion. Various explanations of the same words, epithets and phrases have been used since 1920s until the end of his life. This causes various interpretations of his ideas and thoughts.

I think it is, therefore, more advisable not to base these on all Vernadsky’s statements about living matter but to refer to his last book where, according to his own words, he reviewed his scientific activities (Vernadsky, 1965, p. 245). I refer to his monograph ***"The Chemical Structure of the Earth Biosphere and Its Surroundings."***

From memoirs, letters and other documents, it seems that Vernadsky had been working on this book since 1935 till the end of his life; he did not finish it. He often spoke of this work as “the main book” of his life. After Vernadsky’s death his close colleagues and followers edited the book and it was published for the first time in 1965. It is difficult to overestimate the heuristic importance of the scientist’s thoughts expressed in this book. Of course, I cannot now analyse this outstanding work of V.I. Vernadsky but dwell on living matter, which is of interest to us. In my opinion the chapters XIX and XX of the book are the quintessence of all whatever Vernadsky wrote about living matter. There V.I. Vernadsky defined living matter “both as biological envelopes that are indissolubly tied up with its environment of biosphere and as a powerful geologic factor that forms its integral" (ibid, p. 290). Yet, you can ask me: what is so great about, at first sight, the simple fact that there is a special natural body called “living matter” on earth and consequently in the Cosmos? In fact, it was known before Vernadsky’s discovery, was it not?

Of course it was known, but having evaluated life as a significant feature of the Universe, V.I. Vernadsky was the first to discover the fundamental laws that control the geochemical activities of living organisms in the biosphere. He understood and explained the mechanism, and the laws of function of that definitely unique cosmic system. That is what is very special about his teaching about living matter. The laws discovered by him Vernadsky presented them in the form of his own biogeochemical concepts* mentioned above as biosphere postulates of V.I. Vernadsky (Guegamian, 1981).

They stated the following:

1. The biogenic migration of the atoms of chemical components in the biosphere always tends toward the maximum of its manifestation;

2. The evolution of species throughout a geological time make possible to create certain existing types, which are stable in biosphere and tend to increase the atoms migration in biosphere.

Vernadsky presented the basic functional characteristics of living matter with the help of these postulates. It follows from the first biosphere postulate that among other geospheres that are without life, where domains the chemical balance and "tranquillity", life speeds up reactions in the biosphere and with the help of this disturbs the chemical inertness of the planet; that among other biogenic phenomena the biogenic migration of atoms is dominant in the biosphere. This phenomenon cannot be secondary because it "tends toward the maximum of its manifestation".

The tendency of life to occupy the whole surface of the Earth, according to V.I. Vernadsky, is the statement of the first principle. At the same time he pointed out "vsjudnost" (omnipresence) of life, which can be achieved by "pressure", "load" of life. Vernadsky supposed that the evolutionary process throughout the geological time has and keeps intensifying this "vsjudnost" (omnipresence) and load the of life in the biosphere and that the acme of this process is the existence of *Homo sapiens faber* in biosphere.

V.I. Vernadsky distinguished three types of biogenic migration of the atoms:

---

* In 1928 V.I. Vernadsky (1992, pp. 134–145) stated two mentioned "principles" but later (1965, p.286) he added a third one and noted that this statement can be considered the third principle "in case there would be a need". In fact Vernadsky offered more precise formulation of Charles Darwin's theory of the possibility of densely populated existence during the past geological periods of the Earth.

1 - For microscopic unicellular and microbes (For animalcular monadiforms and microbes);
2 - For multicellular organism and
3 - Biogenic migration of atoms of the third type that has become known to a man not long ago and which during our "Psychozoique" era of the history of the biosphere becomes dominant.

He thought that the first biogeochemical principle was a generalisation of the three types of migrations mentioned above and it demonstrated that " all living matter of the planet in the whole…is the source of an active free energy and can execute work" (Vernadsky, 1965, p.267).

As for the second principle V.I. Vernadsky offered the following additional clarifications: "According to the second biogeochemical principle the evolution of living matter species has to follow a certain direction, which increases the biogenic migration of atoms in the biosphere, i.e. must be directed" (ibid, 272). As we can see V.I. Vernadsky not only confirmed his statement in the first postulate concerning the significant importance of the phenomenon of biogenic migration of atoms in the biosphere functioning but also provided an absolutely new and unexpected explanation of the evolution theory.

According to Vernadsky, in the vital forms stable in biosphere (which is the same at Ch. Darwin as "best adapted") are the forms that increase (i.e. intensify) biogenic migration of the atoms in the biosphere. In other words the second postulate implies that in the struggle for existence those species that increase the biogenic migration of the atoms in biosphere can survive, and that the evolution of living matter has a tendency to accelerate biogenic migration and to intensify biogenic cycles of atoms in the biosphere. Thus a logical conclusion: the ability of every species to "turn over" atoms in "organism-environment" system at any speed should be a principle biosphere feature of this species, should be its biosphere constant. Consequently for the record of biogeochemical work of every species and of every living organism in the biosphere we must measure these constants. The results obtained will allow us not only to understand the mechanism of biosphere functioning but also to select the best adapted species in the biosphere sense to create associations with the bioproductivity of the highest possible level.

That's why, in my opinion, V.I. Vernadsky could believe, indeed, that, first, the teaching of living matter was ***"specific form of environment understanding"*** and, second, ***"the results are significant"***.

And now I would like to add some more words concerning "significant results".

4. The Energy of Living Matter

According to V.I. Vernadsky, living matter together with its habitat form the upper Earth mantle – the biosphere, which is ***energetically*** tied to both the Universe and the lower strata of the planet. I emphasise on the word "energetically" because this point has an exclusive place in the scientific work of V.I. Vernadsky. The issue of biosphere energy and the importance of living matter in this energy runs throughout the teaching of living matter and biospherology founded by him. V.I. Vernadsky believed that together with a human being (who is an integral part of living matter) there is no other such geological force on Earth than living matter. At the beginning of 1920s he reached a significant conclusion about the fact that "the study of lifecycle inconvertibility takes us to the sphere of science, which is in the process of formation and concerns the great aim of life and energy. This science is the domain field of the future i.e. the future *energy of our planet"* (Vernadsky, 1994, p.342). What exactly were his ideas about "future energy of our planet"? Let us deviate slightly from the subject of my report:

V.I. Vernadsky was one of the firsts who took over the radioactivity research with a view to use intra-atomic energy. There were few people at that time who believed in the possibility of such an opportunity but Vernadsky delivered his report at the general meeting of the Academy of Sciences of Russia on December 29, 1910 and stated: "The phenomenon of radioactivity makes it possible for us to use energy sources, which are much more powerful by millions of times than those that can be imagined by a man" (Vernadsky, 1997, p.13).

It is fascinating that in 1922 this eminent person already suspected the danger of nuclear energy and warned us about it: "We approach a great revolution in the life of the mankind, to which nothing experienced earlier can be compared. The time when a man will master atomic energy, a powerful source that will allow him to build up his life in accordance with his desire, is not far. It can happen within the coming years or it can happen in a century. But it is absolutely clear that it will have to happen. Will a man manage to make use of this power, to direct it to good purpose but not at self-destruction? Is a human mature enough to use the power that science will inevitably supply him?" (Vernadsky, 1989, pp. 331–332).

As for the energy of living matter, V.I. Vernadsky came to grips with this for the first time in Paris in 1924. First, he attempted to organise a biogeochemical laboratory but failed. He could only obtain a small subsidy from the Rosenthal Fund and thanks to this subsidy continued his activity in Paris. Vernadsky delivered a report on this work titled «Living matter in the Biosphere» in 1925.

We read the following words in this document: “It is impossible to realise how important living matter is for the chemistry of the earth's crust, for the movement of chemical components in biosphere, if the geochemical energy of life is not measured” (Vernadsky, 1994, p.557).

He decided to organise such work and after returning to his native land (1926) he tried, together with A.P. Vinigradoff, classify living organisms according to chemical composition, according to their weight and way of reproduction and finally, according to their biogeochemical energy. He accomplished a little but on the whole, it was a failure because he did not have the needed entire biometrics database. Nevertheless, V.I. Vernadsky was an optimist and wrote in his “book of life”: “*I have no doubt that there is time for everything and biologists will have to take into account a very important factor – biogeochemical energy in order to solve pure biological issues both of theoretical and applied importance*” (Vernadsky, 1965, p.298).

Unfortunately, nobody has yet taken this important factor into account.

5. Biospherology

In my archive there are faded press cuttings from two March issues of 1968 of a small-town newspaper called “Vperjod” (Forward), which was published in Obninsk. N.V. Timofeeff-Resovsky's lecture on " Biosphere and mankind " read by him in February – March in 1967 in the Moscow House of Scientists and in Institute of Medical Radiology in Obninsk was published in this newspaper for the first time.

It is not incorrect to maintain of this modest article, for the first time after Vernadsky’s death, Russian scientists turned to his teaching of biosphere. The article had been published several times while the author was alive and even after his death, nevertheless, nobody had paid attention to it.

For me among all of the published variants of this article, the most interesting one is that rewritten and expanded variant of the author, which was published in a journal called “Priroda” (Timofeev-Resovsky, 1970). The title of the article was “Biosphere and man” (the editors seem to change “mankind” into “man” but in my opinion Timofeev-Resovsky preferred “biosphere and mankind”).

This article bears certain similarity to a well-known article of Vernadsky “Several words about the noosphere”. First, both articles were written for popular consumption (It should be remembered that V.I. Vernadsky sent his papers to be published in “Pravda” newspaper, as well). Second, in spite of the 23 years difference between them, the authors worried about the same thing - the destiny of humankind. And third, both articles are written in a manner of scientific testament.

V.I. Vernadsky wrote: " *in the geological history of biosphere before the man enormous future is opened, if he understands this and does not use his mind and his labour for self-destruction* ".

We read in the work of N.V. Timofeeff-Resovsky: "*humanity will have to improve its rational mutual relationships* [with the biosphere]. *There is a danger for humanity of many difficulties in the near future without such rational mutual relationships*."

In several pages of the most popular natural-science journal of the time of the Academy of Sciences of the USSR, N.V. Timofeeff-Resovsky in his distinct manner posed the point, as he said, of the " "problem № 1" of natural science of the present and of the near- future."

The end of the article was written in a form of a real program – a testament for succeeding generations. I quote this short but very persuasive program since, in my opinion, it can be included in any national or international program of "Sustainable Development". He wrote:

"...I deem it expedient... to enumerate some (in some instances tightly bounded with each other and interacting) issues, which first of all should attract attention of researchers:

a) quantitative and semi-quantitative inventory of fauna and flora in land, aqueous and pedocenoses, definition of a biomass and connected to it biological productivity of various landscapes and regions of the world;

b) possible full studying of material and energy turnover in various biogeocenoses;

c) fundamental studying of cultivated biogeocenoses (agrocenosis);

d) work in the field of experimental biogeocenology in the open environment and on biocenoses samples;

e) impact assessment of economic and industrial activity of a man and development of rational principles of environment usage;

f) development of rationalisation principles of mutual relationships between developing

economy of mankind and development of biological productivity of natural biogeocenoses;

g) experimental and theoretical study of the dynamic equilibrium and its dysfunction in biogeocenises as the main task associated with understanding of biosphere structure and its biogeochemical operation and with melioration changes of its areas;

h) definition of mathematics and computer models for the process analysis running in populations and biogeocenoses, which is primarily connected with the problem solving of dynamic equilibrium in living organisms associations;

i) development of approach to study the problem of evolution in biogeocenose."

N.V. Timofeeff-Resovsky concluded the article with the following words:

"Vital necessity of mankind makes it absolutely inevitable and necessary to include "the Problem N° 1" in the definite program of the natural science"

It is to be regretted that N.V. Timofeeff-Resovsky for some reason did not resume his work on "The Biosphere and Mankind" (he lived for 10 years after the article was published). Unfortunately, the representative of the USSR at the conference on biosphere organised by UNESCO in September 1968 in Paris was not N.V. Timofeeff-Resovsky. To tell the truth, he was involved in the process of the report preparation, which was presented at the conference by the soviet delegation. Just thanks to that fact the report was made on the basis of Vernadsky's teaching but if N.V. Timofeeff-Resovsky whose opinion was more than highly respected appeared in the world arena he could play a key role in further development of biospherology of V.I. Vernadsky in the world.

6. Final Remarks

In March 1920 V.I. Vernadsky wrote in his diary: "I'm clearly getting on to the fact that I have to tell mankind something new in my teaching about living matter and it is my mission, my obligation imposed on me, which I am to put into practice as a prophet who feels a voice within that encourages him to act. I felt a demon of Socrates inside me. Now I understand that this teaching can have the same influence as Darwin's book…" (Vernadsky, 1997, p.32).

V.I. Vernadsky fulfilled the "obligation imposed" on him and developed the teaching about living matter. Today everything is in the hands of mankind, particularly in the hands of those who shape the destiny of the biosphere.

Will be directed all forces of people for studying of living substance and biosphere?

Has man matured enough to be responsible for the future of the biosphere?

Even as long ago as 1922 V.I. Vernadsky demanded: "Will a human be able...?"

45 years later N.V. Timofeeff-Resovsky warned: "…there is danger for mankind of many difficulties in the near- future"

Only 40 years have passed since then but do our contemporaries understand that a global biosphere crisis is just around the corner and that it concerns the future of his children or, his grandchildren?

We can confidently maintain, unfortunately, that man does not understand, and, as Georg Hutchinson it were right: humanity continues to shorten its life by millions of years.

References

Aksionov G.P., 1997, *Voprosi istorii estestvoznania i tekhniki* [Questions of the history of natural science and technology], **3**: 129–135.

Guegamian G.V., 1981, Lamark, Vernadsky i biosferologia [Lamark, Vernadsky and biospherology]. Priroda, **9**: 78–81.

Guegamian G.V., 1980, O biosferologii Vernadskogo [On Vernadsky's Biospherology]. Zhurn. Obsh. Biol., **41** (4):.581–595.

Kaznacheyev V.P., 1989, *Uchenie V.I. Vernadskogo o biosphere and Noosphere* [V.I. Vernadsky's Study on Biosphere and Noosphere]. Novosibirsk: Nauka.

Kirzhaev S.N., Memelov A.V., Neapolitanskaya V.S. and Sorokina M.Yu (Eds.), 1979, *Perepiska V.I. Vernadskogo s B.L. Litchkovim 1918-1939*. [The correspondence of V.I. Vernadsky with B.L. Litchkov 1918-1939]. Moscow: Nauka: 31–32.

Lapo A.V., 1987, *Traces of Bygone Biospheres. Oracle, Arizona*. Synergetic Press, and Moscow: Mir Publishers.

Lapo A.V., 1977, Zhivoe vechestvo biosferi i formirovanie osadochnikh porod i rud. [Living matter of the biosphere and formation of sedimentary rocks and ores"]. *Izv.AN SSSR, geol. series,* **11**: 121–130.

Mirzoyan E.N., 1994, Teoria zhivoi materii [The theory of the living matter]. *Zhurn. Obsh.Biol.*, **55**(1): 13–29.

Moiseev N.N., 1994, Vernadsky i sovremennost v kn. *Zhivoe vechestvo in biosfera* [Vernadsky and the present in "Living matter and biosphere"],. Moscow: Nauka: 634-647.

Nazarov A.G., 1988, Otkritie biosferi [The discovery of the biosphere]. Moscow: Molodaia Gvardia, *Promotei* **15:**172–181.

Polinov B.B., 1956, *Izbrannie trudi* [Selected Works", Moscow: Publ. House of the USSR Academy of Sciences: 1–652.

Sitnik K.M., Apanovitch E.M., Stoyko C.M., 1988, *V.I. Vernadsky. Zhizn i deyatelnost na Ukraine* [V.I. Vernadsky. Life and activity in the Ukraine]. Kiev: Naukova dumka: 1–25.

Timofeeff-Resovsky N.V., 1970, Biosfera i chelovek [Biosphere and the man"], *Priroda*, **8**: 2–9.

Vernadsky V.I., 1997, *Dnevniki 1917-1921* [The Diaries 1917-1921]. Kiev: Naukova Dumka: 1–55.

Vernadsky V.I., 1965, "Khimicheskoe stroenie biosferi Zemli i ee okruzhenia [The Chemical Structure of the Earth's Biosphere and its Surroundings]. Moscow: Nauka.

Vernadsky V.I. 1989. *Nachalo i vechnost' zhizni* [The Beginning and the eternity of the life]. Moscow, Sovetskaya Rossiya.

Vernadsky V.I., 1994, *Zhivoe Veshchestvo i Biosfera* [Living Matter and the Biosphere]. Moskow: Nauka.

Vernadsky V.I., 1992, *Trudy po Biogeokhimii i Geokhimii Pochv* [Contribution Concerning the Biogeochemistry and Geochemistry of Soils]. Moscow: Nauka: 1–335.

Vernadsky V.I., 1997, *Trudy po radiogeologii* [Contributions Concerning the Radiogeology]. Moscow: Nauka.

Vernadsky V.I., 1994, *Trudy po geokhimii* [Contributions Concerning the Geochemistry]. Moscow: Nauka.

Nuclear accidents and weapons

RADIOECOLOGICAL ASSESSMENT OF THE CHNPP ACCIDENT IN THE WESTERN EUROPE AND ADJACENT AREA, WITH SPECIAL REFERENCE TO THE MODERN PROBLEMS OF RADIOECOLOGY IN THE MEDITERRANEAN

ARRIGO A. CIGNA
Past President Union Internationale de Radioécologie
Strada Bottino 2, Frazione Tuffo
I-14023 COCCONATO (Asti) Italy
E-mail: arrigocigna@tiscali.it

Abstract. The impact on Western Europe of the Chernobyl accidents is assessed. In particular an evaluation of the contamination of air, soil, seawater and food with special reference to geographical and orographical situations.

A simple and reliable method to evaluate the levels of ^{134}Cs, ^{137}Cs and ^{90}Sr in green vegetables, hay and milk on the basis of the soil deposition only is described and experimental results are given. The model appears to be rather conservative and to overestimate the contamination.

The situation prior to Chernobyl with respect to regulations of radiation protection against the consequences of a major accident is considered. The development of the recommendations and regulations issued by the Commission of the European Communities for the Maximum Permitted Levels of different groups of radionuclides in foodstuffs is reviewed.

The different reactions to the accident are examined and some data on the average individual effective dose equivalents estimated in a number of countries are also reported. Also the consequences of the countermeasures are discussed. Some main problems concerning the information of the public and the preparedness for possible future accidents are summarised.

Finally the present status of radioecology in the Mediterranean Area is described pointing out the unjustified reduction of the efforts on the monitoring and research with a special reference to the radiation protection problems in emergency situations.

Keywords: Chernobyl, radionuclides, countermeasures, consequences, radio-ecology, radiation protection.

A.A. Cigna and M. Durante (eds.),
Radiation Risk Estimates in Normal and Emergency Situations, 49–67.

1. Introduction

The environmental behaviour of the radionuclides released by and the possible radiological consequences of the Chernobyl accident have been widely studied throughout the world. On the other hand the problems related to the practical management of such an emergency and, in particular, the transmission of clear and comprehensible information from scientists and laboratories to public authorities and from these to the population, as well as the rationale choice and acceptance of the most effective countermeasures remain a weak link in the preparedness against accidents.

Unfortunately the applications of behavioural sciences are a bit fuzzy because people do not react as computers. The identification of an error cannot automatically ensure its elimination because sometimes the application of the appropriate countermeasure is prevented by a number of factors. Nevertheless it seems convenient to take into considerations these problems also, with the hope that in the future the present situation will be further improved by the experience achieved.

2. The Radionuclides Deposition in Europe

The deposition of radionuclides released by the Chernobyl accident occurred as a combination of two factors, i.e. the presence of the radionuclides in the atmosphere and the occurrence of precipitation. The concentration in soil was higher when a high concentration in the atmosphere occurred at the same time of a precipitation.

In Europe, outside the former USSR, the deposition of radionuclides was obviously rather uneven. Therefore it is not possible to report a detailed view of the contamination measured locally. Nevertheless in order to supply a rough evaluation of the distribution of this contamination in Tab. 1 some data referring to ^{137}Cs are reported. In some instances such data are taken from grey literature, i.e. working documents distributed during a meeting organised by the Commission of the European Community (CEC, 1986).

Notwithstanding its intrinsic uncertainty such data, together with the the "Atlas of caesium deposition on Europe after the Chernobyl accident" (De Cort et al., 1998) give an idea of the pattern of the deposition to soil of ^{137}Cs in most of the European countries.

Table 1. Maximum ^{137}Cs concentrations measured in some European countries.

Country	Air (Bq/m^3)	Deposition (kBq/m^2)	Vegetables (Bq/kg, fresh)	References
Sweden	8	200	-	Moberg, 1991
Finland	10	-	-	Sinkko et al, 1987;
	-	67	-	Saxén etal., 1987
Austria	-	60	-	NEA, 2002
Switzerland	-	>37	-	NEA, 2002
Germany	9	19	1200	CEC, 1986
	-	>37	-	NEA, 2002
Italy	3	10	2500	Giorcelli, 1988
Netherlands	3	3.8	320	CEC, 1986
Belgium	1	2	140	CEC, 1986
Denmark	2	1.2	240	CEC, 1986
United Kingdom	-	0.8	112	CEC, 1986
Ireland	-	0.5	160	CEC, 1986
France	1.5	0.07	-	CEC, 1986
Spain	0.1	-	-	CEC, 1986
Portugal	-	0.02	-	NEA, 2002

3. Limits and Suggested Countermeasures

As it is well known, the first individual limits recommended by ICRP for different radionuclides were expressed as Maximum Permissible Concentrations (MPC) in air and drinking water and Maximum Permissible Body Burdens (MPBB). These quantities have been misused to imply maximum values that should never be exceeded under any circumstances.

For these and other logical reasons MPCs and MPBBs were replaced since 1979 by the Annual Limits of Intake (ALI) and the Derived Air Concentrations (DAC). An ALI is defined as the activity of a radionuclide which taken alone would irradiate a person, represented by Reference Man, to the limit set by ICRP for each year of occupational exposure. A DAC equals the corresponding ALI divided by the volume of air inhaled (2,4OO m^3) by Reference Man in a working year (ICRP, 1979).

Notwithstanding the valuable efforts of ICRP to avoid any misuse of its recommendations any limit is commonly considered as a borderline between "safe" and "unsafe". Therefore any value greater than the corresponding limit is assumed to be instantaneously dangerous, disregarding completely the assumption that the limit itself is averaged over 1 year.

ICRP (1978) stated clearly: "The data provided therein should therefore not be used indiscriminately out of context, e.g. to estimate the risk of cancer in individual cases. The various assumptions made should be regarded as such and not as matters of established fact."

These words leave no room for different interpretations but, outside a limited domain of radiation protection experts, people generally ***do use*** the ICRP data out of context and ***do regard*** the various assumptions (e.g. linearity of dose-effect relationship, absence of a threshold, etc.) as results scientifically sounded and proved.

At the time of the Chernobyl accident both ICRP (1978) recommendations and CEC (1982) regulations were widely used as reference sources for national regulations in many countries. In Table 2 the annual limits of intake for some radionuclides, according to the above-mentioned bodies, are reported.

Table 2. Annual limits of intake for any individual member of the public for some radionuclides in soluble form (ICRP, 1977; 1978).

RADIONUCLIDE	Bq/year	Bq/day over 1 year
^{90}Sr	$1 \cdot 10^5$	300
^{131}I	$1 \cdot 10^5$	300
^{134}Cs	$3 \cdot 10^5$	1000
^{137}Cs	$4 \cdot 10^5$	1000
^{239}Pu	$2 \cdot 10^4$	50

It must be pointed out that the dose-equivalent limits recommended by ICRP (1977, 1984) for members of the public apply to operations involving foreseen radiation exposures. These annual limits are set at a level which, regardless of age or sex, is thought to correspond to a low degree of risk; thus, unless a limit were to be exceeded by a considerable amount, the risk would be still sufficiently low as not to warrant such countermeasures as would themselves involve significant risk or undue cost.

It is therefore clear that in an accident situation it is not obligatory to initiate remedial action if the annual dose-equivalent limit for members of the public has been or might be exceeded. Obviously the same remarks apply to the CEC limits which are directly derived from the ICRP recommendations.

It must be emphasised that ICRP has made a number of recommendations for normal situations when the source of exposure is under control as well as for accidents when the source cannot be controlled and any subsequent exposure may be limited only by some intervention concerning the members of the public. This distinction is crucial and a substantial part of the confusion after the accident originated because the recommendations by ICRP were not properly understood.

Under normal operational conditions the predetermined individual limits for workers are:

50 mSv/year for any year
10 mSv/year for lifelong exposure

The application of these limits to individual members is likely to result in an average dose equivalent of less than, respectively, 5 and 1 mSv/year because only a small number of individuals is exposed to values close to the limit.

For the members of the public the figures are reduced by a factor of ten and, therefore, are as follows:

5 mSv/year for any year (average: 0.5 mSv/year)
1 mSv/year for lifelong exposure (average: 0.1 mSv/year)

After a *severe accident* measures must be taken to minimise the consequences of the accident taking into account the optimisation principle. Certain interventions may have serious health or socio-economic consequences to the individual or society; others have only little impact although there is no intervention, which is not associated with some disadvantages.

4. The Reactions to the Chernobyl Accident

The Chernobyl accident had a completely different impact according to the kind of the people involved. Of course there were many differences in each country, nevertheless it is worthwhile to examine the details of such reactions according a simplified scheme (Cigna, 1989).

4.1 AUTHORITIES

In most countries there were plans to cope with emergency situations. But in the different countries reached by the Chernobyl plume the authorities, both local and national, often preferred to accept opinions and suggestions from their entourage rather than from competent people. In the majority of cases politics overruled science.

For this reason the reactions were, rather frequently, over-protective against the nuclear risk with a consequent high cost quite unjustified within any frame based on the optimisation principle. In many cases to be an "expert" was considered as a firm qualification to be excluded from the group of reliable persons.

Only in France it was clearly stated that the contamination due to the Chernobyl accident had no significant health risks. But this wise approach was often considered to be an untrue statement when compared with the countermeasures taken outside the French borders.

4.2 SCIENTISTS

It was not difficult for the few persons who had some experience from the nuclear weapons fallout era and were still active to start immediately a wide series of measurements in order to obtain the greatest amount of information on the behaviour of radionuclides. To these people it appeared clear from the very beginning that, outside the local area, the accident had a large scientific interest without health implications.

Unfortunately many persons, who can be classified as scientist on account of their job, but without significant competence in the problems involved, spread a large amount of "background noise" which covered the correct information.

4.3 GENERAL PUBLIC

The contrasts among the decision-makers, the lack of clear information and a number of wrong data released by many uncontrolled sources caused a high level of confusion and much fear in the general public. In fact the general public was unable to screen the different sources of information: each of them appeared equally reliable.

For many individuals in the general population such an uncomfortable situation lasted for a rather long time after the contamination released by the Chernobyl accident had decreased to the background values in many compartments.

4.4 THE COMMISSION OF THE EUROPEAN COMMUNITIES

After the Chernobyl accident the Commission of the European Communities had to establish a regulation to lay down the maximum permissible levels of radioactive contamination of foodstuffs and feeding stuffs which may be placed on the market.

Initially a simple "Recommendation of the Commission" (86/156/CEE of the 6 May 1986) was published in the Official Journal of 7 May 1986 (Table 3). These values were quite unrealistic firstly because a limit not related to specified radionuclides in a mixture of radionuclides with a wide range of half lives and energies was meaningless. Secondly, because too large a number of samples already exceeded the limits for the global amount of gamma emitters and therefore it was a nonsense to forbid the consumption of foodstuffs in a very large area without sound health reasons. In addition they were not consistent because the values for the dairy produce (which should be more conservative because this food is considered to be the only food for babies) were higher than those for fruits and vegetables.

Table 3. Maximum permitted levels of radioactivity in foodstuffs, which may be placed on the market (Bq/kg), recommended by the Commission (Recommen-dation 86/156/EEC of 6 May 1986).

From:	Maximum activity (Bq/kg)	
	Dairy produce	Fruits and vegetables
6 May 1986	500	350
16 May 1986	250	175
26 May 1986	125	90

Only on 30 May 1986 (Council Regulation 1707/86 (EEC) this mistake was amended and the levels for Caesium radioisotopes were established at 370 Bq/kg for dairy produce and 600 Bq/kg for other foodstuffs.

Successively two Council Regulations, as summarised in Table 4, were issued with the aim of providing levels to be adopted *"following a nuclear accident or any other case of radiological emergency"*. Unfortunately the risk perception by the public did not allow explaining the change of such limits. Therefore the issue of other Council Regulations solved the problem by confirming the previous values as *"conditions governing imports of agricultural products originating in third countries following the accident at the Chernobyl nuclear power station"*.

Table 4. Maximum permitted levels of Caesium radioisotopes (Bq/kg), in foodstuffs, which may be placed on the market according Council Regulation.

Date	Council Regulation	Application	Baby Foods	Dairy Produce	Othe Foodstuffs
22 Dec. 1987	EURATOM 3954/87	Nucl. accid.	-	1000	1250
18 Jul. 1989	EURATOM 2218/89	Nucl. accid.	100	400	1250
22 Mar. 1990	EEC 737/90	Chernobyl	370	370	600
20 Mar. 2000	EC 616/2000	Chernobyl	370	370	600

It can be recalled here that the best solution, simpler and safer, would have been the implementation of the annual limit of intakes (as indicated in Table 2) immediately at the time of the accident. With some simple and realistic assumptions on the daily food intake and the persistence of the contamination some values for maximum permissible concentrations in foodstuffs would have been immediately available.

In any case the limits were assumed to be a borderline between "safe" and "unsafe" and not at all as a legal value for commercial purposes. Such a problem appears to be very difficult to solve because it cannot be dealt with on a purely logical basis. The risk perception by people does not correspond to facts and an improvement of this risk perception cannot be foreseen in the next future.

What is more unfortunate in this situation is the additional burden to be supported by the public. In fact any action which is over-protective (i.e. which is not optimised by having a negative risk-benefit balance) implies, by definition, a cost which is not covered by any advantage. A dose reduction below values that are already negligible corresponds to a waste of money because no positive gain can be ensured.

Considerable uncertainty and confusion were the main features of the post Chernobyl scenario because: the recommendations on dose intervention levels had not been translated into derived levels of concentration of radionuclides (which is the quantity directly measured in foodstuffs), there were problems in making the public and the authorities understand the rationales for countermeasures, both the "old" units (rad, curie) to the "new" ones (sievert, bequerel) were currently used. Therefore a variety of recommendations resulted not supported by any scientific evidence.

4. The Consequences of the Countermeasures

A distinction between normal and accident situations is essential for the optimisation of countermeasures. In normal situations the restrictions imposed on the doses are applied by restrictions at the source of the radiation exposure. In major accident situations (which are rare, because most of the accidents in nuclear plants will deliver doses to the public well below the dose limits) the source is, by definition, out of control and, therefore, any restrictions have to be applied to the persons involved.

The two categories of situations need different handling at the regulatory level (Dunster, 1987) because in normal situations the restrictions are established in a frame of preventive measures. The application of the corresponding limits can be made at a rather low cost and therefore a certain margin of safety can be maintained.

On the contrary, in accident situations the cost of the countermeasures is generally very high and the balance between such a cost and that attributable to the consequences of the accident should be much more accurate. A margin of safety on the side of the dose reductions might imply a reduction of the global margin of safety for the population on account of an undue health detriment caused by the countermeasures.

Under normal operational conditions the dose limits recommended by ICRP (1991) for individuals of the public is 1 mSv in a year, but in special circumstances a higher value of effective dose could be allowed in a single year, provided that the average over five years does not exceed 1 mSv per year.

In serious accidents some relaxation of the controls for normal situations can be permitted without lowering the long-term level of protection. This is a consequence of two key concept:

1) Deterministic effect are not likely to occur at absorbed dose less than 0.5 Gy.
2) The benefit corresponding to the dose saved by implementing a protective action (averted dose) must be greater than the risk as a consequence of the protective action.

Table 5. ICRP (1993) recommended intervention levels.

Type of intervention	**Intervention level of averted dose (mSv)**	
	Almost always justified	**Range of optimised values**
Sheltering	50	Not more than a factor of 10 lower than the justified value
Administration of stable iodine: equiv. dose to thyroid	500	
Evacuation (<1week):		
- whole body dose	500	
- equivalent dose to skin	5000	
Relocation	1000	For prolonged exposure: 5-15 mSv per month
Restriction to a single foodstuff	10 (in 1 year)	Beta/gamma emitters: 1000-10,000 Bq/kg Alpha emitters: 10-100 Bq/kg

The intervention levels recommended by ICRP (1993) are summarised in Table 5. After a severe accident measures must be taken to minimise the consequences of the accident taking into account the optimisation principle. As it was reported by Ilari & Chamney (1987) a significant effort to develop an international consensus on criteria for the protection of the public in nuclear emergencies had been made during recent years by several international organisations active in this field (ICRP, IAEA, WHO and CEC).

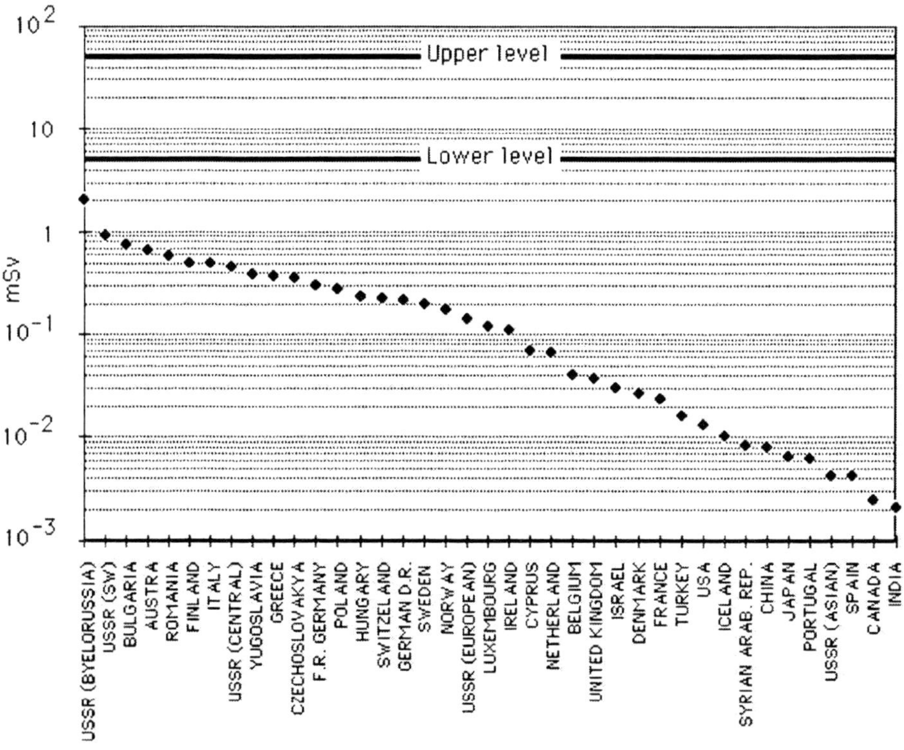

Figure 1. Average individual effective dose equivalents estimates for the first year in different countries; for some of them the ratio of maximum to average values is around one order of magnitude (From data published by: USDOE, 1987; NEA, 1987; Schlesinger *et al.* 1988; UNSCEAR, 1988). Primary intervention levels are also reported: a "Lower level" (below which introduction of countermeasures is not warranted) and an "Upper level" (above which implementation of countermeasures should almost certainly be attempted).

The agreement among the international organisations is virtually unanimous: a few differences concern very minor details only. According to such recommendations a dose range is defined between two levels: a lower level below which introduction of a countermeasure is not warranted and an upper level above which its implementation should almost certainly be attempted.

In Fig. 1 the estimations of the average individual effective dose equivalents delivered by the Chernobyl accident for the first year in different countries are reported. Such estimates have been obtained by calculations carried out by individual countries (USDOE, 1987; NEA, 1987; Schlesinger *et al.* 1988) or (when such calculations are not available) by data reported by UNSCEAR (1988). On account of the size of the USSR and the large difference of the doses over its territory, this country has been divided into five main regions which may be handled more uniformly.

It appears quite clearly that the countermeasures, which have been endorsed in many countries, were not justified because their cost was higher than the advantage obtained by the dose reduction. Only in the vicinity of the plant were the limits substantially exceeded and only there did the countermeasures provide a net benefit for the population.

An extensive account of the response to the Chernobyl accident in OECD member countries was given by NEA (1987). The actions taken in these countries were very different and being completely unrelated to radiological protection criteria such actions were often radically different in two neighbouring countries. This fact was another cause of confusion because the public understood very well that the actions had no logical basis because what was free on one side of an administrative border had to be prohibited on the other side.

To facilitate comparison of alternative control or intervention options with the purpose of optimising protection through cost-benefit analysis techniques, the value of the detriment has to be presented in the same units as the protection efforts which are commonly measured in terms of a cost.

5. The Risk Perception

The risk perception by people does not correspond to facts and an improvement of this risk perception cannot be foreseen in the next future. What is more unfortunate in this situation is the additional burden to be supported by the public.

In fact any action which is over-protective (i.e. which is not optimised by having a negative risk-benefit balance) implies, by definition, a cost which is not covered by any advantage. A dose reduction below values, which are already negligible, corresponds to a waste of money because no positive gain can be ensured.

It must be reminded here that the consequences of excessive countermeasures may result in severe harm also and therefore moral and ethical problems are involved well beyond any economical implication.

It is interesting to report here, as an example, the results of a model to forecast the average levels of contamination of ^{137}Cs e ^{90}Sr in some items of the human food chain, by means of an empirical method established by Van der Stricht et al. (1970, 1971). By studying the radioactive fallout during the sixties at the Joint Research Centre of Ispra (Italy) they found that the concentrations in milk and pastures could be described in terms of the activity deposited onto growing herbage and that cumulated during the previous years. Immediately after the arrival of the contamination released by the Chernobyl accident, such a model was applied to an area in North Italy because the most catastrophic forecast were being spread around ant it was convenient to supply a more realistic view. In Fig. 2 and 3 the prediction of the model for green vegetables and milk is reported.

The results of the measurements obtained successively on pooled samples of green vegetables and milk were at least one order of magnitude lower, notwithstanding the model was considered too optimistic at the moment.

Probably the most widespread health effect in the public is the psychological stress and anxiety which may cause physical symptoms and affect health in a variety of ways, though such symptoms are unrelated to radiation exposure. This fact is a consequence of the perception of radioactivity by the public as great danger without any possibility to protect themselves, with deadly consequences at any level. Obviously it is rather difficult to educate the public by putting radioactivity in right perspective, but there are no other chances.

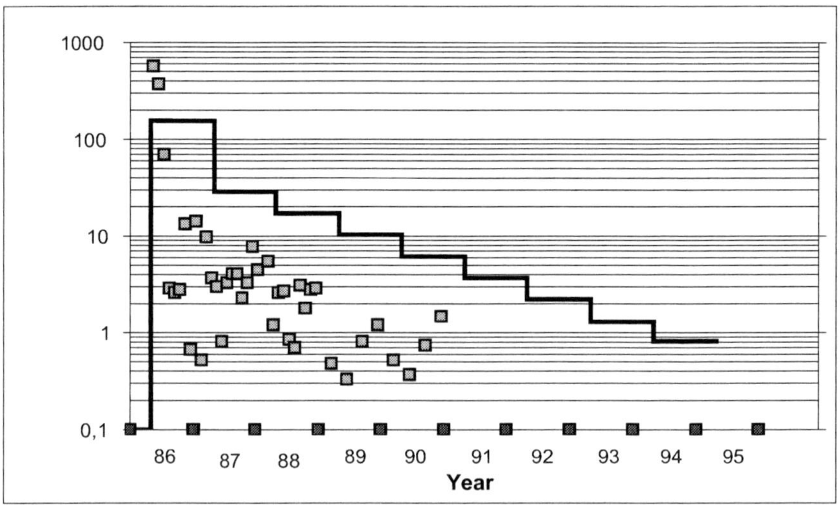

Figure 2. Concentration of ^{137}Cs in green vegetables (line) evaluated after the model by Van der Stricht et al. (1970, 1971). The values measured successively are reported as squares.

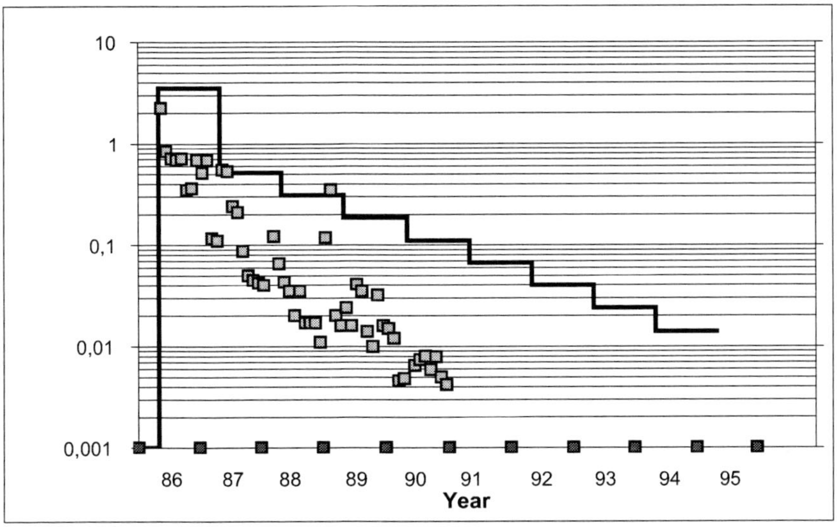

Figure 3. Concentration of ^{137}Cs in milk (line) evaluated after the model by Van der Stricht et al. (1970, 1971). The values measured successively are reported as squares.

The social and psychological effects in other countries were minimal compared with those within the former Soviet Union as it was reported by NEA (2002). Nevertheless, to investigate the occurrence of still more serious health implications, a survey of the induced abortions in some European countries was carried out on the basis of the available literature. A number of confounding factors affects the figures of induced abortions because they vary both during the years and within the month of a given year; in addition in some countries the statistical data concerning abortions differ according to the source and therefore are not reliable.

For this reason in some cases it appeared more useful to consider the data concerning the births, which, on the contrary, are quite reliable, to estimate the increase of induced abortions. In Table 6 the results of the survey are summarised.

In Hungary the percentage of induced abortions was not significantly higher in 1986, nevertheless a decrease of about 8000 births in the first months of 1987 could be correlated to an equivalent increase of induced abortions in September and October 1986.

Table 6. Induced abortions and births in some European countries after the Chernobyl accident. (Cigna, 1997).

Country	Increase in abortions			References
	Measured	Estimated	Decrease in births	
Denmark	230	ND	ND	Knudsen, 1991
Finland	0	ND	350	Harjulehto *et al.*, 1991
Greece	ND	2500	ND	Trichopoulos *et al*, 1991
Hungary	0	?	about 8000*	Czeizel& Billege, 1988 Czeizel, 1991
Italy	ND ND 2600-8000*	about 2000 ND ND	about 3500* 5800* ND	Cigna, 1989 Bertollini *et al.*, 1990 Spinelli & Osborne, 1991
Norway	0*		about 100	Irgens et al., 1991, Egil Skjeldestad *et al.*, 1992
Sweden	500-600*	ND	ND	Källén, 1988

* See text

In Italy the decrease of about 3500 births in the first months of 1987 (Cigna 1989) can be attributed to fewer planned conceptions as well as to an increase of induced abortions in the previous year in agreement with the results of the evaluations carried out by Bertollini et al. (1990) and Spinelli & Osborne (1991). Bertollini et al applied a linear regression model to the monthly birth data from 1977 to 1986 to calculate the expected births during the first months of 1987, while Spinelli & Osborne used four regression models fitted to data of the monthly induced abortions between January 1984 and April 1986 to predict the expected induced abortions in the five months following the Chernobyl accident.

In Norway there were an increase of 192 spontaneous abortions in the year following May 1st, 1986 with respect to the preceding year and a decrease of planned conceptions during the three months after the accident; both occurrences could be attributed to the stress caused by the situation.

In Sweden, according to Odlind & Ericson (1991), the increase reported for 1986 by Källén (1988) is not a consequence of the Chernobyl accident because the increase started before and continued far beyond the time of the accident. There was also an increase in the number of births during the years after the accident. Therefore it seems unlikely that the fear of the consequence of radioactive fallout resulted in any substantial increase in the number of legal abortion in Sweden.

In conclusion, if in some countries induced abortions are not attributable to the concern expressed about the possible damage to the foetuses, in other countries the politicians, ranging from local authorities up to government ministers, had the full responsibility of the confusion of the public (which in some cases was closer to panic rather than to a simple confusion).

When the detriment caused by undue countermeasures taken either by ignorance (because they did not rely on the advice of qualified experts) or by demagogy (because they took decisions turning to account for themselves and not for a net benefit for the public) are beyond any economical implication and results in severe health effects (as the induced abortions reported before) the authorities must respond personally for their decision. Otherwise they will continue to be considered as responsible of "normal" errors which are unavoidable because are strictly connected to any human activity.

On the contrary, an interruption of otherwise wanted pregnancies is a capital crime; therefore, moral and ethical reasons oblige the experts in radiation protection to set up any possible opposition to an obviously wrong decision implying such severe health effects.

6. Present Status of Radioecology in the Mediterranean Area

A steady trend to decrease any radioecological research and also a normal activity of monitoring the environmental radioactivity started in the last decades of last century. At present the laboratories involved in this field could be considered as an "endangered species" and their life is severely threatened.

From this point of view, the case of Italy is emblematic. After the Chernobyl accident the Italian government decided to stop the development of nuclear energy in the country and to shut down the existing nuclear power plants. Also the researches concerning radiation protection declined slowly but steadily and in a few years the large amount knowledge and competence accumulated along tens of years faded away.

Most of the monitoring of the environmental radioactivity disappeared and the results of the few surviving laboratories are no longer published after 1998. This fact was accompanied by the diffusion of unreliable data released by new laboratories without any sound scientific basis.

Since the greatest majority of people with competence in this field is no longer available (they retired or died) it becomes every day harder to counteract the appearance of wrong or biased statements. If at the time of the Chernobyl accident a number of countermeasures set up by the national authorities resulted in a further damage, greater than the direct consequences of the accident. At present also a very minor nuclear accident would imply something close to a disaster for the absence of any scientific support.

Unfortunately the situation reported above extends from single countries also to the European Union. In fact the 7th Framework programme proposed recently by the Commission before the Council and European Parliament for examination has only a small fraction of 6% for EURATOM and radiation protection is not considered at all. The situation is still worst because it results that presently the technical competence of experts who have the task to evaluate the proposal is sometimes poor and the bureaucratic procedure is too long and complicate. The "Management and Co-ordination Advisory Committee" which operated many years ago is a clear recollection of something no longer existing.

Nowadays only the IAEA is still active in the field of research in radiation protection and radioecology, particularly through its excellent Marine Environment Laboratory in Monaco. But notwithstanding the high standard of its scientists they cannot answer any need of the Mediterranean area.

7. Conclusions

The Chernobyl accident provided a unique chance of applying and improving the models set up during the early '60s when the radioactive fallout released by the experimental explosions of nuclear weapons distributed all over the world. Our knowledge of the behaviour of radionuclides in the environment has been extended considerably in recent years.

The Chernobyl accident focused the attention on a problem that seems to exist with any model applied. It appeared that the values frequently used in these models were very conservative, i.e. they were to guarantee a safe assumption, much too pessimistic. Of course it is convenient to use such conservative values in assessment models. However for planning future radioecological countermeasures after an accident, data as close as possible to reality must be used, otherwise the countermeasures applied may imply consequences worse than those produced by the accident itself.

This criterion was clearly stated by ICRP (1978; 1984) but it was frequently neglected nearly all over the world after the Chernobyl accident. Now it is a rather hard task to convince public opinion that the limits adopted by political and administrative authorities were definitely too low and that the much higher limits, proposed by the experts on countermeasures, were chosen in the interest of the population and not for commercial or other interests.

It is worthwhile to realise that radioecological models far better developed than models for other pollutants. Now is the time to yield the experience of many years and to investigate in greater depth the mechanisms, which control the behaviour of other pollutants. The extent to which the chemical industry influences environmental quality is well recognised, whereas radionuclides have, in general, a negligible effect, the associated exposures being less than the fluctuations in natural background.

Research methods using radioactive tracers must not be regarded as a threat to mankind but as a tool, which can play a major role in improving the quality of life.

References

Bertollini R., Di Lallo D., Mastroiacovo P. and Perucci C.A., 1990, Reduction of birhts in Italy after the Chernobyl accident, *Scand J Work Environ Health* **16**: 96–101.

CEC, 1982, Radiological Protection Criteria for Controlling Doses to the Public in the Event of Accidental Releases of Radioactive Material, *A Guide on Emergency Reference Levels of Doses from the Group of Experts Convened Under Article 31 of the Euratom Treaty*. Luxembourg, V/5290/82.

CEC, 1986, *Unedited working papers prepared by participants to a meeting held in Bruxelles*, 25/26 June 1986.

Cigna A.A., 1989, Lesson learned and evaluation of the impact from the Chernobyl accident, *Proc. Standing Conf. Health and Safety in the Nuclear Age, 2nd meeting, Brussels 5-6 Dec. 1989*, Report EUR 12682 EN: 163–177.

Cigna A.A., 1997, Interactions with human nutrition and other indices of population health.*Health impacts of large releases of radionuclides. Wiley, Chichester (Ciba Foundation Symposium 203):* 141–154.

Czeizel E. and Billege B., 1988, A csernobili atomerömü-baleset teratológiai értékelése Magyarországon. *Orv Hetil* **129**: 457–462.

Czeizel A.E., 1991, Incidence of legal abortions and congenital abnormalities in Hungary, *Biomed & Pharmacother* **45**: 249–254.

De Cort M., Dubois G., Fridman Sh.D., Germenchuck M.G., Izrael Yu.A., Fanssens A.; Jones A.A., Kelly G.N., Kvasnikova E.V., Matveenko I.I., Nazarov I.M., ôkumeiko Yu.M., Sitak V.A., Stukin E.D., Tabachny L.Ya., Tsaturiv Yu.S. and Avdyushin S.S., 1998, *Atlas of Caesium deposition on Europe after the Chernobyl accident*. European Communities, Report EUR16733.

Dunster H.J., 1987, The Place of Optimisation in the Setting of Action Levels for Intervention Following an Accident, *Proc. Int. Scient. Seminar on Foodstuffs*

Intervention Levels Following a Nuclear Accident, CEC, Luxembourg 27-30 April 1987, EUR 11232: 25–32.

Egil Skjeldestad F., Storm Munch J. and Magne Madland T., 1992, Tsjenobyl-ulykken påvirket den svangerskapsutfall i Norge ? *Tidsskr Nor Lægeforen* **112**: 1278–1281.

Giorcelli F.G., 1988, *Misure effettuate nell'anno 1986 dai laboratori dell'ENEA su campioni ambientali e della catena alimentare in seguito all'incidente di Chernobyl.* NEA, Roma.

Harjulehto T., Rahola T., Suomela M., Arvela H. and Saxén L., 1991, Pregnancy outcome in Finland after the Chernobyl accident, *Biomed & Pharmacother* **45**: 263–266.

Källén B., 1988, *Pregnancy outcome in Sweden after Chernobyl-A study with Central Health Registries*, The Swedish National Board of Health, Stockholm.

Knudsen L.B., 1991, Legally-induced abortions in Denmark after Chernbyl. *Biomed & Pharmacother* **45**: 229–231.

ICRP, 1977, *Recommendations of the International Commission on Radiological Protection. Publication 26*, Pergamon Press, Oxford and New York.

ICRP, 1978, *Limits for Intakes of Radionuclides by Workers. Publication 30*, Pergamon Press, Oxford.

ICRP, 1979, Limits for Intakes of Radionuclides by Workers. Publication 30, *Annals of the ICRP*, **2**(3/4), Pergamon Press, Oxford.

ICRP, 1984, Protection of the Public in the Event of Major Radiation Accidents: Principles for Planning. Publication 40, *Annals of the ICRP,* **14**(2) Pergamon Press, Oxford.

ICRP, 1993, Principles for Intervention for Protection of the Public in a Radiological Emergency ICRP Publication 63 *Annals of ICRP* **22**(4): 1–30.

Ilari O. and Chamney L., 1987, Review of existing international recommendations on intervention levels of dose. *Proc. Int. Scient. Seminar on Foodstuffs Intervention Levels Following a Nuclear Accident,* CEC, Luxembourg 27-30 April 1987, EUR 11232: 1/16.

Irgens L.M., Lie R.T., Ulstein M., Skeie Jensen T., Skjærven R., Sivertsen F., Reitan J.B., Strand F. and Egil Skjeldestad F., 1991, Pregnancy outcome in Norway after Chernbyl. *Biomed & Pharmacother* **45**: 233–241.

Moberg L. (Ed.), 1991, *The Chernobyl fallout in Sweden.* SSI, Stockholm.

NEA, 1987, *The radiological impact of the Chernobyl accident in OECD countries* . OECD, Paris.

NEA, 2002, *Chernobyl: assessment of radiological and health impacts. 2002 uodate of Chernobyl : Ten years on.* NEA/OECD, Paris.

Odlind V. and Ericson A., 1991, Incidence of legal abortion in Sweden after the Chernobyl accident. *Biomed & Pharmacother* **45**: 225–228.

Saxén R., Taipale T.K. and Aaltonen H., 1987, Radioactivity of wet and dry deposition and soil in Finland after the Chernobyl accident in 1986. *Säteilyturvakeskus*, Helsinki, Report STUK-A57.

Schlesinger T., Biran T., Even O., Dukhan R., Shamai Y., Tal A., Friedman K., Koch J. and Israeli M., 1988, Environmental contamination related to the Chernobyl accident as measured in Israel during May-June 1986. IRPA Abstracts, *Health Physics,* **54**(3):361/362.

Sinkko K., Aaltonen H., Mustonen R., Taipale T.K. and Juutilainen J., 1987, Airborne radioacivity in Finland after the Chernobyl accident in 1986. *Säteilyturvakeskus*, Helsinki, Report STUK-A56.

Spinelli A. and Osborne J.F., 1991, The effects of the Chernobyl explosion on induced abortion in Italy. *Biomed & Pharmacother* **45**: 243–247.

Trichopoulos D., Zavitsanos X., Koutis C., Drogari P., Proukakis C., and Petridou E., The victims of Chernobyl in Greece: induced abortions after the accident. *Br Md J* 295: 1100.

UNSCEAR, 1988, *Sources, effects and risks of ionizing radiation.* New York.
USDOE, 1987, *Health and Environmental Consequences of the Chernobyl Nuclear Power Plant Accident*. Report DOE/ER-0332.
Van der Stricht E., Gaglione P. and de Bortoli M., 1970 – *Valutazione della contaminazione dell'erba e del latte dovuta al ^{137}Cs presente nelle ricadute radioattive.* Giorn. Fis. San. e Protez. Contro le Radiaz., Torino, **14**(2):53–60.
Van der Stricht E., Gaglione P. and de Bortoli M., 1971 – *Predictions dell'erba e del latte of ^{90}Sr levels in milk on the basis of the deposition values.* Health Phys., **21**(2):217–226.

DOSE RATES AND EFFECTS OF CHRONIC ENVIRONMENTAL RADIATION ON HYDROBIONTS WITHIN THE CHERNOBYL EXCLUSION ZONE

DMITRI GUDKOV*, NATALIA SHEVTSOVA, OLENA DZYUBENKO, MIKHAIL KUZMENKO
Institute of Hydrobiology,
12 Geroyev Stalingrada Avenue, 04210 Kiev, Ukraine
E-mail: digudkov@svitonline.com

ALEXANDR NAZAROV
Chernobyl Radioecological Centre, 6 Shkol'naya Street, 07270 Chernobyl, Ukraine

* To whom correspondence should be addressed

Abstract. The rate of chromosome aberrations in cells of freshwater snail (*Lymnaea stagnalis* L.) embryos and in the apical meristem of roots of common reed (*Phragmites australis* (Cav.) Trin. ex. Steud.) and arrowhead (*Sagittaria saggitifolia* L.) from reservoirs within the Chernobyl exclusion zone have been analyzed. The absorbed dose rate for hydrobionts living within the littoral zone of water bodies, due to external irradiation and radionuclides incorporated in tissue was in the range from 0.0018 to 3.4 Gy /year. The highest value was measured for hydrobionts from lakes within the dammed territory on the left-bank flood plain of the Pripyat River. The lowest doses were measured for specimens from the water courses. The hydrobionts from the heavily contaminated lakes are characterized by the maximal rate of chromosome aberration – about 20–25%. The chromosome aberration rate of hydrobionts from conditionally clear was on average 1.5%, and the maximal rate does not exceed 2.5%.

Keywords: Chernobyl exclusion zone; chromosome aberration rate; dose rate; chronic environmental radiation; hydrobionts.

A.A. Cigna and M. Durante (eds.),
Radiation Risk Estimates in Normal and Emergency Situations, 69–76.

1. Introduction

The radioactive contamination of the environment as a result of the Chernobyl accident produced a substantial increase in the radiation background, accumulation of radioactive substances by living organisms and, accordingly, an excess radiation dose to biota. Only a few relevant studies deal with radiation impact on hydrobionts from water bodies within the Chernobyl exclusion zone. The lack of accurate dosimetry measurements creates difficulties in the interpretation of observed (or missing) effects, and their correlation with dose rate. Our research aimed to the assessment of radiation dose resulting from external irradiation and incorporated radionuclides in tissue of hydrobionts of different taxonomy. In addition, we measured the frequency of chromosome aberration in cells of freshwater snail's embryos and in apical meristem of roots of higher aquatic plants.

2. Materials and Methods

Our researches were carried out during 2000–2004 in water bodies within the 30-km exclusion zone of the Chernobyl nuclear power plant (ChNPP). External gamma-background measurement and sampling of hydrobionts for radionuclide content and cytogenetic analysis were made in the littoral zone of Azbuchin Lake, Yanovsky (Pripyatsky) Crowl, the cooling pond of the ChNPP, the lakes of the left-bank flood plain of the Pripyat River – Glubokoye Lake and Dalekoye 1 Lake, and in the Uzh (Cherevach village) and Pripyat (Chernobyl town) river (Figure 1).

External gamma irradiation dose rate was measured by DKS 01 dosimeter and by NaI field radiometer SRP-68-03. The ^{137}Cs content was measured by gamma spectrometry complex: PGT IGC-25 detector (France), “Nokia LP 4900 B” analyzer ("Nokia", Finland), low-volt feeding source – crate NIM BIN, amplifier NU-8210 ("Elektronicus Merokeszulekek Gyara", Hungary) and 100 mm thickness leaden protection. The ^{90}Sr content was measured on low-background NRR–610 β-radiometer (“Tesla”, Czech). Minimal detectable activity was 0.04 Bq under 1000 s sample exposition. ^{238}Pu and $^{239+240}$Pu content in electrolytic samples was determined by α-spectrometric tract by NUC-8192 impulse analyzer ("Elektronicus Merokeszulekek Gyara", Hungary). The ^{241}Am content was measured by x-ray-spectrometric line including x-ray detector EG&G Ortec LOAX-51370/20 CFG-SU-GMX ("EG&G Ortec", USA) and analyzer “Nokia LP 4900 B”. The estimation of the absorbed dose rate for hydrobionts was carried out according to the methods described elsewhere (Amiro, 1997; Brown *et al.*, 2003).

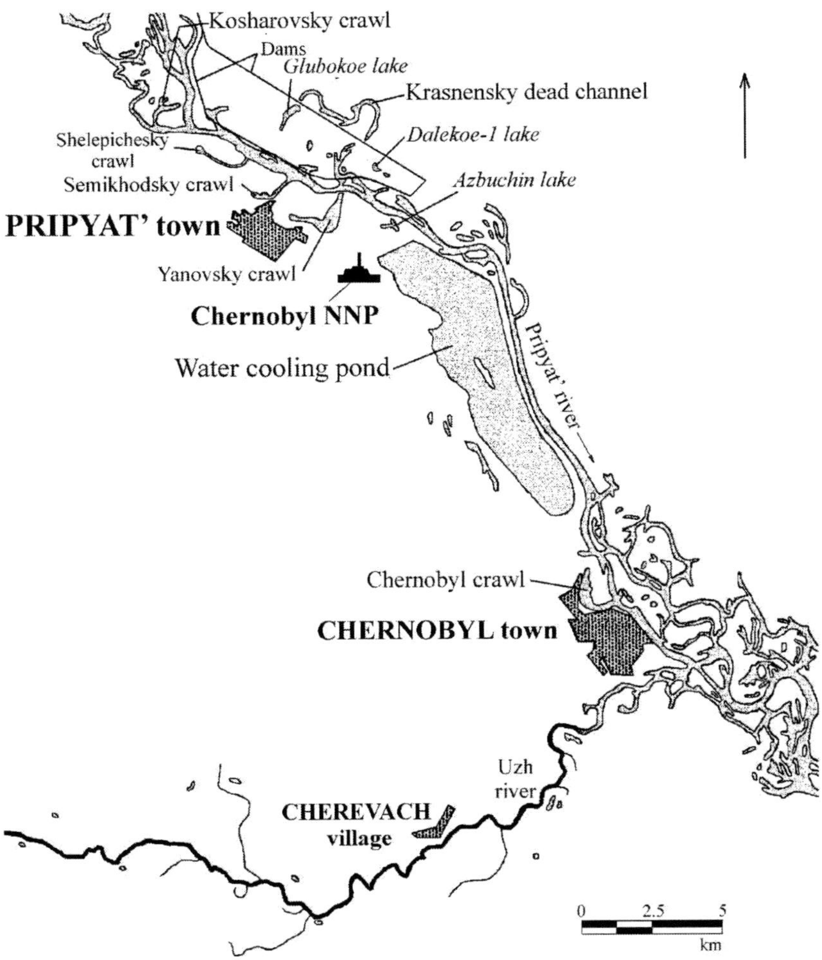

Figure 1. Map of reservoirs within the Chernobyl NPP exclusion zone.

We studied the rate of chromosome aberrations in cells of freshwater snail (*Lymnaea stagnalis* L.) embryos and in the apical meristem of roots of the higher aquatic plant common reed (*Phragmites australis* (Cav.) Trin. ex. Steud.) and arrowhead (*Sagittaria saggitifolia* L.). The samples were collected in different seasons from 2002 to 2004. The chromosome aberration rate was measured by standard (for snails) and modified (for macrophytes) (Shevtsova *et al.*, 2005) anaphase methods. The results of the analyses compared to the data received for hydrobionts from conditionally clear Goloseevo lakes, located within the Kiev City area.

3. Dose Rates for Hydrobionts

The dose rate for hydrobionts from reservoirs of the ChNPP exclusion zone were found to be in the range from 0.0018 to 3.4 Gy·year^{-1}. The highest value was found for hydrobionts from water ponds within the embankment territory on the left-bank flood plain of Pripyat River – lakes Glubokoe and Dalekoe-1. The lowest dose was observed for specimens from the water courses – Uzh and Pripyat river (Table 1).

Table 1. Absorbed dose rates for hydrobionts of littoral zone from different water objects within the sampling sites, in Gy year^{-1}.

Dose		Water object						
		Uzh River	Pripyat River	Cooling pond of the ChNPP	Yanovsky Creek	Azbuchin Lake	Dalekoye-1 Lake	Glubokoye Lake
Internal dose from incorporated radionuclides	^{90}Sr	9.9E–06 1.8E–04	1.5E–05 8.6E–05	2.2E–04 2.1E–03	7.7E–04 8.1E–04	4.9E–03 5.1E–02	1.8E–03 3.2E–02	2.2E–03 6.1E–02
	^{137}Cs	1.0E–04 1.8E–04	1.2E–04 4.5E–04	5.7E–03 2.0E–02	3.2E–03 4.2E–02	8.3E–03 1.5E–02	6.2E–03 8.2E–03	3.6E–02 4.9E–02
	^{238}Pu	×*	×	5.2E–06 7.9E–05	5.4E–05 2.5E–04	4.2E–05 3.3E–04	6.5E–05 3.7E–04	1.5E–04 1.3E–03
	$^{239+240}$Pu	×	×	8.3E–05 1.6E–04	1.3E–04 5.9E–04	1.3E–04 7.2E–04	1.6E–04 7.9E–04	3.2E–04 2.3E–03
	^{241}Am	×	×	2.3E–05 3.6E–04	4.5E–05 6.3E–04	6.7E–05 7.5E–04	9.7E–05 8.0E–04	7.2E–04 2.5E–03
	Total	1.1E–04 3.6E–04	1.4E–04 5.4E–04	6.0E–03 2.3E–02	4.2E–03 4.4E–02	1.3E–02 6.8E–02	8.5E–03 4.2E–02	3.9E–02 1.2E–01
External water dose	^{90}Sr	4.9E–08 9.8E–08	6.8E–08 1.5E–07	5.2E–07 5.8E–07	2.1E–05 2.8E–05	4.4E–05 7.0E–05	2.8E–05 3.3E–05	3.2E–05 4.0E–05
	^{137}Cs	2.7E–07 3.8E–07	2.9E–07 4.3E–07	7.2E–06 8.3E–06	1.3E–05 1.6E–05	4.5E–05 1.3E–04	2.7E–05 4.0E–05	3.8E–05 4.5E–05
	Total	3.2E–07 4.8E–07	3.6E–07 5.8E–07	7.7E–06 8.9E–06	3.4E–05 4.4E–05	8.9E–05 2.0E–04	5.5E–05 7.3E–05	7.0E–05 9.5E–05
External γ-dose		1.7E–03 2.9E–03	2.3E–03 3.6E–03	7.4E–03 7.9E–03	3.3E–03 6.2E–03	4.8E–03 1.2E–02	4.3E–02 5.0E–02	1.5–3.3
Total dose		1.8E–03 3.3E–03	2.4E–03 4.1E–03	1.3E–02 3.1E–02	7.5E–03 5.0E–02	1.8E–02 8.0E–02	5.2E–02 9.2E–02	1.6–3.4

* The measurements were not carried out

The ratio of external and internal doses varied considerably for hydrobionts from different reservoirs and depended on the contents of γ-emitting radionuclides in bottom sediment of littoral zone, as well as in soils close to the riverbank. Thus, in Glubokoye Lake, which contains a so-called abnormal contamination strip at the shoreline border, about 95% of absorbed dose results from external exposure, and only about 5% is due to internal exposure by radionuclides incorporated in tissues. A similar ratio was observed for Uzh and Pripyat rivers; to the lower values measured therein are caused by the high flow rate and to the relatively low radionuclide content in water and, consequently, in hydrobionts tissues.

In Azbuchin Lake and Yanovsky Crawl, at a relatively low external radiation dose, the main contribution to the absorbed dose is caused by radionuclides incorporated in hydrobionts tissues. It is linked to the high radionuclide content in water and at the same time to the low contamination level of the bottom sediment within littoral zone and the soils of nearby areas (with sandy soils showing low levels of radionuclide fixation). In this respect, the cooling pond of the ChNPP is in a mid-way position.

According to the G. Polikarpov's classification (1978, 1999), the studied littoral sites of Uzh and Pripyat river belong to the radiation safety zone; the sampling stations of Azbuchin Lake, Yanovsky Crawl, the ChNPP cooling pond and Dalekoye-1 Lake are zones of physiological and ecological disguise, where the radiobiological effects are difficult to detect due to the influence of other ecological factors. Glubokoye Lake approaches to the ecosystem of effect zone where reduction in aquatic organism numbers and loss of radiosensitive species can be observed.

4. Chromosome Aberrations

Radiation-induced DNA damage represent the main cause of reproductive cell death and chromosome aberrations. The analysis of cytogenetic effects of radiation on biosystems is extremely important to predict the late consequences of the Chernobyl NPP accident

4.1 FRESHWATER MOLLUSCS

Gastropod molluscs, especially freshwater snail *Lymnaea stagnalis,* are the prevailing invertebrates in the littoral zones of water bodies within marshy woodlands in Europe. These snails have a high bioproductivity and their embryonated eggs could be collect easily on the back of leafs of water soldier (*Stratiotes aloides*), white (*Nuphar lutea*) and yellow (*Nymphaea candida*) water lilies, and frogbit (*Hydrocharis morsus-ranae*) during April– November.

We have analyzed 7995 cells from 106 snail's laying. We found 947 aberrant cells and 985 aberrations. Single bridges were the most abundant in aberration spectrum, representing 56.4% of all aberrations (Figure 2). Other chromosomal rearrangements included single fragments (39.7%), pair bridges (3.1%) and pair fragments (0.8%).

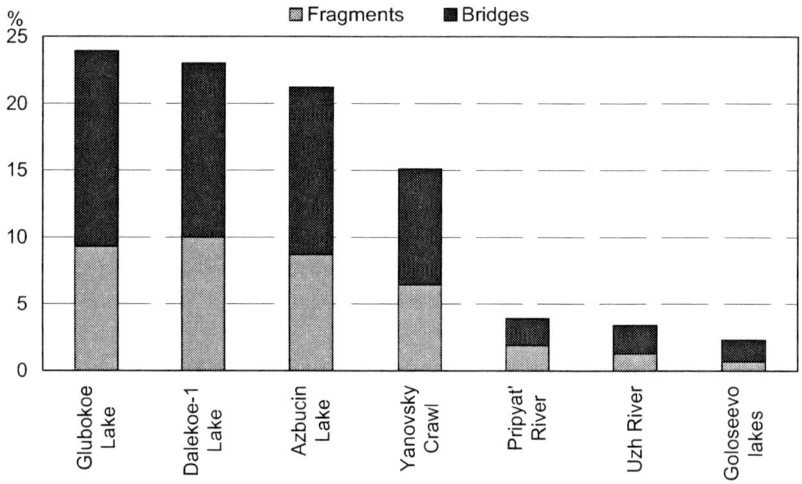

Figure 2. Rate and spectrum of the main chromosome aberration in cells of snails (*Lymnaea stagnalis*) from water bodies within the Chernobyl exclusion zone and Goloseevo lakes.

The highest value was found for hydrobionts from lakes within the dammed territory on the left-bank flood plain of the Pripyat River (Dalekoye-1 Lake and Glubokoye Lake). The lowest frequency was measured in specimens from Uzh and Pripyat rivers. The molluscs from Dalekoye-1 Lake and Glubokoye Lake were characterised by the maximum measured frequency of chromosome aberration – about 20–25%, 10 times exceeding the level of spontaneous mutagenesis for hydrobionts, which is around 1.5–2.0% (Polikarpov, Tsytsugina,1993). A little bit less rate is registered for snails from Azbuchin Lake and Yanovsky Creek. The chromosome aberration rate of hydrobionts from Goloseevo lakes was about 1.5%, and the maximum rate was lower than 2.5%.

4.2 HIGHER AQUATIC PLANTS

The reed is the most widespread species of the helophytes (aerial-aquatic plants) within the Chernobyl exclusion zone. We have analyzed 7525 root cells from 83 plants and measured 177 aberrant cells and 181 aberrations.

As shown in Figure 3, single bridges were the most frequent aberration in reed's meristem cells (57.1% of all aberrations). Frequency of single fragments was 40.7%, pair bridges were 1.7 %, and pair fragments 0.6%. The minimal rate of chromosome aberrations was found in control plants from Goloseevo lakes and was always lower than 1.9%.

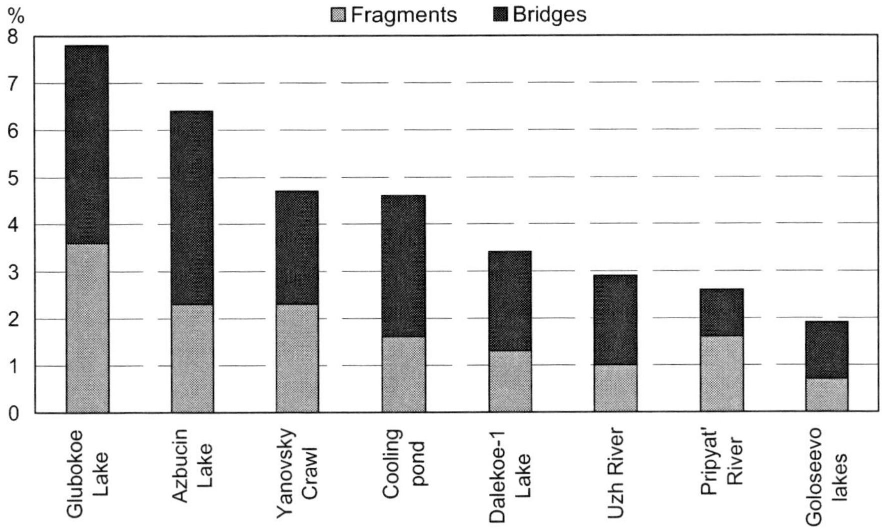

Figure 4. Rate and spectrum of the main chromosome aberrations in cells of reed (*Phragmites) australis*) from water bodies within the Chernobyl exclusion zone and Goloseevo lakes.

We found 2.6% aberrant cells in common reed of Pripyat' River , and 2.9% in Uzh River . Higher frequencies were observed in cooling ponds of ChNPP and Yanovsky Crowl –4.6% and 4.7%, respectively. Maximum rate of chromosome aberration was registered for plants from Azbuchin and Glubokoe lakes –6.4% and 7.8%, respectively.

The arrowhead was sampled in Glubokoe Lake and Pripyat' River (near Chernobyl town) only. The rate of chromosome aberrations in arrowhead's cells did not exceed 2.5% in Pripyat' River and was about 6.5% in plants from Glubokoe Lake.

5. Conclusion

Different radiation effects in hydrobionts within the Chernobyl exclusion zone have been registered in the post-accident period. Some of these effects disappear shortly, while an increasing importance is expected by the remote consequences – genetic damages induced by a long-term irradiation. These remote consequences are a delayed manifestation of changes in germ cells, where the initial molecular damages have a latent period, without any evident effect, but can be transferred through many generations of cells to trigger genome instability in future. The long-term impact of low dose irradiation in aquatic ecosystems, especially in closed water bodies within the Chernobyl exclusion zone, is shown by the increased level of chromosome aberrations and, connected with it, reproductive death of cells.

It seems extremely important to continue studies connected to reconstruction of dose rate on natural populations and organization of regular cytogenetic monitoring on contaminated territories, necessary for forecasting and prevention of the negative remote consequences of radiation impact.

Acknowledgements

This study was supported by the Ministry of Ukraine on the Emergency and Affairs of Population Protection Against the Consequences of the Chernobyl Catastrophe and by the National Academy of Sciences of Ukraine as part of the state budgetary projects 0101U004987, 0102U004665 and the project of young scientists 0102U003541. The authors wish to thank the personnel of Radioanalytical Laboratory of the Chernobyl Radioecological Centre for the radionuclides measuring procedure.

References

Amiro, B.D., 1997, Radiological Dose Conversion Factors for Generic Non-Human Biota Used for Screening Potential Ecological Impacts, *Journal of Environmental Radioactivity*. **35**(1): 37–51.

Brown J., Strand P., Hosseini A. and Borretzen P. (Eds.), 2003, Handbook for Assessment of the Exposure of Biota to Ionising radiation from Radionuclides in the Environmen*, A project within the EC 5th Framework Programme*: 1–101.

Polikarpov, G.G., 1978, Effects of ionizing radiation upon aquatic organisms (Chronic irradiation*), Atti della Giornata sul Tema "Alcuni Aspetti di Radioecologia". XX Congresso Nazionale, Associazione Italiana di Fisica Sanitaria e Protezione contro le Radiazioni, Bologna, 1977.* Parma Poligrafici: 25–46.

Polikarpov, G.G., 1999, Effects of nuclear and non-nuclear pollutants on marine ecosystems, *Marine Pollution. Proc. of a Symp. held in Monaco, 5–9 October 1998*. IAEA-TECDOC-1094. IAEA-SM-354/22: 38–43.

Polikarpov, G.G., Tsytsugina, V.G., 1993, Regularities of distribution of chromosome aberrations on hydrobionts' cells caused by radiation and chemical mutagens of environment, *Radiobiology*: **33**(5): 205–213 (in Russian).

Shevtsova, N.L., Gudkov, D.I., Stoyko, U.A., Syvak, E.V., 2005, To the method of determination of chromosome damages of higher aquatic plants at the example of common reed and arrowhead, *Scientific Acta of the Ternopil State Teacher's Training University*. Series: Biology, Special issue: Hydroecology: **3**(26)**:** 478–479 (in Russian).

EFFECTS OF CONTAMINANT EXPOSURE ON PLANTS

STANISLAV A. GERAS'KIN*
VLADIMIR G. DIKAREV
ALLA A. OUDALOVA
DENIS V. VASILIEV
TATYANA A. BAYKOVA
NINA S. DIKAREVA
Russian Institute of Agricultural Radiology and Agroecology, 249032, Obninsk, Russia

*To whom correspondence should be addressed. Stanislav Geras'kin, Laboratory of Plant Ecotoxicology, Russian Institute of Agricultural Radiology and Agroecology, Kievskoe shosse, 109 km, 249032 Obninsk, Kaluga region, Russia; e-mail: stgeraskin@gmail.ru

Abstract. The results of long-term field experiments in the 30-km Chernobyl NPP zone, in the vicinity of the radioactive wastes storage facility (Leningrad Region), and in Bryansk Region affected by the ChNPP accident that have been carried out in our laboratory on different species of wild and agricultural plants are discussed. These findings indicate that plant populations growing in areas with relatively low levels of pollution are characterized by the increased level of both cytogenetic disturbances and genetic diversity. The seeds from plant populations experiencing a man-caused impact showed a higher radioresistance than the reference ones. Therefore, the chronic low-dose exposure appears to be an ecological factor creating preconditions for possible changes in the genetic structure of a population. These processes have a genetic basis; therefore, understanding changes at the genetic level should help in identifying more complex changes at higher levels. Presented findings add to filling a major gap in our knowledge on remote effects of man-made impact on plant populations and ecosystems.

Keywords: radioactive and chemical contamination; plant populations; cytogenetic disturbances

A.A. Cigna and M. Durante (eds.),
Radiation Risk Estimates in Normal and Emergency Situations, 77–83.

1. Introduction

Contamination of the environment has become a worldwide problem. A clear understanding of all the dangers posed by environmental pollutants to both human health and ecological systems are needed. A key question in dealing with contaminated sites is whether, and to what extent ecotoxical effects occur. There are acute effects at severely contaminated sites, but the main problem refers to possible long-term effects of low doses and multi-pollutant exposure. These types of effects are of special concern because they can manifest themselves long after the source of contamination has been eliminated.

Interaction of contaminants with biota takes place first at the cellular level making cellular responses not only the first manifestation of harmful effects, but also suitable tools for an early and sensitive detection of exposure. It is becoming increasingly clear (Theodorakis, Blaylock, and Shugart, 1997) that cellular alterations may in the long run influence biological parameters important for populations such as growth, health and reproduction. Therefore, just genetic test-systems should be used for an early and reliable display of the alterations resulting from human industrial activity.

2. Results

An important gap in our knowledge is information about long-term ecotoxical effects induced by chronic low dose-rate radiation exposure and multi-pollutant exposure at contaminated sites. Indeed, there are few studies directly relevant to revealing responses of plant and animal populations to radionuclides in their natural environments. Although radioactive and chemical contaminants cause primary damage at the molecular level, there are emergent effects at the level of populations, which are not predictable solely from the knowledge of elementary mechanisms of single pollutant action. The use of data gathered from both laboratory-based assays and field-based monitoring may therefore be significantly affected by our present lack of knowledge in this area of environmental research.

Previously work in our laboratory and ongoing field studies of biological effects in different species of wild and agricultural plants are briefly summarized in Table 1. In 1987-1989, an experimental study on the cytogenetic variability in three successive generations of winter rye and wheat, grown at four plots with different levels of radioactive contamination, was carried out within the 10-km ChNPP zone. In autumn of 1989, aberrant cell frequencies in intercalary meristems of winter rye and wheat of the second and third generations significantly exceeded frequencies for the first generations (Fig. 1).

Table 1. Field studies on wild and agricultural plants.

Species	Site	Contamination	Assay
Winter rye and wheat, spring barley and oats	10-km ChNPP zone (11.7-454 MBq/m^2), 1986-1989	Radionuclides	Morphological indices of seeds viability, mitotic index, cytogenetic disturbances in intercalar and seedling root meristems (Geras'kin *et al.*, 2003a)
Scots pine, couch-grass	30-km ChNPP zone, (250-2690 μR/h), 1995	Radionuclides	Cytogenetic disturbances in seedling root meristem (Geras'kin *et al.*, 2003b)
Scots pine	Radioactive waste storage facility, Leningrad Region, 1997-2002	Mixture	Cytogenetic disturbances in needles intercalar and seedling root meristems (Geras'kin *et al.*, 2005)
Wild vetch	Radium production industry storage cell, Komi Republic, (73-3300 μR/h), 2003	Heavy natural radionuclides	Embryonal lethals, cytogenetic disturbances in seedling root meristem
Scots pine	Sites in Bryansk Region radioactively contaminated in the Chernobyl accident (451-2344 kBq/m^2), started in 2003	Radionuclides	Cytogenetic disturbances in seedling root meristem, enzymatic loci polymorphism analyses
Scots pine	10-km ChNPP zone (1100 μR/h), 2004	Radionuclides	Morphological modifications in pine needles, cytogenetic disturbances in seedling root meristem
Feather-grass, fescue	Semipalatinsk Test Site, (74-2050 μR/h), started in 2005	Radionuclides	Developmental instability in plants, mitotic index, cytogenetic disturbances in seedling root meristem

In 1989 plants of all three generations were developing in the identical conditions and were exposed to the same doses, which is why the most probable explanation of the observed phenomenon relates to a genome destabilization in plants grown from radiation-affected seeds. This finding relates to higher-order ecological effects, as well as to contaminant-induced selection of resistant phenotypes. From these viewpoints, the results observed in this study, indicating a threshold for genetic instability induction may be a sign of an adaptation processes beginning, that is, the chronic low-dose irradiation appears to be an ecological factor creating preconditions for possible changes in the genetic structure of a population.

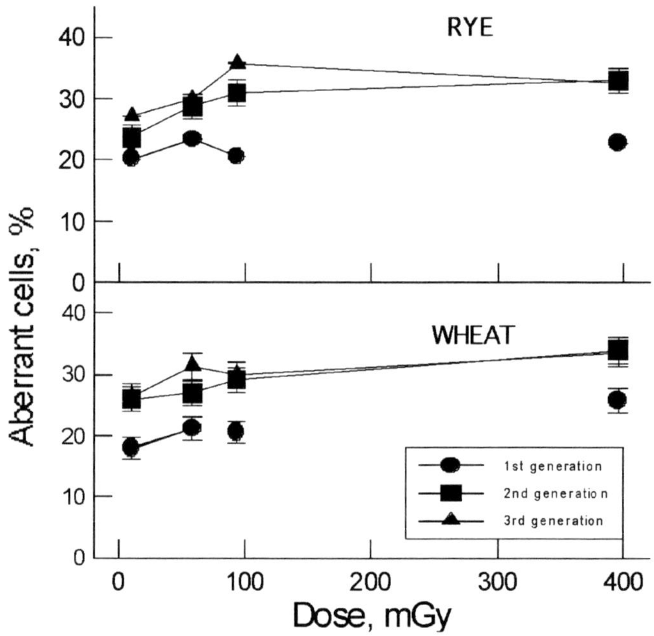

Figure 1. Aberrant cells frequency in three successive generations of winter rye and wheat, grown on contaminated plots (Geras'kin *et al.*, 2003a).

In frames of other studies, adaptation processes in impacted wild plant populations were investigated. The results of these experiments indicate (see, for example, Table 2) that an increased level of cytogenetic disturbances is a typical phenomenon for plant populations growing in areas with relatively low levels of pollution. Such information can be used to identify cellular mechanisms responding to environmental stress, which in turn may lead to a better understanding of the consequences of contaminant exposure.

Table 2. Aberrant cell frequency in seedling root meristem of Scots pine growing in Bryansk Region of Russia, radioactively contaminated in the Chernobyl accident (preliminary data).

Test site	^{137}Cs contamination (kBq/m^2)	Estimated absorbed dose rate mGy/year§	Aberrant cells (mean±se) (%)
Reference	-	0.14	0.90±0.09
VIUA	451	7.4	1.47±0.15*
Starye Bobovichy	946	15.3	1.32±0.12*
Zaborie1	1730	28.3	1.69±0.17*
Zaborie2	2340	37.8	1.63±0.15*

Note: Seeds were collected in 2003;

§ - absorbed doses are calculated for β- and γ-radiation.

* - difference from the reference population is significant, p<5%;

In the field study (Geras'kin *et al.*, 2005), Scots pine populations were used for an assessment of the genotoxicity originating from the operation of a radioactive waste storage facility. Specifically, the frequency and spectrum of cytogenetic disturbances in reproductive (seeds) and vegetative (needles) tissues sampled from Scots pine populations were studied to examine whether Scots pine trees have experienced environmental stress in areas with relatively low levels of pollution. The temporal changes of the cytogenetic disturbances in seedling root meristem from 1997 to 2002 are shown in Fig. 2. There are essential differences between these dependencies for the reference and impacted Scots pine populations. Statistical analysis revealed (Geras'kin *et al.*, 2005) that cytogenetic parameters at the reference site have a trend towards cyclic fluctuations in time, whereas in affected populations these peculiarities could not be revealed with confidence.

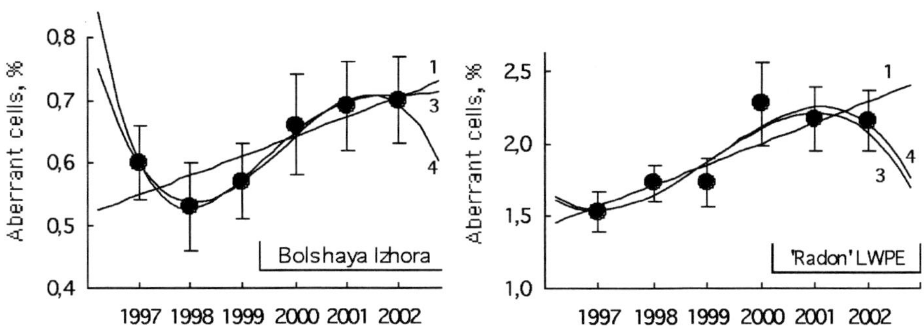

Figure 2. Aberrant cells percentages in seedling root meristem of Scots pine trees in dependence on year and their approximation by the best models. 1 – linear model, 3 and 4 – polynomial models of 3rd and 4th degree, correspondingly.

Thus, man-caused impact in this region is strong enough to destroy natural regularities. But the mechanisms involved in this plant response are still unclear.

To study possible adaptation processes in impacted plant populations, a portion of the seeds collected were subjected to an acute γ-ray exposure (Geras'kin *et al.,* 2005). The seeds from the Scots pine populations experiencing a man-caused impact showed (Fig. 3) a higher resistance than the reference ones. There is a convincing proof (Shevchenko, Pechkurenkov, and Abramov, 1992) that the divergence of populations in terms of radioresistance is connected with a selection for changes in the effectiveness of the repair systems.

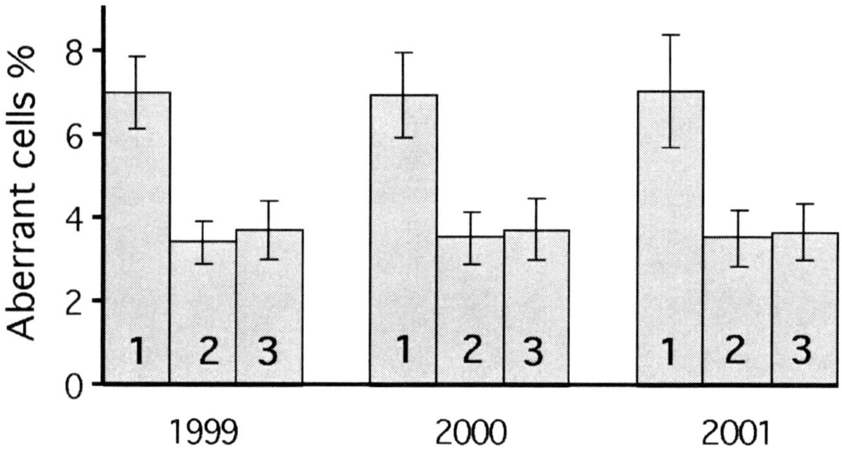

Figure 3. Aberrant cell frequency in root meristem of Scots pine seedlings grown from seeds sampled in Leningrad Region in 1999-2001 and exposed to an acute γ-ray dose of 15 Gy.

(1 – Bolshaya Izhora (control), 2 - Sosnovy Bor town, 3 – 'Radon' LWPE)

To find out whether genetic differentiation had occurred between the reference and impacted populations, genetic structure of five Scots pine populations growing in Bryansk Region under conditions of radioactive contamination from the Chernobyl accident has been evaluated, using gel electrophoresis for 4 enzymatic loci (Lap, Est, Mdh, Gdh). Radionuclide-affected populations demonstrate higher genetic diversity than those from the reference site. These results provide evidence that a general pattern of adaptation strategy of populations to pollution stress involves an increase in genetic variation.

3. Conclusions

A better understanding of genetic aspects of population, community and the whole ecosystem responses to toxic agent's exposure is vital to future environmental management programs. The genetic nature of such effects as well as their dynamics in subsequent generations remains inadequately explored up to now. The presented findings taken together suggest that a long-term existence of some factors (either of natural origin or man-made) in the plants environment activates genetic mechanisms, changing a population's resistance to exposure. These processes have a genetic basis; therefore, understanding changes at the genetic level should help in identifying more complex changes at higher levels. Studies briefly illustrated here add to filling a major gap in our knowledge on remote effects of man-made impact on plant populations and ecosystems and make a vital contribution to scientific and public understanding of the environmental risks of radiation and to debates on the environmental costs and benefits of nuclear energy. The further evolution of investigations in the field would issue in a development of theoretical bases and practical procedures for environmental protection against radioactivity. This should be addressed in the future.

References

Geras'kin S.A., Dikarev V.G., Zyablitskaya Ye.Ya., Oudalova A.A., Spirin, Ye.V. and Alexakhin R.M., 2003a, Genetic consequences of radioactive contamination by the Chernobyl fallout to agricultural crops, *J. Environ. Radioactivity* **66**: 155–169.

Geras'kin S.A., Zimina L.M., Dikarev V.G., Dikareva N.S., Zimin V.L, Vasiliyev D.V., Oudalova A.A., Blinova L.D. and Alexakhin R.M., 2003b, Bioindication of the anthropogenic effects on micropopulations of *Pinus sylvestris,* L. in the vicinity of a plant for the storage and processing of radioactive waste and in the Chernobyl NPP zone, *J. Env. Radioactivity* **66**: 171–180.

Geras'kin S.A., Kim J.K., Oudalova A.A., Vasiliyev D.V., Dikareva N.S., Zimin V.L. and Dikarev V.G., 2005, Bio-monitoring the genotoxicity of populations of Scots pine in the vicinity of a radioactive waste storage facility, *Mutation Research* **583**: 55–66.

Shevchenko V.A., Pechkurenkov V.L. and Abramov V.I., 1992, *Radiation Genetics of Natural Populations: Genetic Consequences of the Kyshtym Accident*, Nauka, Moscow.

Theodorakis C.W., Blaylock B.G. and Shugart L.R., 1997, Genetic ecotoxicology I: DNA integrity and reproduction in mosquitofish exposed *in situ* to radionuclides, *Ecotoxicology* **6**: 205–218.

VARIABILITY AND VIABILITY OF SEED PLANT POPULATIONS AROUND THE NUCLEAR POWER PLANT

V.L. KOROGODINA*, V.P. BAMBLEVSKY,
B.V. FLORKO, V.I. KOROGODIN[1]
Joint Institute for Nuclear Research (JINR),
141980 Dubna, Moscow region,
Russia
E-mail: *korogod@jinr.ru*

*To whom correspondence should be addressed

Abstract. Earlier was shown that in populations around the Nuclear Power Plant, the seed survival can decrease by up to 20%, statistical simulations showed that CAs can appear independently and Poisson-distributed (P) as well as correlative enhanced and geometrical-distributed (G) (Korogodina *et al.*, 2004). Our aim was to determine the regulatory mechanisms of viability and variability of seed plant populations. We showed that damage transmission in meristem (bystander effect) is described by relay-race scheme and corresponds to G-distribution. We showed that P-subpopulation did not change significantly, whereas value of G-fraction did with dose/dose rates. We conclude that G-machinery plays an adaptive role.

Keywords: low dose irradiation; seeds; cells with abnormalities; survival; adaptation

1. Introduction

It is often believed that the only environmental factors that are dangerous for an organism are those that exceed the usual limits of its ecological niche. We studied plantain seeds of populations growing around the Nuclear Power Plant (NPP) and observed that the seed survival can decrease by up to 20%, whereas the frequency of rootlet meristem cells with abnormalities (CAs) increases (or not) depending on the seeds'

[1] Dr. Korogodin past away shortly after the meeting.

A.A. Cigna and M. Durante (eds.),
Radiation Risk Estimates in Normal and Emergency Situations, 85–93.

antioxidant status (AOS), and the mitotic activity (MA) increases up to threefold (Korogodina *et al.*, 2004). The distributions of CAs in root meristems had "tails", and consisted of two separate components, they were Poisson (P) and geometrical (G) ones (Korogodina *et al.*, 2004, 2005). The CAs appearance can be described by the formula

$$T_{CA} = P_{CA} + G_{CA}, \qquad (1)$$

where T_{CA} is the total probability of CA appearance; P_{CA}, G_{CA} are probabilities of CA appearance in P- and G- subpopulations, respectively;

$$P_{CA}(n) = P \cdot e^{-m} \cdot m^n / n!$$

$$G_{CA}(n) = G \cdot \theta(1-\theta)^n ,$$

where $P_{CA}(n)$, $G_{CA}(n)$ - probabilities of n CAs appearance in P- and G- distributions; m – average number of CAs in meristem in P-subpopulation; θ - quota of meristems without CAs in subpopulation G; P and G – values of subpopulations.

We tested γ-irradiation, when a hit of one γ-quanta induces cells' absorbed energy of 1.4 KeV/cell. Other authors suggested dynamics models of bystander effect when absorbed cell energy would exceed ~0.3 MeV/cell [α-irradiation (Khvostunov & Nikjoo, 2002)] and ~40 MeV/cell [γ-irradiation (Khvostunov & Nikjoo, 2002)]. All these modeling showed a signal transmission from one damaged cell to several dozens of others.

Our aim was to determine the regulatory mechanisms of viability and variability in the presence of oxidizing factors such as low doses of ionizing radiation and high temperatures, neither of which exceeded the norms of the tested plant habitats. We studied the features of P- and G- parameters in dependence on AOS and dose irradiation.

2. Materials and Methods

2.1 SEEDS

The plantain seeds (*Plantago major*) were used. Seeds were collected from 20–30 plants at the end of August in 1998 and 1999. Seeds storage was standard (Korogodina *et al.*, 2004). Seeds were germinated in petri dishes at 23°C until seedling roots reached 3.5 ± 2 mm, a length corresponding to the first mitoses. Seedlings less than 1.5 mm were scored as non-surviving (1-S) after 13 days. The methods of seed sprouting and rootlet fixing have been described in Korogodina *et al.* (2004, 2005). AOS were studied with a photochemiluminescence method. The procedure was described in

Korogodina *et al.,* (2004). We find an amount of seed infusion that inhibited chemiluminescence by 50 % ($C_{1/2}$).

Ana-telophases were scored for CAs containing chromosome bridges and acentric fragments. The mitotic activity (MA) of seedling meristem cells was scored as the number of ana-telophases in apical meristem.

2.2 CHARACTERISTICS OF SITES AND RADIOACTIVITY

The plantain populations were located at sites within 80 km of the Balakovo NPP (figure 1, P2 – P10) and in Chernobyl trace area (Saratov region, P11), and in JINR territory (Moscow region, P12). Mostly, the NPP atmospheric fallouts influence on populations P2 – P6 resulting from the direction of the winds in summer. In 1999, the summertime high temperatures in the European part of Russia averaged 2-3°C above normal (MAPRF, 1998).

Figure 1. Locations of the selected plantain populations in the vicinity of the Balakovo NPP: 2 - 10 (P2 - P10). Populations P1, P11 are located 80 km and 100 km from NPP; P12 one is placed in the Moscow region.

For sites within a 100 km radius of the NPP, the annual γ-radiation dose rates (DR) and ^{137}Cs soil concentrations (C_{Cs}) varied little from the ranges ~0.10-0.15 μSv/h and ~5-10 Bq/kg; in site P11 DR is ~0.10-0.15 μSv/h and C_{Cs} is 30 Bq/kg (State Committee, 2000; Min. Atomic Power, 1998), DR is ~0.10-0.12 μSv/h and C_{Cs} is ~5-10 Bq/kg in site P12 (JINR) (Alenitskaja et al., 2004). These values (excluding C_{Cs} in site P11) do not exceed the average values over the Saratov and Moscow regions (Alenitskaja *et al.*,

2004). The ^{137}Cs and ^{40}K soil contaminations were 5-10 and 400-500 Bq/kg, respectively (Korogodina *et al.,* 2004), that agree with Alenitskaja *et al.* (2004).

Plantain seeds experienced the NPP fallout irradiation (annual NPP fallouts on isotopes: Kr ~2.5 TBq; Xe ~2.5 TBq, and I ~4.4 TBq) (Min. Atomic Power, 1998) Distribution of the particulate emissions and gases were estimated according to the Smith-Hosker model based on NPP characteristics (Min. Atomic Power, 1998), and winds in summer near the ground in the NPP region (State Committee, 2000). The isotopes' fallouts result in γ-emitters mainly, and each seed gets no less than one γ-quantum in summer additionally to the natural ground. The calculated mean γ-quanta energy deposition per plant cell nucleus is 1.4 keV. The DR values were calculated in the ratio to the dose in site P1. In P12 site, the expected irradiation dose was calculated after the JINR facilities operation in 1998. The neutron dose exceeded the background ~ twofold (1 mSv) and dose rate ~eightfold (0.8 μSv/h). In 1999, the neutron irradiation did not increase. The averaged neutron energy was 5.5 MeV.

2.3 STATISTICS AND MATHEMATICAL PROCESSING

In the experiment with plantain seeds we used about 200 (1998) and 500 (1999) seeds and analyzed about 400-1000 (1998) and 900-3800 (1999) ana-telophases for each population. The statistical simulation was conducted by the methods of χ^2-minimization, minimization of standard deviation, and maximum likelihood ones (Korogodina *et al.*, 2004, 2005). The results obtained by each method agreed within the standard errors.

3. Results and Discussion

3.1 AVERAGE CA FREQUEBCY, MA AND FRACTION OF NON-SURVIVED SEEDS

It was shown ones (Korogodina *et al.*, 2004) that seed survival was not low as well as cytogenetic values were not unusual in 1998. A strong correlation between the (1-S) value and the CA frequency was demonstrated ($|r| = 0.92$, df=9, $p < 0.001$). The increase of the MA in P12 population (Korogodina *et al.*, 2004) may be the result of neutron irradiation in this year.

In 1999, for the populations located near the NPP, the average (1-S) value grew high as well as the average CA frequency was elevated in comparison with 1998 (Korogodina *et al.*, 2004). Average value of the MA is higher than the same value over all studied populations in 1998

(~ threefold, $p < 0.05$), and higher than in P11, P12 plants (~ twofold, $p < 0.05$) in 1999. The data on P11 did not differ both years. It was expected because the population located in chronicle radiation conditions becomes more resistant in some generations. For seeds of P12 population, (1-S) value and CA frequency are in the ordinary ratio with reasonable MA elevation. According to all 1999' data, the correlation between the (1-S) value and the CA frequency disappeared ($|r| = 0.02$), and we can suspect induction of some regulatory mechanisms around the NPP. Bystander mechanisms can be involved in CAs appearance and (1-S) value that could disturb their correlation.

To understand changes in these values, we examined scheme of bystander mechanism.

2.4 THE "RELAY-RACE" SCHEME

We consider a damage transmission that could be induced by primary cell damage in N proliferated meristem cells of rootlets (Figure 2, A). A running of cells with mutations finishes at an appearance of adaptive mutation ("relay-race" mechanism). Let us an adaptive cell appearance is characterized by a frequency $R_{ad..}$ At high AOS, non-targeted CA number (N) increases up to the appearance of adaptive cell ($N/N \approx R_{ad}$) that provides cells' dividing (1). At low AOS, both apoptosis and cell differentiation frequencies are high, and damage transmission leads to death of rootlets (2). A "relay-race" mechanism is described mathematically by geometrical distribution (Feller, 1957). We can presuppose that higher doses will induce an appearance of some CAs originated from one damaged cell, which would be described mathematically by a distribution with maximum.

Figure 2, B shows CAs appearances in P- and G- subpopulations. In P-subpopulation, number of CAs $N \approx R_{ad} \cdot N$ can't be originated by low-dose irradiation. In G-subpopulation, a threshold of an adaptive cell emerging is reached by bystander mechanism. In P-subpopulation, cell's systems aimed to protect old genotype. Contrarily, the mechanisms "looking-for adaptive genotype" (at high AOS) and "abolishing of the old one" (at low AOS) act in G-subpopulation. Figure 2, C show flow block of CAs appearance as well as death/growth of seeds rootlets. The flow block can be described with formula 1.

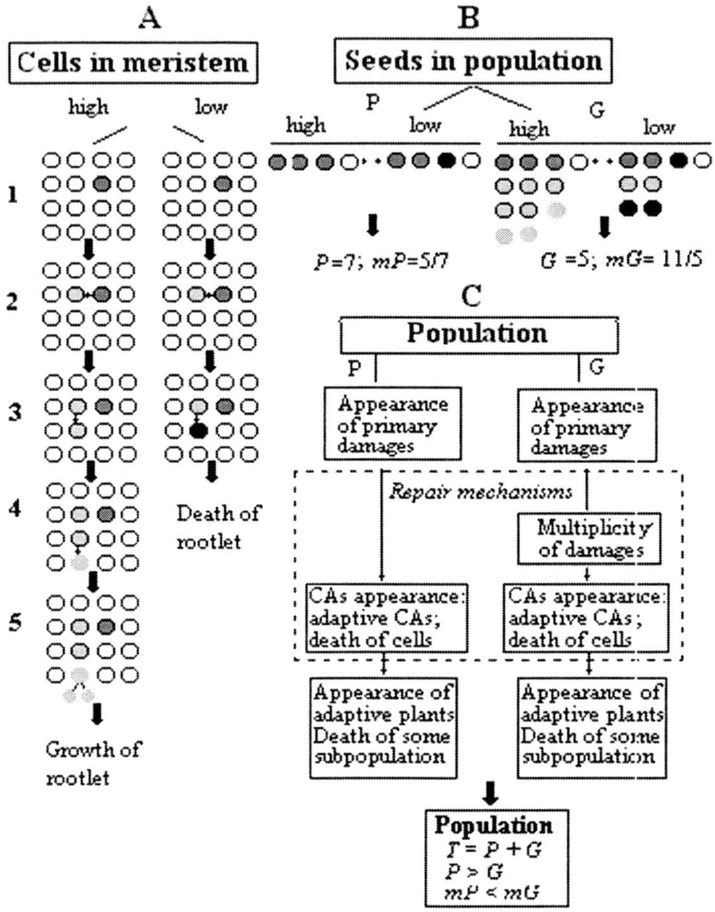

Figure 2. Schemes of relay damages transmission in rootlet meristem cells. Undamaged cells: white; with abnormalities: gray; dead: black. A: Scheme of non-targeted damages transmission in meristem of seeds with high (h) and low (l) AOS value. A numeration of cells is from left to right and down. The event 1 (h, l): primary damage of cell no.7. The events 2, 3, 4 (h): a damage transmission leads to CAs appearance (cells no.6, 10, 14). The event 5 (h): an adaptive CA (no.14) starts to divide that stimulates growth of rootlet. The event 3 (l) induces cell death (no.10) that leads to death of rootlet. B: Scheme of CAs appearance and cell death in P- and G- subpopulations. P-subpopulation: induction of primary CAs (h) with dead cell (l). G-subpopulation: multiplicity of damaged cells and appearance of adaptive (h) and dead (l) ones. C: Flow block of appearance of damaged, adaptive and dead cells, as well as adaptive and dead plants in P- and G- subpopulations.

2.5 AOS- AND DOSE- DEPENDENCE OF P- AND G- PARAMETERS

Sample mean of P-subpopulation is not changed significantly with AOS (figure 3a, A, C), excluding the decreased level of *mP* at $AOS_{50} \sim 0.16$ in 1999. It is expected because both the quantity and quality of antioxidants influence the radiosensitivity (Zhuravskaya *et al.*, 1998). In 1999, N_P increases with AOS (figure 3a D) that can be clarified by increasing of cell proliferation. Parameters of G- subpopulation *mG*, N_G increase with AOS both years (figure 3a, A – D). A distance to NPP influences seeds because AOS-dependence of G-parameters demonstrate two approximately parallel graphs respected to 10 km (P4, P6) and 20 km (P2, P3) from the NPP. Parameter *mG* increases, and N_G one decreases with dose irradiation (Figure 3b, C, D, graphs 2).

The population P5 experienced the most intensive irradiation (Korogodina *et al.,* 2004), which damaged each cell nucleus during vegetation period. At this irradiation the parameter *mG* decreases dramatically, as the N_G value is low (Figure 3b, B). It can be evidence that cells' elimination dominated at this dose irradiation. This conclusion agreed with data that apoptosis could be induced by irradiation with doses higher than mutations (Mothresill *et al.*, 2000). In father populations P4, P6, irradiation induced mutagenic bystander effect.

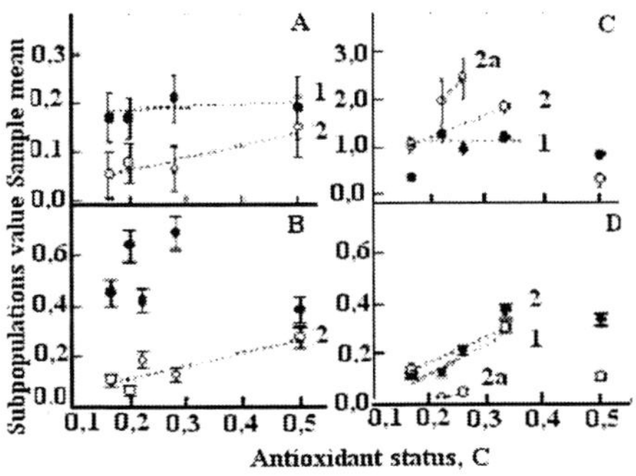

Figure 3a. The dependence of parameters of subpopulation with damage transmission on AOS in 1998 (A, B) and 1999 (C, D). The parameters: P – ● (1), G – ○ (2). The regressions, A: y= 0.17+0.08·x ($p > 0.001$) (1), y= 0.01+0.26·x ($p < 0.05$) (2); B: y= -0.02+0.56·x ($p < 0.05$) (2); C: y= 1.16-0.11·x ($p > 0.001$) (1), y= 0.25+4.80·x ($p < 0.05$) (2), y= -0.92+12.96·x ($p < 0.05$) (2a); D: y= -0.15+1.43·x ($p < 0.05$) (1), y= -0.04+1.02·x ($p < 0.05$) (2), y= -0.10+0.53·x ($p < 0.05$) (2a).

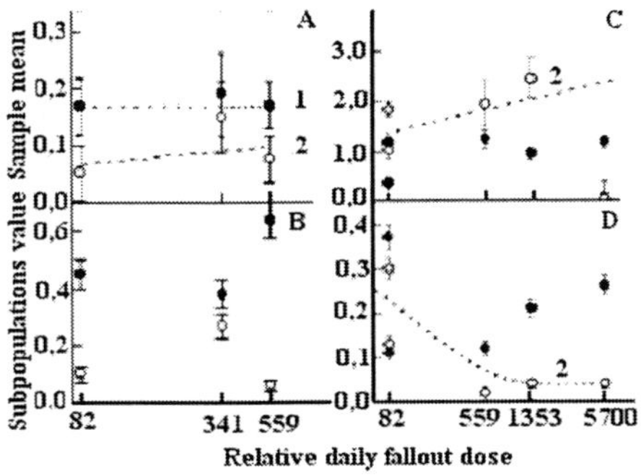

Figure 3b. The dependence of parameters of subpopulation with damage transmission on the relative fallout dose in 1998 (A, B) and 1999 (C, D). The parameters: P – ● (1), G – ○ (2). The regressions: **A**: $y= 0.17+2\cdot10^{-6}\cdot x$ ($p > 0.05$) (1), $y= 0.07+4\cdot10^{-5}\cdot x$ ($p > 0.001$) (2); **C**: y=1.34+0.001·x ($p < 0.05$); **D**: $y=100/x^{1.4}+0.027$.

2.6 MANIFESTATIONS OF ADAPTATION OF SEED POPULATIONS

We examined a correlation between (1-S) values and P- and G- parameters. Both years, (1-S) value was connected with either N_P and N_G (1998: $|r_{S,P}|=0.74$, $df=7$, $p < 0.05$, $|r_{S,G}|=0.67$, $df=7$, $p < 0.05$; 1999: $|r_{S,P}|=0.94$, $df=5$, $p < 0.001$, ($|r_{S,G}|=0.75$, $df=5$, $p < 0.05$). P-parameters are independent on dose irradiation ($|r_{mP,D*}| = 0.22$, $|r_{P,G*}| = 0.14$, $df = 4$); contrarily, G-parameters correlate with dose ($|r_{mG,D*}| = 0.82$, $|r_{G,D*}| = 0.68$, $df = 4$). G-mechanisms increase a number of CAs and provide selection of seeds according their AOS value.

N.W. Timofeeff-Ressovsky (1939) pointed at two requirements of evolution: sharp increasing of evolution material coupled with partial death of population because new genotype would be advantaged in this case only. All these are complicated by G-mechanisms, and formula 1 describes seeds adaptation.

3. Conclusions

A correlation between the (1-S) values of seeds and CA frequencies of meristem cells can be disturbed in plantain populations in 30-km zone of NPP. We presuppose that bystander mechanism can be induced by NPP irradiation.

It is shown that cell damages transmission can be presented by "relay-race" scheme. These mechanisms are "looking-for an adaptive genotype" and "abolishing of the old one". This scheme is described by geometrical distribution.

Findings early formula $T_{CA} = P_{CA}+G_{CA}$ described CA appearance. In P-subpopulation, cell systems aimed to protect the old genotype. "Looking-for an adaptive genotype" and "abolishing of the old one" act in G-subpopulation, these are bystander mechanisms.

Acknowledgements

The authors would like to thank LRB JINR Direction for investigation conditions, E.B. Burlakova, J.W. Drake, V.B. Priezzhev for helpful discussions.

References

Alenitskaja, S.I., Bulah, O.E., Buchnev, V.N., Zueva, M.V., Kargin, A.N., Florko, B.V., 2004, Results of long-term supervision over the environment radioactivity in area of JINR arrangement. *PAPAN, Lett. 1,* no.**5** (122), 88–96.

Feller W., 1957, *An introduction to probability theory and its applications,* New York: John Wiley & Sons, London: Chapman & Hall, Limited, 2nd ed.

Khvostunov, I.K., Nikjoo, H., 2002, Computer modelling of radiation-induced bystander effect. J. Radiol. Prot.. **22**, A33–A37.

Korogodina, V., Bamblevsky, V., Grishina, I., Gustova, M., Florko, B., Javadova, V., Korogodin, V., Lozovskaya, E., Malikov, A., Maslov, O., Melnikova, L., Shlyakhtin, G., Stepanchuk, V. and Zueva, M., 2004, Evaluation of the consequences of stress factors on plant seeds growing in a 30-km zone of Balakovo NPP, *Radiats. Biologija. Radioecologija* **44**(1), 83–90 (in Russian).

Korogodina, V.L., Florko, B.V., Korogodin, V.I., 2005, Variability of seed plant populations under oxidizing radiation and heat stresses in laboratory experiments. *IEEE Trans. Nucl. Sci.* **52**(4).

MAPRF, 1998, *The General Information on Balakovo NPP*, The Ministry of Atomic Power of the RF, Rosenergoatom concern, Balakovo NPP: 3–5 (in Russian).

Mothersill, C., Kadhim, M.A., O'Reilly, S., Papworth, D., Marsden, S.J., Seymour, C.B., and Wright, E.G., 2000, Dose- and time-response relationships for lethal mutations and chromosomal instability induced by ionizing radiation in an immortalized human keratinocyte cell line. *Int J Radiat Biol.* **76**(6): 799–806.

State Committee on the Environment Protection in the Saratov Region, 2000, *Conditions of the environment in the Saratov region in 1999.* Acvarius, Saratov: 1–193 (in Russian).

Timofeeff-Ressovsky, N.W., 1939, Genetik und Evolution. *Z. und Abst. Vererbl.,* **76**(1-2): 158–218.

Zhuravskaya, A.N., Stognij, V.V., Kershengolts, B.M., 1998, A dependence of radiosensitivity of plant seeds on ecological conditions of vegetation. *Radiats. Biologija. Radioecologija* **38**(5):531–539 (in Russian).

RADIOSENSITIVITY OF CHROMOSOME APPARATUS OF VOLES FROM ALIENATION ZONE OF CHERNOBYL'S ACCIDENT

OXANA A. KOVALOVA*, TATIANA T. GLAZKO
Institute of Agriecology and Biotechnology
UAAS, Kiev, Ukraine.
E-mail: glazko@biotech.relc.com

*To whom correspondence should be addressed

Abstract. The mutation occurrences at 9 cytogenetic characters in bone marrow cells in different vole's species which were trapped in Chernobyl's zone with various levels of radionuclide pollution were analyzed. The spontaneous spectra of cytogenetic anomalies in bone marrow cells are characterized by the species-specific traits in the voles both on predominance of cytogenetic anomalies and also on evolving into anomalies of the individual chromosomes. In conditions of chronic ionizing irradiation speeding cell proliferation in different vole species and also increasing the frequency of those cytogenetic anomalies, which had species-specific particularities is investigated in vole species.

Keywords: vole, Chernobyl, cytogenetic anomalies,

1. Introduction

Investigations of cytogenetic anomaly frequencies in somatic cells in connection with ionizing irradiation in the last 50 years have been widely conducted on humans, plants, small-sized *Rodentia* species and others. The small-sized rodents are the traditional object for bioindication of environmental pollution by different genotoxic agents. However, the high individual variability in tested species in the same conditions, and also the absence of a precise correlation between the quantity of cytogenetic damage in somatic cells and doses of ionizing irradiation was found from the accumulated data. It is not excluded, that one of causes of this result may relate to both the peculiarities of species karyotype and/or the morphology of individual chromosomes. Thus, the goal of our work was the study of mutation occurrences in bone marrow cells in different vole species. These were trapped in Chernobyl zones with various levels of radionuclide pollution ranging from 20 to 100 - 1000 Ci/km^2.

A.A. Cigna and M. Durante (eds.),
Radiation Risk Estimates in Normal and Emergency Situations, 95–100.

2. Materials and Methods

The different vole species were trapped in Chernobyl zones with various levels of radionuclide pollution ranging from 20 – 100 - 1000 Ci/km^2. The sites included: Razezzhee, Nedanchichi: <5 Ci/km^2; Lelev: ~ 20 Ci/km^2; Like Deep: 500 Ci/km^2; Chistogalovka: >500 Ci/km^2; Red forest: ~1000 Ci/km^2. Species differed by the number of acrocentric and metacentric chromosomes in the karyotypes: *Microtus arvalis* (2n=46, Fna=84), *Microtus subarvalis* (2n=54, Fna=54), *Clethrionomus glarealus* (2n=56, Fna=56), *Microtus oeconomus* (2n=30, Fna=56), *Microtus agrestis* (2n=50, Fna=54). The preparations of bone marrow cells of representatives of these vole species were obtained by a standard technique without colchicine. Nine cytogenetic characters in bone marrow cells were included in the analysis, such as the frequency of aneuploidy. This was evaluated in two ways: general aneuploidy (A1) and aneuploidy (A2) on one chromosome (2n±1)). Other characters were polyploidy (PP), the frequency of metaphase plates with chromosome aberrations (CHA), interchromosome fusion on a type of Robertsonian translocation (RB), with asynchronous separation of centromere chromosome region (ASCR) (in%). Quantity of metaphase plates (MI), binuclear leukocytes (BL) and leukocytes with the micronuclei (LM) in 1000 cells were calculated on the same preparations in cells with saved cytoplasm (in%). Statistical reliability of between group differences was evaluated with use of the Student t-test (tS).

3. Results and Discussion

In the zones with low-level radioactive pollution the spectra of cytogenetic parameters had significant species variability. For example, in *Microtus arvalis* the frequency of aneuploidy cells was higher than in *Microtus oeconomus* ($P<0{,}05$) or the other species. In the voles with acrocentric autosomes in the karyotype (*Clethrionomus glarealus* and *Microtus subarvalis)* the frequency of metaphases with interchromosome fusion of the type of Robertsonian translocation were seen more often than in the voles with metacentric autosomes in the karyotype (*Microtus arvalis* and *Microtus oeconomus)* (Table 1).

In the zones with high-level radio pollution in all species of voles an increase of the mitotic activity ($P<0.001$ in *Microtus arvalis* and *Microtus oeconomus,* $P<0.05$ in *Clethrionomus glarealus*) was seen. Correlated with this was the small decrease of metaphase frequency with asynchronous fusion of centromeres (ASCR) ($P<0.05$ at the *Microtus arvalis* and *Microtus oeconomus)* (Table 2).

Table 1. Spontaneous mutation spectra in different species of voles.

N. anim	N. met.	Frequency of metaphases on 1000 lymphocytes								
		A1	A2	PP	RB	CHA	ASCR	MI	BL	LM
Razezzhee (<5 Ci/km²)		*Microtus arvalis*								
15	948	44,4 ±5,1	8.6 ±2.8	0.9 0.5	0.1 ±0.5	2.5 ±0.6	16.5 ±4.9	4.5 ±0.9	5.0 ±0.8	3.0 ±0.4
Lelev (~ 20 Ci/km²)		*Microtus oeconomus*								
4	370	21.2 ±6.4	4.2 ±14	1.7 ±0.8	0	2.7 ±0.9	12.7 ±3.3	3.7 ±0.2	8.0 ±2.1	4.5 ±0.9
Lelev (~ 20 Ci/km2)		*Microtus subarvalis*								
3	170	36.3 ±13.9	5.0 ±34	3.4 ± 3.4	31.6 ±25.9	2.0 2.0	0.7 ±0.7	3.6 ±2.8	3.4 ±1.3	2.4 ±0.8
Nedanchichi (<5 Ci/km²)		*Clethrionomus glarealus*								
4	97	33.7 ±6	9.0 ±3.5	14.0 ±3.5	0.5 ±0.5	1.2 ±0.7	6.2 ±3.6	3.2 ±0.6	3.5 ±0.6	5.5 ±1.5

In the cytogenetic anomaly spectra of animals from Chernobyl zones with high-level radiopollution an increased frequency was observed in of those cytogenetic anomalies which were unstable in the zones with low-level radio pollution. (*Microtus arvalis* – aneuploidy. the voles with acrocentric autosomes – centric fusion of the chromosomes). A high frequency of chromosome inversions in the group of small-sized chromosomes in *Microtus agrestis* (about 30% of metaphases) was also revealed.

Table 2. Induced mutation spectra in different species of voles.

N anim.	N met.	Frequency of metaphases on 1000 lymphocytes								
		A1	A2	PP	RB	CHA	ASCR	MI	BL	LM
Chistogalovka (>500 Ci/km²)		*Microtus arvalis*								
9	784	52.7 ±8.3	17.9 ±4.4	0	0.4 ±0.4	3.6 ±0.8	3.7 ±0.7	10.0 ±0.6	7.9 ±0.3	6.8 ±0.5
Like Deep (500 Ci/km²)		*Microtus oeconomus*								
6	579	23.0 ±5.0	12.5 ±4.7	0	0	5.0 ±09	1.8 ±0.7	9.8 ±06	7.2 ±04	5.5 ±0.6
Red forest (~1000 Ci/km²)		*Clethrionomus glarealus*								
3	252	33.7 ±0.9	5.0 ±2.1	3.7 ±3.7	5.7 ±3.3	7.3 ±3.4	2.3 ±1.9	10.3 ±1.9	7.0 ±1.0	9.3 ±1.9
Red forest (~ 1000 Ci/km²)		*Microtus agrestis*								
2	124	25 ± 25	3.9 ± 3.6	0	16.1 ±11.9	9.7 ±14	21.7 ±11.6	1.5 ±05	1.8 ±02	2.8 ±0.2

Individual chromosome mutation spectra of some vole species were also analyzed. The group of biggest chromosomes evolved into aneuploidy with high frequency in the *Clethrionomus glarealus* (Table 3) but the same chromosomes participated with low frequency in asynchronous separation of centromere regions (Table 4).

Table 3. Participation of individual chromosomes in aneuploidy in the *Clethrionomus glarealus*.

Place of trapping of animals	**Groups of different chromosomes %**				
	NN 1-9	**NN 10-19**	**NN 20-26**	**N 27**	**N 28**
< 5 Ci/km^2 % metaphases	15.8 ±5.8	26.8 ±1.8	37.5 ±7.6	12.5 ±7.5	7.5 ±6.8
120 – 200 Ci/km^2 % metaphases	27.2 ±6.8	16.1 ±3.9	39.5 ±6.3	9.7 ±3.1	7.1 ±4.1
500 – 1000 Ci/km^2 % metaphases	35.3 ±2.2**	18.2 ±3.1**	30.5 ±4.9	6.4 ±2.1	9.7 ±3.5
Theoretical expected results	32.1	35.7	25	3.6	3.6

** P<0.01

Table 4. Participation of individual chromosomes in asynchronous separation of centromere regions in the *Clethrionomus glarealus*.

Place of trapping of animals	**Groups of different chromosomes %**				
	NN 1-9	**NN 10-19**	**NN 20-26**	**N 27**	**N 28**
< 5 Ci/km^2 % metaphases	29.6 ±4.6	8.9 ±3.4	28.2 ±8.7	27.1 ±22.9	6.3 ±6.3
120 – 200 Ci/km^2 % metaphases	14.7 ±3.2*	30.5 ±5.8*	31.8 ±4.1	17.2 ±2.6	5.3 ±2.7
500 – 1000 Ci/km^2 % metaphases	17.4 ±3.8*	35.6 ±6.5*	21.1 ±4.0	19.9 ±5.3	6.0 ±3.3
Theoretical expected results	32.1	35.7	25	3.6	3.6

* P<0.05

In contrast. in *Microtus arvalis* the group of small chromosomes evolved into aneuploidy with high frequency (Table 5) however the group of biggest chromosomes evolved with high frequency into asynchronous separation of centromere regions (Table 6).

Table 5. Participation of individual chromosomes in aneuploidy in the aneuploidy in the *Microtus arvalis*.

N metaphases	M= 46%	M= 45%	M= 47%	Chromosome loss				
				NN 1-5	NN 6-18	NN 19-22	X	Y
43	69.8	9.3	4.7	2	8	13	0	0
42	52.4	11.9	9.5	0	26	3	2	0
56	67.9	8.9	12.5	0	8	8	2	-
40	72.5	7.5	12.5	1	9	7	1	1
30	73.3	3.3	10.0	1	7	5	0	-
43	41.9	11.6	9.3	2	21	7	0	-

Table 6. Participation of individual chromosomes in asynchronous separation of centromere regions in the *Microtus arvalis*.

N metaphases	Metaphases with ASCR %	Chromosomes with ASCR				
		NN 1-5	NN 6-18	NN 19-22	X	Y
43	51.2	10	32	4	3	1
42	31.0	2	23	6	2	0
56	58.9	18	82	17	9	-
40	45.0	7	24	12	3	0
30	33.3	6	16	0	2	-
43	49.5	3	37	7	1	-

However. in cells of *Microtus oeconomus* chromosomes 10 and 14 preferably evolved into aneuploidy with asynchronous separation of centromere regions (Table 7, 8).

Table 7. Participation of individual chromosomes in aneuploidy in the aneuploidy in *Microtus oeconomus*.

N metaphases	M= 30%	M= 31%	M= 29%	Chromosome loss					
				N 1	NN2-9. X	N 10	NN 11-13	N 14	Y
24	70.8	8.3	16	-	5	1	3	1	-
46	69.6	0	15.2	-	12	4	9	3	-
49	61.2	10.2	10.2	1	7	2	4	3	-
52	69.2	19.2	3.8	-	5	2	6	-	-
29	72	0	13.8	-	2	2	1	3***	4***

*** P>0.001

Table 8. Participation of individual chromosomes in asynchronous separation of centromere regions in the *Microtus oeconomus*.

N metaphases	Metaphases with ASCR %	Chromosome with ASCR						
		On the 1 metaphase	N 1	N 2-9. X	N 10	NN 11-13	N 14	Y
24	16.7	0.13	-	-	1	1	1	-
46	24.5	0.39	-	8	1	6	3	-
49	30.6	0.59	1	7	6***	6	9***	1
52	21.2	0.42	2	11	4	3	2	-
29	10.3	0.14	-	2	1	1	-	-

*** P>0.001

4. Conclusion

Our data show that spontaneous spectra (at the low-dose radiopollution zone) of cytogenetic anomalies in bone marrow cells are characterized by the species-specific traits in voles both in the predominance of cytogenetic anomalies (centric fusion – for species with predominance of acrocentric chromosomes. aneuploidy – for *Microtus arvalis*. asynchronous separation of centromere regions – for *Microtus arvalis* and *Microtus oeconomus*) and also in evolving into anomalies of the individual chromosomes.

In the Chernobyl zones animals with constitutive chromosome aberrations did not occur.

Chronic ionizing irradiation did not lead to appearance of the unusual cytogenetic damages. In these conditions speeding up of cell proliferation in different vole species was marked and also the increase of the frequency of those cytogenetic anomalies. which had species-specific particularities in investigated vole's species was observed.

STRATEGY OF BIOLOGICAL RADIATION PROTECTION OF BIOTA AT THE RADIONUCLIDE CONTAMINATED TERRITORIES

IGOR N. GUDKOV
National Agricultural University, Geroev oboroni 15, 03041 Kiev, Ukraine,
E-mail: ingudkov@i.com.ua

Abstract. Resident population in territories contaminated by long-lived radionuclides receive most of the radiation dose by internal irradiation. Under these circumstances, the main radiation protection measure is minimization of radionuclide transfer into organisms through food and water. First, this can be achieved by food processing, using special technologies reducing the content of radionuclides in food; second, substances accelerating excretion of radionuclides from the organism and stimulation of recovery processes can be exploited.

Keywords: radionuclides, contaminated territories, radiation protection, radioprotectors, growth factors

1. Introduction

Biological or pharmacological radiation protection, as a preventive method, is generally associated to the use of radioprotectors – definite drugs or dietary supplements, given prior or during ionizing radiation exposure. Such interpretation follows the practice of radiation protection in the period of the threat of mass acute external irradiation as a consequence of a nuclear conflict.

However, besides direct protection from lethal and sublethal action of ionizing radiation in the high dose range, this class of substances finds an application for radiation protection in such spheres of the human activity as space flights, mining industry, energy radiation productions, biological and medical technologies.

N.V. Timofeeff-Ressovsky is the coauthor of several scientific works, conducted in his laboratory together with N.V. Luchnic, N.A. Izmozherov, L.S. Tsarapkin and N.A. Porjadkova (Luchnic *et al.*, 1964; 1969), devoted to radiation protection. He is also author of fundamental researches about

A.A. Cigna and M. Durante (eds.),
Radiation Risk Estimates in Normal and Emergency Situations, 101–108.

prevention of radioactive contamination and decontamination of the environment (Timofeeff-Ressovsky, 1957; 1962; Timofeeff-Ressovsky *et al.*,1960).

2. Results

Radiation protection measures are determined by the radiation exposure scenario. In radioecology, the natural habitat where radiation damage of organisms is observed occupies significantly smaller areas than contaminated territories, i.e. the regions where permissible concentration of radionuclides in food-stuff is registered (Alexakhin, 1993). After the Chernobyl accident in 1986, millions of residents in contaminated territories received up to 90-95% of the dose by internal irradiation. The main countermeasure is minimization of radionuclides uptake into an organism by food. This goal can be achieved by food processing, using special technologies, decreasing the content of radionuclides in food stuffs. On the other hand, chemicals limiting (blocking) absorption of radionuclides into an organism (Gudkov, 2002; Gudkov & Lazarev, 2003) can be exploited. Thus, such blockage can be effective not only in a link "food – man", but also in much earlier links of trophic chains: "soil – crops", "forage - agricultural animals", "crop and animal productions - food stuffs". Among agents which block radionuclide transfer into alive organisms, it is important to point out three basic groups: radionuclide antagonists, enterosorbents and complexones (substances forming an insoluble complex with radionuclide).

In a link "soil-crop" the applications of lime or others lime materials (calcium is an antagonist of strontium) and potassium fertilizers (potassium is an antagonist of cesium) belong to the first group, whereas the addition of phosphoric fertilizers (phosphorus forms with strontium the hardly soluble phosphates) belong to the second group. Finally, the application of zeolites, montmorilonits, bentonites and some other natural, specially prepared, minerals, belong to the third group. The application of these measures allows reducing radionuclide transfer into crops, and chronic radiation dose (Table 1).

In the link "crop - agricultural animals" it is also necessary to consider the preparations containing calcium, potassium and phosphorus. They provide additional minerals to fodder, and saturate a ration of animals by plants-calciphillics (leguminous), potassiphillics (maize, potato, beet). An important role rely in the immobilization of radionuclides, such as for ferrocyne. Having a specific crystal lattice, ferrocyne selectively immobilizes isotopes of cesium, forming an insoluble complex, which transits into the gastrointestinal path. The addition of ferrocyne to a ration allows to reduce the ^{137}Cs transfer to milk and meat by 6-8 times or more (Table 2).

However, nowadays one of the basic measures, promoting decrease of radionuclide transfer in animal production, is insufficient and radical improvement of meadows and pastures, including a complex of agrotechnical and agrochemical measures (providing good herbage with minimal content of radionuclides) is necessary.

Table 1. Efficiency of agrochemical measures in reducing ^{137}Cs and ^{90}Sr concentration in plant-breeding production.

Method	Type of soil	Reduction factor compared to control	
		^{137}Cs	^{90}Sr
Liming of soil	soddy-podzol light grey forest, peat bog	1.5-4 1.5-2.5	1.5-2.5 -
Applying higher (1.5-2-fold) norms of phosphoric and potassium fertilizers	soddy-podzol, grey forest peat bog	1.5-2 1.8	1.2-1.5 -
Applying of organic fertilizers, 40 t ha.$^{-1}$ and higher	soddy-podzol, grey forest peat bog	1.5-3	1.5-2
Combined applying of lime, mineral and organic fertilizers	soddy-podzol, grey forest	2.0-5	2-4
Applying of natural mineral sorbents (zeolites, vermiculites, bentonites, paligorskit)	soddy-podzol	1.5-2.5	1.5-2

Table 2. Efficiency of measures to reduce ^{137}Cs and ^{90}Sr concentration to cattle-breeding production.

Measure	Reduction factor compared to control	
	^{137}Cs	^{90}Sr
Superficial improvement of meadows and pastures	1.5-6	1.5-5
Radical improvement of meadows and pastures	1.5-10	1.5-5
Addition to fodder of ferrocyne	2-8 (to 20)	-
Addition to fodder of zeolites	2-4	-
Addition to fodder of mineral salts	1.5-2	2-3
Transfer of animal before slaughter during one month on clean fodder	2-4	-

A series of the common and special technologies of food processing allows to reduce the content of radionuclides in food stuffs (Table 3).

Table 3. Efficiency of technological measures to reduce ^{137}Cs and ^{90}Sr content in food-stuff.

Measure	Reduction factor compared to control	
	^{137}Cs	^{90}Sr
Cleaning of seeds of cereals	1.5-2	
Processing of potato to starch	15-50	
Processing carbohydrate-containing production to sugars	60-70	
Processing carbohydrate-containing production to ethyl alcohol	Up to 1000	
Processing of milk to cream	6-12	5-10
Processing of milk to butter	20-30	30-50
Culinary treatment of meat	2-4	-

Eventually, in a link "food stuff - man" the alginates and pectin can be considered the radioblockators.

The majority of radioprotectors can be divided into four groups: antioxidants, stabilizers of DNA and membranes, inhibitors of metabolism and adaptogenes, and vitamins-antioxidants and metals-microelements. Some of them, based on pharmaceutical preparations of the natural origin, can be considered more or less effective in conditions of chronic irradiation. The mechanisms of their effect, apparently, is related to adaptogenous properties.

An important role belongs to radiodecorporants – substances which intensify the radionuclide excretion from organism. Radiodecorporants are specific complexes, which bind to radionuclides (isotopes) in tissues and increase their solubility, thus accelerating their transport and excretion. The best are synthetic preparations: pentacyne, zinkacyne, tetacyne. Pentacyne (calcium trisodium pentenate) increases the excretion of plutonium, cerium, yttrium, lead and zinc isotopes from organism. Unfortunately, it does not show noticeable influence on the velocity of cesium and strontium excretion. Therefore, pentacyne does not influence the content of potassium and calcium, which are usually excreted by sorbents and complexones.

Antocyans, pigments of flavonoid group, which contribute to the coloration of plant organs (mainly flowers and fruits) from pink to black-violet, also have the ability to decorporate cesium and strontium isotopes. Antocyans have the unique ability to form water-soluble complexes with ions of one- and divalent metals. The complex with potassium correspond to dark red color, whereas with calcium and magnesium generates a dark blue compound. Analogous complexes are formed with cesium and strontium. Some other flavonoids - flavones, catechols - have similar properties.

These measures are represented in figure 1. The safety of this protection by separate remedies is presented in table 4.

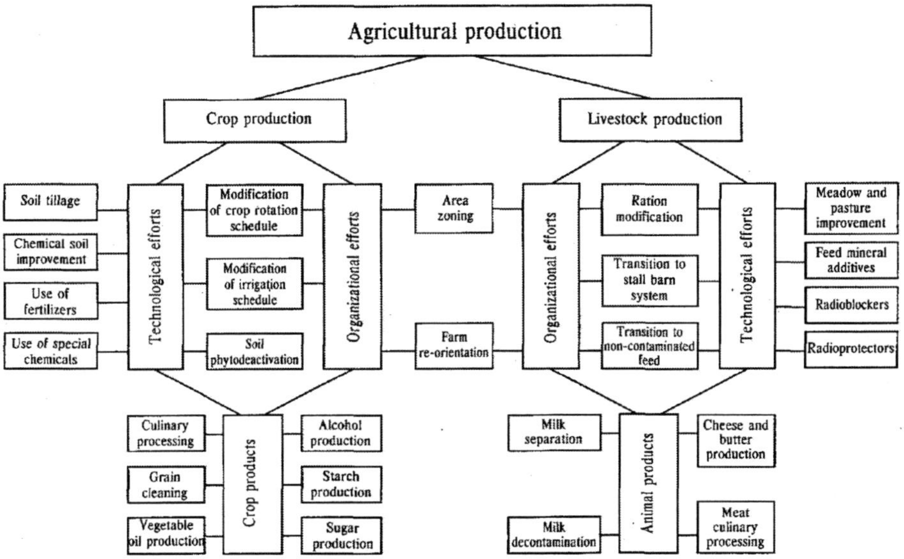

Figure 1. Radioprotective methods in crop and animal agricultural production.

It appears to be necessary to include other preparations, accelerating processes of post-irradiation recovery, to anti-radiation agents. First of all, we have the group of nonspecific physiologically-active substances: the growth factors, stimulating processes of division of surviving cells and processes of repopulation and regeneration.

The strategy of radiation protection may be schematically represented as in Fig. 2 and Table 3. These measures are represented in figure 1. The safety of this protection by separate remedies is presented in Table 4.

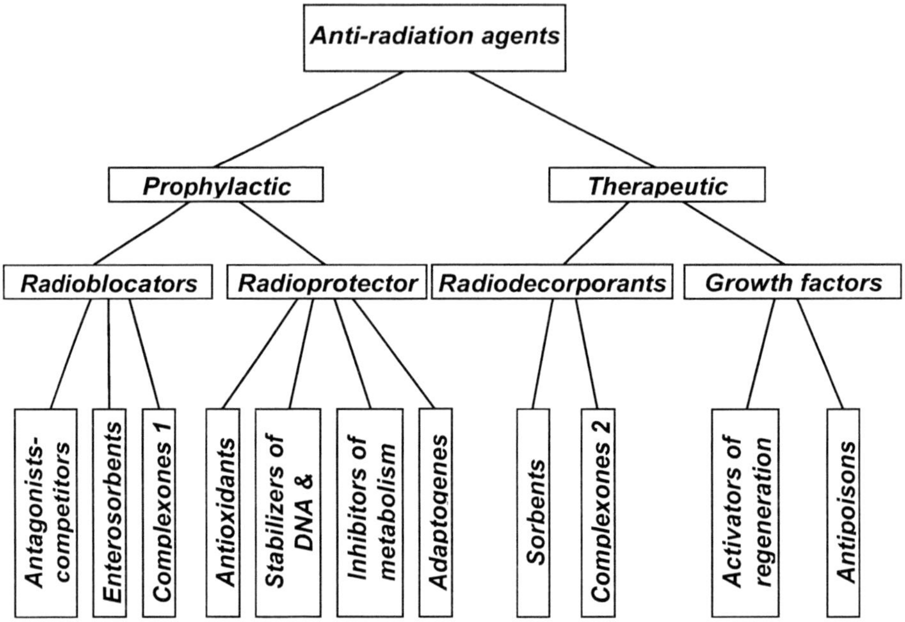

Figure 2. Anti-radiation agents.

3. Conclusion

That the arguments above suggest that radioprotective preparations should be created by the following compounds: radioprotector + radioblockator + radiodecorporant + growth factor. Such complex should provide the maximal degree of radiation protection of alive organism.

Thus, at present time the strategy of chemical/pharmacological protection of biota, including men, should include a multi-step complex of measures: blocking of radionuclides transfer into an organism in all links of trophic chain; protection from irradiation, both exterior and interior; accelerated excretion of radionuclides from organism; and activation of the recovery processes.

Table 4. Anti-radiation remedies.

Anti-radiation agents	Type of action	Remedy
Radioblocators	Antagonists-competitors	Stable isotopes
		Elements-chemical analogues
	Enterosorbents	Activated charcoal
		Natural (zeolite, montmorilonite, humolite)
		Synthetic ("Vokacit", "Sorbex", "Sorboces")
	Complexonates 1 (preparations formed insoluble complexes)	Ferrocyanides
		Alginates
		Pectins
		Phosphates
Radioprotectors	Antioxidants	Aminothiols
		Aminophosphorthioates
		Disulphides and thiosulphates
		Vitamins E, A, C,U
	Stabilizers of DNA and membranes	Metal ions
		Chelate-formed agents
		Flavonoids
	Inhibitors of metabolism	Cyanides
		Nitrils
		Azides
		Endotoxins
	Adaptogenes	Immunostimulators
		Vitamins
		Peptides
		Nucleotides
		Phytomixture
		Microelements
Radiodecorporants	Sorbents	«Algisorb»
	Complexonates 2 (preparations formed soluble complexes)	Natural (flavonoids: flavones, anthocyans, cahetins)
		Synthetic ("Zinkacyne", "Pentacyne", "Tetacyne")
Growth factors	Activators of regeneration	Cytokines
		Hormones
		Lipopolysaccarided
		Vitamins
		Microelements
		Food additions
	Antipoisons	Sterils
		Prostoglandins
		Detergents

References

Alexakhin R.M., 1993, Radioecological lessons of Chernobyl. *Radiobiology*. **33**(1): 3–14 (in Russian).

Gudkov I.N., 2002, The strategy of radioprotection of agricultural produce on radionuclide contaminated territory. *Problems of Ecological Security in Agriculture.*:Rus. Acad. Agr. Sci. Moscow, **5**: 5–14.

Gudkov I.N., Lazarev N.M., 2003, Peculiarities of agricultural keeping on radionuclide contaminated territories of Forest-stepper region. *Scientific provision of sustainable development in Forest-stepper region of Ukraine.* Kiev: Alefa,. **1:** 747–775 (in Ukrainian).

Luchnik N.V., Izmozherov N.A., Porjadkova N.A., Tsarapkin L.S., Timofeeff-Ressovsky N.V., 1969 , Reversibility of cytogenetic damages induced radiation *(Preprint). Acad. Sci. of USSR*,. Moscow: 1–16 (in Russian).

Luchnik N.V., Porjadkova N.A., Tsarapkin L.S., Timofeeff-Ressovsky N.V., 1964, About mechanism of repair of radiation damages in cell level. *Recovery processes in radiation damages*. Atomizdat, Moscow: 5–14 (in Russian).

Timofeeff-Ressovsky N.V., 1957, Use of radiations and radioactive sources in experimental biogeocenology. *Bot. Zh.* **42**(2):161–194 (in Russiun).

Timofeeff-Ressovsky N.V., 1962, Some Problems of Radiation. *Biogeocenology. Dissertation*. Sverdlosk, (in Russian).

Timofeeff-Ressovsky N.V., Luchnik N.V., 1960, Collected Works of Biophysical Laboratory of Ural Branch of Acad. Sci. Sverdlovsk, No. 3 (in Russian).

POPULATION-GENETIC CONSEQUENCES OF THE ECOLOGICAL CATASTROPHE (CHERNOBYL'S EXAMPLE)

VALERIY I. GLAZKO
Institute of agriecology and biotechnology uaas, Metrologicheskaja St. 12, Kiev, 03143, Ukraine
glazko@biotech.relc.com

Abstract. Analysis of the population-genetic consequences of technogenic catastrophes, e.g., Chernobyl, is of special interest in connection with the development of global ecological changes and rising technogenic contamination. Research on the dynamics of cytogenetic anomalies in bone marrow cells of different *Rodentia* species (trapped in the alienation zone near Chernobyl's NPP in 1994-2001 in places with different levels of contamination from 5 up to 1000 Ci/km^2) and in peripheral blood cells of cattle generations of experimental herd (Pripjat, 200 Ci/km^2) were carried out. The changes of genetic structure in cattle generations were analyzed employing family analysis of allele's transfer in structural genes and ISSR-PCR markers. Increases of mutant animals were not detected, but a reversal of genetic structure in cattle generations from an initial breed to ones, typical for more primitive breed was revealed. Our results indicate that ionizing radiation does not induce new genetic anomalies but allows realization of inherently unstable species- and individual-specific genetic characteristics.

Keywords: population-genetic adaptation, ionizing irradiation, protein polymorphism, ISSR-PCR, cytogenetic anomalies

1. Introduction

The life on the Earth has arisen and developed in the presence of the natural radioactive background (RB), a constant external abiotic factor. RB for

A.A. Cigna and M. Durante (eds.),
Radiation Risk Estimates in Normal and Emergency Situations, 109–126.

human, calculated for altitude of 1 meter above ground, fluctuates in world-wide average annual exposure near 3.5 mSv (Masse R., 2000). Exposure to natural sources is characterized by very large fluctuations, not excluding a range covering two orders of magnitude. The availability of the territories on Earth, in which RB differs dozens and hundred times from world average (Masse R., 2000), testifies the relativity of concepts of injuring dozes of ionizing radiation.

Genetic consequences of residing in radioactive provinces have been investigated for many years. The broad research of the human populations in the areas with increased RB (for example, several states in India, Ramsar in Iran) where the exposure dose of the population for one year measures from 35 to 260 mSv, have not revealed an increased level of hereditary diseases in the populations (Jaikrishan et al., 1999; Ghiassi-nejad et al., 2002).

Sudden significant elevation and persistence of sub-lethal radioactive exposure caused by human error as in Chernobyl can be considered a "changed" ecological condition. Not the new local elevated RB, but abruptness of the change, is a real long-term problem related to Chernobyl accident. The "novelty" of ecological conditions caused by human activity is the specific trait of modern time. Chernobyl serves as invaluable model for the study of the effects of abrupt fast changes in the environment on the well-being of humans and ecosystems.

Investigations of cytogenetic anomaly frequencies in somatic cells in connection with ionizing irradiation in the last 50 years were widely conducted on the people, plants, small-sized *Rodentia* species and others. The high individual variability in tested species in the same conditions, and also the absence of precise relations between quantity of cytogenetic damages in somatic cells and dozes of ionizing irradiation were the singularity of the accumulated data. Usually it is supposed that the increase of accuracy of methods will allow a more precise detection of correlations between low dozes of irradiation and induced mutation events. The first data, confirming this hypothesis were obtained by Dubrova et al. (1996) and concerned the occurrence of new minisatellite loci mutations in children of the liquidators. However, the increase of mutation frequencies was observed only in 3 of 8 investigated microsatellite loci (Dubrova et al., 1996). In other research (Weinberg et al., 2001) the increase of new mutations was detected by RAPD-PCR, but not by ISSR-PCR DNA markers. So, revealing of mutation events on DNA level in these investigations determined by specificity of variability in investigated minisatellite loci (Dubrova et al., 1996) and in DNA fragments, flanking by particular decanucleotide or microsatellite invert repeats (RAPD-PCR, ISSR-PCR) (Weinberg et al., 2001). Hence, as in cytogenetic investigations, DNA marker analysis fails to provide unambiguous data about genetic effects of low dozes of ionizing irradiation, which would not depend on specificity of initial-variability of separate character.

We propose several methods to solve this problem. First, we need to search for the markers of the individual "resistance" to ecological changes, evaluate the changes in the fitness of genotypes facing changed conditions. This could be done by analyzing generation-to-generation changes in allelic composition of populations of several species reproducing in the conditions of ecological catastrophe (it is impossible in human populations because of large life duration). It is reasonable to expect that a set of molecular genetic markers distributed across the genome whose variability has nonrandom character in generations under conditions of increased ionizing radiation can be identified.

Thus, we propose our model of the control of population-genetic changes in Chernobyl's zone using several species, for example, species *Rodentia.* Among the advantages, are fast change of generations, these animals are convenient for direct cytogenetic analysis of dividing cells in bone morrow and for population genetic research. Among the deficiencies of this model system are short cycle of reproduction, complex migration processes, and impossibility of the family analysis of animals in field conditions.

Another good object for such investigations is the cattle of the experimental farm Novoshepelichi (Pripjat). The advantages of this model is: each animal has strictly determined genealogy and is useful for family analysis; all generations born after Chernobyl's catastrophe in conditions of a zone of alienation of Chernobyl's accident are available. The cattle of the same breed in rather pure regions is available as control. Large quantity of molecular genetic markers in cattle is available and their localization in chromosomes in cattle is well characterized. In addition, similarity between cattle and human in gene syntheny is noted. deficiencies of this model could be low number of offspring and special service required.

2. Materials and Methods

2.1 ANIMALS

Mice of laboratory lines BALB/c (35 animals), C57BL/6j (35 animals) and CC57W/Mv (27 animals) were investigated in two age groups (2-3 month and 12-18 month) in control conditions (vivarium of Institute in Kiev, Ukraine (Glazko et al., 1996) and their sibs in experimental vivarium near Chernobyl's NPP, in which they exposed to chronic low ionizing irradiation (mean dose approximately 0.6 Sv,). The investigation was carried out in 1993 – 1999 in different seasons.

Representatives of a *Microtus oeconomus* (10 animals), *Clethrionomys glareolus* (29 animals) and *Microtus arvalis* (43 animals) were trapped in places of zone alienation of Chernobyl's NPP, distinguished by radio nuclide pollution from high (Red Forest, 1000 Ci/km^2), middle (Janov, near 200 Ci/km^2) and to low (Ivankov, Nedanchichi 1-3 Ci/km^2) levels.

Cytogenetic and population-genetic investigation were performed on Holstein cattle at Novoshepelichi farm located within a 5 km zone surrounding the Chernobyl nuclear power plant (near 200 Ci/km^2). As result of the accident the zone showed dramatically increased ionizing radiation (~200 Ci/km^2). These cattle are further referred as "exposed" group. Parent's generation (FP), born in "pure" zones and founded the experimental herd in farm "Novoshepelichi", included two subgroups.

(1) Bull *Uran* and three cows -*Alfa, Beta* and *Gamma* trapped in 1987 near the Chernobyl reactor when the accident in 1986 happened. These animals (FP) were founders of F1, F2 and F3 generations, born in "Novoshepelichi".

(2) Another subgroup of cows (FP) brought to Novoshepelichi from "pure" zones of Ukraine in the years 1990-1993 being founders of another Fl and F2 generations, born in experimental farm near the Chernobyl's reactor.

In both subgroups, FP as well as F1 and F2 cows were mated exclusively to only one bull -*Uran* belonging to FP. In summary, the experimental exposed herd included 17 parents, 96 F1 and 50 F2 (first and second generations born in conditions of chronic influences of low doze of ionizing irradiation).

Cattle of the same breed kept in an uncontaminated region (Dnepropetrovsk, Ukraine) served as control, (a total of 46 animals). The cattle group (36 animals) of Grey Ukrainian breed (from "pure" Kherson region, Ukraine) was included in analysis as example of more primitive breed in comparison of Holstein ones.

2.2 CYTOGENETIC INVESTIGATIONS

The preparations of bone marrow cells of representatives of *Rodentia* species and peripheral blood cells of cattle (with the use of short cell cultivation throughout 72 hours with phytohemagglutinin) were obtained by standard technique without colchicine. In *Rodentia* species the bone marrow from back legs was washed away by hypothonic solution of KCl (0.54%), fixed by a mix of methanol spirit and ice acetic acid (3:1), three times changing a fixing solution, then the cells were spread out the cold glass slides, dried and stained using Giemsa dye ("Merck", Germany). They were analyzed a binocular microscope (Karl Zeiss) at a magnification of 1000. Metaphase plates were photographed on a film "Micrat-300". The frequency of metaphase plates (in %) with following cytogenetic characteristic were counted (in %): metaphase with aneuploidy, polyploidy (PP), chromosome aberrations (CHA), interchromosome associations on a type of Robertsonian translocation (RB), with asynchronous separation of centromere chromosome region (ASCR). Aneuploidy was evaluated in two variants: general aneuploidy (A1) and aneuploidy (A2) on one chromosome (2n±1).

The quantity of metaphase plates (MI) in 1000 cells, binuclear leukocytes (BL) and leukocytes with the micronuclei (LM) calculated on the same preparations in cells with saved cytoplasm (in %). Additionally the smears of cattle peripheral blood were done and frequency of MI, BL, LM and the erythrocytes with micronuclei (EM) in them was analyzed. Statistical reliability of between group differences was evaluated with use of a Student T-test (t_S).

2.3 POPULATION-GENETIC ANALYSIS

In cattle, we used an electrophoretic method of protein separation in vertical PAGE gels on the according to the modified technique of Gahne (Gahne et al., 1977) for the analysis of the polymorphism of transferrin (TF), posttransferrin-2 (PTF-2) and receptor of vitamin D (GC) loci. The analysis of hemoglobin (HB), ceruloplasmin (CP), amylase-1 and purin nucleoside phosphorylase (PN) loci was carried out by a method of horizontal starch gel electrophoresis with subsequent histology-chemical staining using the standard techniques (Harris H., Hopkinson D. 1976).

Polymorphism kappa-casein genes were investigated with the use of PCR-RFLP analysis. For PCR-amplification of a fragment of a kappa-casein gene used the following primers:

Bocas A: 5 ' - ATGTGCTGAGCAGGTATCCTAGTTATGG - 3 '

and

Bocas B: 5 ' - CCAAAAGTAGAGTGCAACAACACTGG - 3 ',

picked up so that the DNA fragment between them included Pst I site specific for A and B allelic variants (Zadworny D., Kuhlein U., 1990). PCR-amplification carried out in the following mode: denaturation - 60 sec at + 92 C, subsequent 35 cycles - 60 sec at + 62 C, 90 sec at + 72 C. 5 mcl of amplified product was used for restriction analysis which was carried out within 4 hours at + 37 C with restrictase Pst I in the buffer of firm Sibenzime in volume 15 mcl. Allele CSN3 A contains the site to restrictase Pst I, and B – don't contain. The restriction products divided by a method of electrophoresis in 1.5%-s' agarose gels with addition of ethidium bromide and testing under ultra-violet light. The mix for PCR contained in all cases 50 ng of DNA, 15 pmol of each primer, 2.5 mcl 10 x buffer (700 mmol/l TRIS-HCL, pH 8.8 at +25° C, 170 mmol sulfate ammonium, 1.7

mg/ml BCA, 0.3 mmol/l Mg_2Cl), on 200 mcmol/l desoxinucleoside triphosphates, and also on 1,5 U of Taq polymerase ("Bion", Moscow). PCR was carried out in volume 25 mcl in thermocycler PTC-100 MJ Research, Inc. (USA).

We used also the method proposed by Zietkiewicz E. et al., 1994 for PCR-amplification of DNA fragments, flanked by microsatellite repeats (ISSR-PCR), with the use as primers dinucleotide repeats - $(CA)_{10}G$, (CG)9G and three nucleotide repeats - $GT(CAC)_7$, $(CAC)_7T$, $(AGC)_6C$, $(AGC)_6G$.

Statistical analysis (accounts of allelic and genotype frequencies, genetic distances on M.Nei's method, estimation of gene balance according to the Hardy-Weinberg's law, cluster analysis) was carried out with use of the standard computer program "BIOSYS-I", the statistical reliability between frequencies of allelic variants and phenotypes on various loci paid off with use of Fisher's criterion.

3. Results and Discussion

3.1 CONSTITUTIVE (INHERITED) MUTATIONS

Our results indicated the absence of constitutive mutations in the zone of alienation of Chernobyl's NPP in *Rodentia* species and increased resistance to radiation from 1994 to 2001 in the frequency of cells with cytogenetic anomalies in bone marrow cells. The constitutive mutations were not detected following exposure of mice lines (C57BL/6, BALB/c, CC57W/Mv) to increased (approximately 100 times) level of ionizing radiation in special vivarium (Glazko et al., 1996), in species of red and common voles, and in oeconomus voles, surprisingly, including those, trapped in the Red Forest.

In cattle, in one animal (from 160 investigated), the mutation in transferrin gene was revealed and only in the second animal generation, which was born in conditions of increased contamination by radio nuclides (200 Ci/km^2).

Carriers of Robertsonian translocation were not detected in mice and cattle in Chernobyl's zone, in spite of the presence of such mutation quite often was observed even in "pure" zones in the genomes of species with acrocentric autosomes.

3.2 CYTOGENETIC ANOMALIES IN SOMATIC CELLS.

3.2.1 *Laboratory lines of mice*

In laboratory experiments on mice lines (in special vivarium near Chernobyl's reactor), we observed an increase of cytogenetic anomalies in the bone marrow cells subjected to ionizing radiation (approx 0.6 Sv). However, only those types of cytogenetic anomalies were increased which were spontaneously highly variable in an age- or season-dependent manner in the same mice lines not subjected to radiation. For example, from 8 investigated cytogenetic characters only the frequency of binucleated leukocytes (BL) and leukocytes with micronuclei (LM) varied in relation with season of analyzing (winter, summer and autumn) and age of BALB/c mice in control conditions, and only BL and LM were increased in BALB/c exposed experimental population. In control conditions only aneuploidy (A1 and A2 types) was varied in relation with investigation season and age in C57BL/6j mice and only aneuploidy increased in exposed population. The increase of LM and metaphase plates with chromosome aberration (CHA) in old mice and in winter in comparison with summer was revealed in CC57W/Mv mice, and CHA and LM were increased in CC57W/Mv mice in special Chernobyl's vivarium. Moreover, in the group of "old" linear mice CC57W/Mv (aged 16-18 months), some cytogenetic anomalies (LM) were less frequent (5.0±0.8%) than in the mice of the same age in the control group (9.0±1.2%). This corroborates the findings of an increased rates of cell division (MI, updating of cell populations in bone morrow and elimination of defect cells) in Chernobyl's animal populations (7.0±1.8%) in comparison with control group (4.0±0.7%).

Therefore, our results indicate that ionizing radiation does not induce new anomalies in laboratory mice lines, but strengthens expression of inherently unstable line-specific cytogenetic characteristics in the investigated lines of mice.

3.2.2 *Species of voles*

In other experiments, three species of voles (*Microtus arvalis, Microtus oeconomus, Clethrionomys glareolus*) were investigated. Among them, the evolutionary youngest species of common vole (*Microtus arvalis*) characterized by comparative high karyotype instability in area was the most sensitive to ionizing radiation (Kostenko et al., 2001).

This interspecies comparison thus confirmed that an increase of ionizing radiation does not induce new genetic damages, but destabilizes the preexisting genomic "hot spots" that are either species-specific (and more characteristic to evolutionary young species) or genotype-specific (for example, different laboratory lines of mice).

We have revealed selection of radio resistant animals in environments with a high level of radio nuclide contamination. Among red and common voles trapped in Chernobyl contaminated zone in places with high (Red Forest, 1000 Ci/km^2) and middle (Janov, near 200 Ci/km^2) levels of radiation, in 1994-1996, or 16-20 generations after explosion, increased frequency of cytogenetic anomalies in bone marrow cells was revealed. In bone marrow cells of *Microtus arvalis,* trapped in contaminated zones in 1996 y, the frequencies of aneuploid metaphases (A2, 2n±1) and LM were 17.9±4.4% and 6.8±0.5% in comparison with A2=8/6±2.8% and LM=3.0±0.4% in control group (trapped in "pure" zones). In voles of *Clethrionomys glareolus* in these zones in 1996 y the frequency of metaphases with chromosome aberration was 7.3±3.4% in comparison with 1.2±0.7% in animals from "pure" zones. In 1999-2001, after 26-30 generations, no distinguishes from control groups on the frequency of cytogenetic anomalies in bone marrow cells were revealed in animals, trapped in Red Forest. So, the frequencies of A2 and LM in *Microtus arvalis* were 3.1±0.8% and 3.1±0.5%; CHA in *Clethrionomys glareolus* – 0.9±0.3%.

The intensity of selection for radio resistance was mostly expressed in the Red Forest (1000 Ci/km^2). The slower speed of such selection was observed under middle level of radio nuclide contamination. In 1999, in locations with radioactive contamination level of 200 Ci/km^2 (Janov), higher individual variability and increased frequency of cytogenetic anomalies in comparison with the control group from the "pure" zone (lower than 5 Ci/km^2) and population from a Red Forest in 1999-2001 year were revealed. For example, the frequency of metaphases with CHA in bone marrow cells of *Clethrionomys glareolus,* trapped in 1999 y in Janov, was CHA=8.1±4.0%.

3.2.3 *Cattle*

In the parent cattle generation in the experimental economy "Novoshepelichi", frequency of leukocytes with the micronuclei (LM) in blood smears was significantly higher, than in the first, in the second and in the third generations of animals that were born in the zone of increased radio nuclide contamination. This characteristic in the cattle of the third generation was significantly lower ($t_S = 3.00$; $P < 0.02$) than in the second generation. 6 animals from parent generation (LM=4.5±0.3%), 15 cattle from F1 generation, born in experimental farm (LM=1.1±0.8%), 12 animals form F2 generation (3.0±0.3%) and 3 animals from F3 generation (1.5±0.4%) were included in analysis. Frequency of binuclear leukocytes (BL) in smears of peripheral blood also was significantly higher in the parent generation than in first and in the second generations of animals.

That is, on the frequencies of cytogenetic anomalies in smears of peripheral blood in generations of cattle, which were born in conditions of increased ionizing radiation, the clear increase of radio resistance of animals was observed also.

We investigated also the fertility of cows (in average number of calves, born by one cow in one year) in parent's and F1 generations. Fertility of cows in the first generations after Chernobyl explosion on the experimental farm located in contaminated zone was reduced approximately 5 times in comparison with the parent generation (on average, from 0.93 up to 0.12 calves per cow per year).

Table 1. The allele distribution in parent's generation (FP), first (F1) and second (F2) generations of cattle born in experimental farm «Newshepelichi» in zone of alienation of Chernobyl's NPP (cattle group exposed to ionizing irradiation), in ancestor cattle breed Grey Ukrainian (from "pure" zone, Kherson, Ukraine) and in Holstein "control" group (from farm in "pure" zone, Dnepropetrovsk region, Ukraine) on TF, AMI, CP, GC, PTF-2 and CSN3 loci.

	CATTLE'S GROUPS				
	EXPOSED CATTLE's HERD IN EXPERIMENTAL FARM "NOVOSHEPELICHI"			GREY UKRAINIAN	HOLSTEIN ("control" group)
	Parent's generation (sum of 13 genotypes of mothers and 13 the same genotypes from one bull Uran)	F1 off spring	F2 off spring		
Locus					
TF (N)	26	34	21	30	45
A	.423	.412	.333	.167	.411
D1	.346	.279	.476	.000	.267
D2	.231	.309	.190	.667	.300
E	.000	.000	.000	.117	.022
F	.000	.000	.000	.050	.000
AM-1 (N)	26	34	21	30	46
A	.000	.000	.000	.000	.000
B	.038	.103	.143	.933	.283
C	.962	.897	.857	.067	.717
CP (N)	26	34	21	30	46
A	.577	.412	.524	.733	.435
B	.423	.588	.476	.267	.565
GC (N)	26	34	21	30	35
A	.327	.324	.405	.183	.057
B	.673	.676	.595	.817	.943
PTF2 (N)	26	34	21	30	35
F	.596	.529	.357	.850	.529
S	.404	.471	.643	.150	.471
CSN3 (N)	23	15	8	36	43
A	.848	.733	.625	.653	.895
B	.152	.267	.375	.347	.105
The averege hetero-zygosity	.575 (S.E. .144)	.474 (S.E. .086)	.440 (S.E. .081)	.379 (S.E. .080)	.398 (S.E. .087)

(N) - quantity of investigated animals

So, 16 cows of parent's population produced 96 calves (0.93±0.03 calf per cow per year); 20 of them (21%) died before 3 month age. However, sterile cows were absent in parent's population. In F1, of 36 cows, 21 (58%) were sterile; only 15 cows of F1 born animals of F2 generation had calves 50 calves (27♀ and 23♂) in 8 years. 13 of them died before 3 month age (26%).

If calculated for all 36 cows of F1, the cow's fertility decreased from 0.93 in parent's cow generation to 0.12 calves per cow per year in F1 cows. If the number of born calves is calculated only on 15 fertile cows of the F1 generation, the decrease would be less, to 0.73±0.06 calves per cow per year, but some fertility decrease (t_S=2.86; $P<0.01$) was revealed.

Four cows of the F2 generation in gave birth to 10 calves (F3) over 2-4 years, average, 0.94±0.06 calf per year per cow. This allowed us to suppose, that the fertility of F2 cows could increased in comparison with F1 cows (t_S=2.67; $P<0.02$).

The data obtained can result from selection pressure in F1 generation, related with new conditions of cattle reproduction (the increase level of ionizing irradiation), which lead to elimination of some genotypes.

4. Population-genetic Changes in Cattle's Generations

Analysis of the allele inheritance in different genes and DNA fragments, flanked by microsatellite loci (ISSR-PCR) in cattle's generations that were born under increased radio nuclide contamination was carried out. Data on allele frequencies observed in control and exposed groups of cattle are presented in Table 1. The homozygosity of HB locus and low level of polymorphism in PN locus are the specific traits of Holstein breed in different countries. It was true for the exposed experimental herd also and hence the data of HB and PN loci was not included in the following comparative analysis of the genetic structures of cattle groups. It is interesting that the mean heterozygosity in the exposed group (by one sire, bull *Uran)* was not lower than that in the control group sired by a number of bulls (Tab.1). The investigations covered allelic variants of the following polymorphous loci traced in the exposed group: *TF, CP, GC, AM-I, PTF-2,* and *CSN3*. At the *TF locus* three allelic variants -A, D1 and D2 were found. The rare allele Tf E and specific for ancestor breed Grey Ukraininan allele Tf F were not revealed in experimental herd. Two allelic variants were revealed at the *CP* locus - A and B. Polymorphism at the GC locus was due to two alleles: A and B. *AM-I* was represented by variants B and C. *PTF-2* locus showed fast and slow allelic variants - F and S, respectively. *CSN3* had two variants -A and B. For the first time an animal was revealed with a constitutive mutation, the carrier of a unique variant at the *TF* locus, having electrophoretical mobility different from the other five *TF* variants, including the parental and rare ones, revealed in Holstein from "pure" zone (Tf E) and in Grey Ukrainian breed (Tf F). The mutated allele (mut) had faster electrophoretic mobility than allele E, but slower than allele D2. Its genetic nature was confirmed by data on its inheritance (Tab. 2). Neither in literature nor in the control animals (bred in a relatively clean environment) was a similar allelic variant found, thus confirming the uniqueness of the mutation. Basing on the available data, it is not possible to establish precisely whether the mutation came from the dam or the sire. One may only assume, that the mutation had appeared in cow No.49 (FI. the daughter of cow Beta) and next was inherited by No. 113 (F2) and her daughter No.155 (F3), but did not appear in any other progeny of Beta and Uran (Tab. 2).

The distribution of allelic variants at the loci studied in parents and their progeny of different generations is presented in Tables 2 and 3. An analysis of the transfer of allelic variants from heterozygous parents to their progeny was carried out. Theoretically, both alleles have an equal chance of being transferred from the parents to the offspring.

In parent generation (FP) genetic structure was described as the sum of 13 different mother's genotypes and 13 identical genotypes of bull Uran, which was the father in all cases (Tab. 1). Excess of heterozygosity in some loci was observed. However, among parents excess of heterozygosity in TF locus was revealed on Uran's Tf AD1 genotype (Chi-square=15.384; p=0,002), but in F1 – on Tf AD2 genotype (Chi-square=8.975; p=0,030). In four out of the investigated six loci deviations were found from the expected parent → offspring transfer values in two generations, being significant, however only for AMI and *CP* loci (t_S=2.0, $P<0.05$ and t_S=2.8, $P<0.01$, respectively) and only in FI derived from cows *Alfa* and *Gamma* the preference of Tf D2 allele transferring from mothers to offspring was obvious (Tab. 2, 4). In general, in the case of the *TF* locus in FI, allele A was more often transferred to the offspring from the sire (AD1 genotype) while allele D2 - from the dams. In F2 the allelic transfer from *Uran* was closer to that expected (Tab. 2, 3, 4). In case of CSN3 locus the increase of B allele frequency in F1 in comparison with FP was observed also, but the differences were more small (t_S=1.77, $P<0.10$). It is very important to note that the comparative high frequencies of allele D2 in TF, B in AMI, B in CSN3 loci are the specific traits of gene pools of ancestor Grey Ukrainian breed (Tab.1). So, the data obtained demonstrated some shift of gene pools in F1 offspring, born in conditions of ionizing irradiation, from parent generation to ones, typical for more primitive cattle breed, Grey Ukrainian.

An analysis of the changes in allele frequencies occurring over the two generations demonstrated a clear disturbance in their distribution in F1, as compared to that characteristic for the parental (FP) generation. In generation F2, the observed frequencies were close to those expected.

Table 2. The transferring of allelic variants of TF, CP, AM-I, GC, PTF2, CSN3 loci from cows-mothers Alpha, Gamma, Beta and bull Uran to offspring of 1, 2. 3 generation (F1, F2, F3), which were birthing in experimental economy of alienation zone of Chernobyl's accident "Novoshepelichi".

Mother name	Mother genotype TF	Allele Tf from mother	to off-spring	TF genotype of offspring	Allele Tf from bull Uran	CP genotype of offspring	Cp allele from father	AM-I genotype of offspring	GC genotype mother-offspring		PTF2 genotype mother-offspring		CSN3 genotype mother-offspring	
Bull Uran					AD1		AB	CC	AB		FS		AB	
F1 from cow Alpha														
Alpha	AD2	D2	Rosa	AD2	A	AB	B	CC	BB	AB	FS	FF	AB	BB
Alpha	AD2	D2	Galka	AD2	A	AB	B	CC	BB	BB	FS	SS	AB	AA
Alpha	AD2	D2	105	AD2	A	AB	B	CC	BB	BB	FS	FF	AB	AA
Alpha	AD2	D2	120	AD2	A	AB	B	CC	BB	AB	FS	SS	AB	-
Alpha	AD2	D2	167	AD2	A	AA	A	CC	BB	BB	FS	FS	AB	-
F2 from cow Alpha														
Rosa	AD2	A	83	AD1	D1	AA	A	CC	AB	AA	FF	FF	BB	-
Rosa	AD2	A	116	AD1	D1	BB	B	CC	AB	AB	FF	FS	BB	-
Galka	AD2	D2	80	AD2	A	AA	A	CC	BB	AB	SS	FS	AA	AB
Galka	AD2	A	95	AD1	D1	BB	B	CC	BB	AB	SS	SS	AA	AB
Galka	AD2	A	11	AD1	D1	AB	?	CC	BB	AB	SS	FS	AA	AB
120	AD2	D2	152	D1D2	D1	BB	B	CC	AB	AA	SS	SS	-	-
F1from cow Gamma														
Gamma	AD2	D2	Maika	AD2	A	AB	B	CC	BB	BB	FS	FS	AA	-
Gamma	AD2	D2	32	AD2	A	AA	A	CC	BB	BB	FS	FF	AA	AA
Gamma	AD2	D2	15	AD2	A	AB	B	CC	BB	BB	FS	FF	AA	AB
F2 from cow Gamma														
Maika	AD2	D2	85	D1D2	D1	BB	B	CC	BB	BB	FS	FS	-	-
Maika	AD2	A	99	AD1	D1	AB	?	CC	BB	AB	FS	FS	-	AB
32	AD2	D2	93	AD2	A	AA	A	CC	BB	AB	FF	FF	AA	AA
15	AD2	A	81	AD1	D1	BB	B	CC	BB	BB	FF	FF	AB	AA
15	AD2	A	100	AA	A	AB	?	CC	BB	BB	FF	FF	AB	BB
F1 from cow Beta														
Beta	AD1	A	103	AA	A	BB	B	BC	BB	BB	FF	FF	AB	-
Beta	AD1	?	49	AD1	?	AB	A	BC	BB	BB	FF	FS	AB	-
F2 from cow Beta														
49	AD1	?	92	AD1	?	AB	?	CC	BB	BB	FS	SS	-	AA
49	AD1	?	113	**Amut**	?	AA	A	CC	BB	BB	FS	SS	-	-
49	AD1	D1	144	D1D1	D1	AA	A	BC	BB	BB	FS	SS	-	-
49	AD1	D1	158	D1D1	D1	AA	A	BC	BB	AB	FS	FS	-	-
F3 from cow Beta														
113	**Amut**	**mut**	155	**Amut**	A	AB	B	CC	BB	AB	SS	SS	-	-
113	**Amut**	A	168	AD1	D1	AB	B	CC	BB	BB	SS	FS	-	-

The notes:
"?" - cases, in which it is impossible to establish, what allele is received from mother, and with what - from the father; "-" - the data are absent. **«mut»** – mutation in transferrin's locus

Table 3. The transferring of allelic variants of TF, CP, AM-I, GC, PTF2, CSN3 loci from cows-mothers, introduced in experimental economy «New-Shepelichi» of alienation zone of Chernobyl's accident in 1990-1993 years to offspring of 1 and 2. generation (F1, F2), which were birthing in condition of increased ionizing pollution.

Mother name	Mother's genotype TF	Allele Tf from mother	to offspring	TF genotype of offspring	Allele Tf from bull Uran	CP genotype of offspring	Cp allele from father	AM-I genotype of offspring	GC genotype mother-offspring		PTF2 genotype mother-offspring		CSN3 genotype mother-offspring	
Bull Uran					AD1		AB	CC	AB		FS		AB	
without name	-	D2	4768	AD2F1	A	AB	?	CC	??	AB	??	FS	-	-
without name	-	D2	4776	D1D2F1	D1	AB	?	CC	?B	BB	?S	SS	-	-
without name	-	D2	4789	AD2F2	A	AB	?	CC	??	AB	?F	FF	-	-
6843	AD2	A	91	AA F1	A	AA	A	CC	AB	AB	FF	FF	BB	BB
6843	AD2	D2	160	D1D2 F1	D1	AB	?	CC	AB	AB	FF	FS	BB	-
6841	D1D2	D2	40	AD2 F1	A	BB	B	CC	BB	AB	FF	FS	-	AB
6827	D1D2	D2	42	AD2 F1	A	AB	B	CC	AB	AB	SS	FS	BB	AB
6827	D1D2	D1	88	AD1 F1	A	BB	B	CC	AB	AA	SS	FF	BB	-
6827	D1D2	D1	104	AD1 F1	A	AB	B	CC	AB	AB	SS	FS	BB	AB
6827	D1D2	D1	166	D1D1 F1	D1	BB	B	BC	AB	BB	SS	SS	BB	-
6824	D1D2	D1	162	D1D1 F1	D1	BB	B	BC	BB	AB	FF	FS	AA	-
6824	D1D2	D2	149	D1D2 F1	D1	AB	?	BC	BB	BB	FF	FS	AA	-
6824	D1D2	D1	79	AD1 F1	A	AB	B	BC	BB	BB	FF	FS	AA	AA
6824	D1D2	D1	7Golk	AD1 F1	A	BB	B	BC	BB	BB	FF	FS	AA	AA
6803	AD2	D2	102	D1D2 F1	D1	BB	B	CC	AB	BB	FS	FS	-	AA
6803	AD2	D2	150	D1D2 F1	D1	AB	?	CC	AB	AB	FS	SS	-	-
4789	D1D2	D2	86	AD2 F1	A	AA	A	CC	AA	AB	FF	FS	-	AA
4789	D1D2	D1	94	AD1 F1	A	BB	B	CC	AB	AA	FF	FS	-	AB
4789	D1D2	D1	169	AD1 F1	D1	BB	B	CC	AB	AA	FF	FS	-	-
4776	D1D2	D2	87	D1D2 F1	D1	BB	B	CC	BB	AA	FS	FF	-	-
4776	D1D2	D2	101	AD2 F1	A	AB	?	CC	BB	BB	FS	SS	-	AA
4776	D1D2	D1	147	D1D1 F1	D1	AA	A	CC	BB	BB	FS	SS	-	-
4768	AD2	D2	84	AD2 F1	A	BB	B	CC	AB	BB	FS	FS	AB	-
4768	AD2	A	97	AA F1	A	AB	B	CC	AB	AA	FS	FS	AB	AB
118	?D2	D2	151	AD2 F2	A	BB	B	BC	??	AB	?S	SS	-	-
118	?D2	D2	157	D1D2 F2	D1	BB	B	BC	?B	BB	?S	SS	-	-
42	AD2	D2	136	D1D2 F2	D1	AA	A	CC	AB	AA	FS	SS	AB	-
42	AD2	D2	154	D1D2 F2	D1	AA	A	CC	AB	AA	FS	SS	AB	-
7 Golka	AD1	?	122	AD1 F2	?	AB	A	BC	BB	AB	FS	FS	AA	-
7 Golka	AD1	D1	161	D1D1 F2	D1	AB	A	BC	BB	BB	FS	SS	AA	-

The notes:
"?" - cases, in which it is impossible to establish, what allele is received from mother, and with what - from the father;
"-" - the data are absent.

The reason for the inheritance disequilibrium observed in F1 may be related to a change in selection, as, also, the fertility decrease of F1 cows. Effects of an abiotic stress may lead to selection in different stages of offspring forming from parental generation which was subjected of increased level of ionizing irradiation in the zone of alienation around Chernobyl.

Thus, the change of environment could directly affect the preferable reproduction of some genotype combination of the parental animal gametes, and also change the pattern of genetic structure in F1. The effect of the environment on preferable genotype reproduction is most clearly demonstrated in the disruption of the expected allelic transfer.

Table 4. The frequency of allele transfer in transferrin (TF) and ceruloplasmin (CP) loci from cows *Alpha, Gamma* and bull *Uran* to offspring in first (F_1) and second (F_2) generation.

Allele		F1		F2	
sex	TF	offspring /parent	frequency of transfer	offspring /parent	frequency of transfer
Uran	A	8/8	1.00	3/11	0.27
cows	D2	8/8	1.00	4/11	0.36
sex	CP	offspring /parent	frequency of transfer	offspring /parent	frequency of transfer
Uran	B	6/8	0.75	5/8	0.62
cows	B	0/0	0.00	5/8	0.62

Despite close inbreeding in the exposed herd (one bull serving several generations), the heterozygosity at the loci analyzed in generation F2 was close to the mean heterozygosity in Fl. In the case of *PTF-2* and *CSN3* loci the mean heterozygosity in F2 was even higher than that in FP. Thus, over the two progeny generations examined the principal effect of inbreeding (increase in the number of homozygotes) was not observed. This phenomenon may possibly be explained by the involvement of mechanisms preserving a stable heterozygosity level, expressed as a disrupted allelic transfer from the parents to the offspring.

Thus, under changed environmental conditions the disrupted transfer of certain alleles from the parents to the offspring leads to a significant difference in allele frequencies from those expected in F1 and to a stabilization of these differences in F2.

The comparison of genetic structure of the same molecular-genetic markers between Holstein and Grey Ukrainian cattle breeds from "pure" zones, parent's generation of experimental herd and their children (born under influences of ionizing irradiation) demonstrated the shift of the offspring's genetic structure in some loci from typical ones for parents (belonging to Holstein breed) to more primitive, ancestor breed, Grey Ukrainian cattle.

Early on we revealed the increase of frequencies of cytogenetic anomalies which did not lead to cell death (such as inversions, inserts, reciprocal translocations) in blood cells of children (14 – 16 years age) who received the doze of ionizing radiation (0.3-0.4 Sv) in utero (Nastiukova et al., 2002). Obviously, it related to the clonal cell expansion after acute ionizing irradiation. Our data about the cattle fertility decrease in F1, allowed to suppose, that children exposed to low dozes of ionizing radiation in utero, could face reproductive problems in the future.

The analysis of polymorphism and heritability of some molecular genetic markers of anonymous sequences in DNA (ISSR-PCR) was carried out using the method described by Zietkiewicz E. *et al.* (1994). The following sequences were used as primers: dinucleotide repeats - (CA)10G, (CG)9G and trinucleotide repeats - GT(CAC)7, (CAC)7T, (AGC)6C and (AGC)6G.

The amplification spectra were analyzed for two families of the exposed group -from cows *Alfa* and *Beta* mated, as always, to *Uran.* In the case of dinucleotide repeat (CA)10G, in the amplification spectra of all investigated animals, eight distinct DNA fragments were observed -from 750 to 1900 bp. These fragments were identified in all animals -both in the parents and in the offspring, and neither individual variation nor the occurrence of new variants was observed in the progeny. The same pattern was observed when (CG)9G primer was used -the amplification spectra had six precisely identified fragments 650-1500 bp long. Neither individual variation nor differences of amplification spectra from parental variants were found when using the trinucleotide repeat (CAC)7T as a primer, and ten fragments were recorded in its amplicon spectrum -400-1600 bp long. Thus, considering each fragment of amplified DNA as a separate locus, it was possible to conclude that in all 24 DNA *loci* and in all animals analyzed no individual variation appeared and no new mutation variants occurred in the progeny born under increased ionizing radiation.

The use of three other primers, consisting of trinucleotide microsatellite repeats -GT(CAC)7, (AGC)6C and (AGC)6G -resulted in the formation of polymorphous spectra of amplification products. Using the GT(CAC)7 primer produced ten fragments 650-2000 bp long, in the spectra of amplification products. The 2000 bp fragment was absent in cow *Alfa* and in her FI offspring (Nos. 120 and 105). The progeny of cow *Beta* (No.113 and 155, respectively) were also lacking this fragment. Fragments 1800 and 1700 bp long were absent in F2 and F3 (No.113 and 155, respectively) of cows *Alfa* and *Beta.* The 1500 bp fragment was absent in cows Nos. 49, 113, 144 and 155 *(Beta's* progeny), while fragment 1400 bp -in No.105 *(Alfa's* progeny). Thus, among ten fragments (loci) of GT(CAC)7 primer, five appeared polymorphous as they were present or absent from the amplification spectra. Basing on the data obtained one may conclude that

Uran was heterozygous with respect to the 2000 bp fragment, *Alfa* was homozygous as regards the absence of the fragment, while *Beta* homozygous as regards its presence. The distribution of this fragment in the progeny was close to that expected.

The use of the (AGC)6G primer resulted in a wide spectrum of amplification products, consisting in total of 27 fragments (600-2600 bp) from different animals. Again no amplification products which would point to the occurrence of mutations were found. Altogether, with the existing polymorphous spectra in different animals the use of these two trinucleotide primers resulted in 49 amplification products.

The most complex spectrum of amplification products was observed with (AGC)6C primer. Two amplification products were revealed in *Alfa's* daughter No.105 that were absent in both parents. A poor reproducibility of the amplification spectra in this case was also marked. One may assume that such complexity of the amplification spectra and poor reproducibility were caused by the primer itself, in particular by the presence on its 3' end of the nucleotide combination GCC, which promotes formation of a "mini-pin". It could result in a poor annealing accuracy. For this reason it proved necessary to make a special family investigation establishing whether the new bands in animal No.105 were a result of mutation events or artefacts of an inaccurate reannealing of the primers.

In total, using two dinucleotide and three trinucleotide primers, 73 amplification products were found in the progeny of cows *Alfa* and *Beta* mated to the bull *Uran.* No changes were observed which could be interpreted as a new mutation. Two unique bands found in one offspring of *Alfa* with the (AGC)6C primer could be the effect of the reduced accuracy of annealing, and thus requires further research. Interesting is the high heterozygosity of *Uran,* shown by the analysis of polymorphous spectra of anonymous DNA fragments and by the polymorphism of structural genes. It is possible that the prolonged and successful fertility of the bull under both increased ionizing radiation and inbreeding was caused by his high heterozygosity.

It is important to note that observed shift of a genetic structure in cattle generations in the direction of the less specialized forms was in agreement with the literature data about a decrease of a number of behavioral specialized functions in voles (more primitive relatives of burrows) in conditions of increased radio nuclide contamination (Maslov B.I., 1972), and also with the data of the Danish investigators about disturbance of functions of associative thinking in Danish children after the first air explosions of nuclear bombs and after Chernobyl accident (Sankaranarayanan K., 1991).

All these appearances corresponded to a rule of I.I. Shmalgauzen (1983) that any change of the environment lead to preferable reproduction of the more primitive forms within a species. Thus, the main problem after the Chernobyl's catastrophe, as well as other ecological changes, lies not in the occurrence of the new mutant organisms, but in the long-term changes of the genetic structure of populations and, accordingly, in the appearance of the new interspecies interactions between the less specialized (marginal) representatives of each species in species communities.

4. Conclusions

1. Problem of Chernobyl's catastrophe is that the populations of different organisms were subjected to dozes of ionizing radiation which were new to them.

2. Increased level of ionizing radiation did not induce qualitatively new damages of the genetic material, but increased chromosomal instability in those species or at those cytogenetic anomalies that have been determined to be *a priori* more prone to appearance of cytogenetic defects than others. This provides a basis for the hypothesis that evolutionary younger species are more sensitive to the change of ecological conditions at the chromosome rearrangement level in comparison with older ones.

3. Effects of detrimental environmental changes have deferred realization in generations. Decrease of the reproduction function in cattle was observed in cows, which were born in first generation in a zone of alienation of Chernobyl's NPP, possibly, in connection with the particularities of mammalian oogenesis (maturation of ooblasts to meiotic stage before birth). Strong selection for radio nuclide resistance in voles emerged through approximately 26 generations after the beginning of the ionizing radiation exposure and it was dose-dependent.

4. No increase in the quantity of constitutive mutations in investigated genes, ISSR-PCR markers or chromosomes in analyzed species (cattle and *Rodentia* species) was observed.

5. In generations of cattle, disturbance of equiprobable transmission of alleles of a number of molecular genetic markers, increase of heterozygosity and radio resistance were observed.

6. In family analysis the changes of genetic structure in cattle generations of experimental farm "Novoshepelichi", the shift of gene pool from typical for specialized parent dairy breed Holstein to that characteristic for the less specialized breeds was revealed (decrease in level of specialization).

References

Dubrova Y.E., Nesterov V.N., Krouchinsky N.G., Ostapenko V.A., Neumann R., Neil D.L., Jeffreys A.J., 1996, Human minisatellite mutation rate after the Chernobyl accident. Nature, 380: 683–685.

Gahne B., Juneja R.K., Grolmus J., 1977, Horizontal polyacrylamide gradient gel electrophoresis for the simultaneous phenotyping of transferrin, post-transferrin, albumin and post-albumin in the blood of plasma of cattle. *Animal Blood Groups and Biochemical Genetics*, **8:** 127–137.

Ghiassi-nejad M., Mortazavi S.M., Cameron J.R., Niroomand-rad A., Karam P.A. (2002) Very high background radiation areas of Ramsar, Iran: preliminary biological studies. Health Physics, **82:** 87–93.

Glazko T.T., Saphonova N.A., Stolina M.R., Maluta S.S., Glazko V.I. (1996) Cytogenetic variability in laboratory mice lines C57BL/6, BALB/c and CC57W/Mv in 30-km zone of Chernobyl's NPP. Dokladi Nacionalnoi Akademii Nauk Ukraini (Proceedings of National Academy of Science of Ukraine), **57:** 137–144.

Harris H., Hopkinson D.A., 1976, *Handbook of enzyme electrophoresis in human genetics.* North-Holland Publ. Comp., Amsterdam.

Jaikrishan G., Andrews V.J., Thampi M.V., Koya P.K., Rajan V.K., Chauhan P.S., 1999, Genetic monitoring of the human population from high-level natural radiation areas of Kerala on the southwest coast of India. I. Prevalence of congenital malformations in newborns. *Radiation Research*, **152:** 149–153.

Kostenko S.A., Buntova E.G., Glazko T.T., 2001, Species specificity of karyotype instability under conditions of radionuclide contamination (the Chernobyl Nuclear Power Plant) in the voles, *Micrototus oeconomus, Micrototus arvalis,* and *Clethrionomys glareolus. Tsitology and Genetics*, **35**: 11–18.

Maslov B.I., 1972, Radiation situation in dwellings and shelters of mice-like *Rodentia* in contitions of biogeocenosis with increased natural radiation. – In: Verkhovskaya I.N. (Ed.) *Radioecological investigation in natural biogeocenosis*. Nauka, Moskow: 216–226,

Masse R., 2000, Ionizing radiation. *Comptes Rendus de l'Academie des Sciences - Series III.*, **323:** 633–640.

Nastiukova V.V., Stepanova E.I., Glazko V.I., 2002, Cytogenetic effects in children exposed to low doses of ionizing radiation. *Tsitology and Genetics*, **36**: 38–45.

Sankaranarayanan K., 1991, Ionizing radiation and genetic risks. IV. Current methods, estimates of risk of Mendelian disease, human data and lessons from biochemical and molecular studies of mutations. *Mutation Research,* **258**: 99–122.

Schmalgauzen I.I., 1983, *The ways and lows of evolution process*. Moscow: Nauka. 1–360.

Weinberg H.S., Korol A.B., Kirzhner V.M., Avivi A., Fahima T., Nevo E., Shapiro S., Rennert G., Piatak O., Stepanova E.I., Skvarskaja E., 2001, Very high mutation rate in offspring of Chernobyl accident liquidators. *Proceeding of Royal Society, London, B Biological Science,* **268:** 1001–1005.

Zadworny D., Kuhlein U., 1990, The identification of the kappa-casein genotype in Holstein dairy cattle using the polymerase chain reaction. *Theoretical and Applied Genetics*, **80:** 631–634.

Zietkiewicz E., Rafalski A., Labuda D., 1994, Genome fingerprinting by sequence repeat (SSR) - anchored polymerase chain reaction amplification. *Genomics,* **20:** 176–183.

PRINCIPLES AND RESULTS OF GENETIC MONITORING OF CHEMICAL MUTAGENS AND RADIATION EFFECTS IN ARMENIA

ROUBEN AROUTIOUNIAN
Department of Genetics and Cytology, Yerevan State University, 375025, Yerevan, Armenia
E-mail: genetik@ysu.am

Abstract. The genetic effects of environmental pollution in Armenia are presented. Most of them are underway at Yerevan State University in cooperation with other institutions in Armenia, Germany, France and USA. The results of genetic monitoring of effects of environmental pollution obtained by different methods are presented. The routine analyses of chromosomal aberrations in groups of genetic risk can give data only about nonspecific action of environmental pollutants. The detection of genotoxic effects became more differentiated after studying –clastogenic factors (ultrafiltrates of blood plasma). FISH analysis was used in groups of patients with leukemia and inborn defects. We applied the Comet Assay to study DNA damage and repair in leukocytes of Chernobyl accident liquidators and patients with familial Mediterranean fever. The results showed an increased sensitivity or changed repair capacity of DNA of their cells exposed to UV-C. Analysis of micronuclei induction in exfoliated cells revealed the significant increase of chromosomal damage and their nondisjunction in the groups exposed to mutagens. For the investigation of groups of genetic risk in human populations we apply programmes with combination of presented methods. Cytogenetic programmes are usually combined with epidemiological data. Our research group is responsible for the genetic monitoring of the Armenian nuclear power plant and its environment. The results of chromosomal and point mutations monitoring in plants are presented.

Keywords: genetic monitoring, genotoxins, environmental pollution, Armenia

A.A. Cigna and M. Durante (eds.),
Radiation Risk Estimates in Normal and Emergency Situations, 127–136.

The ecological situation in Armenia, and in the capital city Yerevan, has changed dramatically especially over the past 15 years. During this time many industrial plants have closed down whilst others have reduced operations, but the number of road vehicles is increased three-fold. Because of private building constructions the open areas covered by gardens in Yerevan decreased to around one third. Yet, the pollution of surface water was greatly increased. Thus, the structure of environmental pollution is drastically changed.

Detecting effects of these radical changes has been undertaken by a number of basic techniques of biomonitoring, using different genetic and cytogenetic markers that assist in identification of risk factors.

The following groups at genetic risk in Armenia were investigated: subjects in contact with various pollutants (mainly industrial and agricultural) were analyzed by the chromosomal aberration (CA) test in blood cells and the micronuclei (MCN) test in buccal smears; patients with familial Mediterranean fever - by the Comet-assay, CA, and clastogenic factor test (CF); cancer patients treated with cytostatics – by MCN; cancer patients treated with radiotherapy - by Comet-assay and MCN; and finally Chernobyl accident liquidators- by MCN, Comet-assay, CA, and CF.

Increased levels of genetic and cytogenetic changes in most of the investigated groups were observed, but in general they were not drastic.

We have to state that the analysis of CA alone can not be used as a comprehensive parameter of genetic damage, but it needs to be included within wider frameworks of genetic monitoring.

Conventional cytogenetic methods were therefore enlarged in our research with the molecular-cytogenetic method FISH (fluorescence in situ hybridization). It allowed us both to improve molecular-cytogenetic diagnostics and monitor secondary cytogenetic anomalies including effects of treatment and management of patients.

For FISH diagnostics a DNA probes library was created at our Department (Arutyunyan et al., 2003). Conventional and molecular cytogenetic investigations of more than 400 patients with leukemia were performed. In 30% of them FISH results were compared with data obtained by conventional cytogenetic analysis (Kasakyan et al., 2003).

One of the most effective methods of cytogenetic research is the investigation of clastogenic factors (CF) - cell substances that induce chromosomal aberrations. CF are found in blood plasma of persons from different groups at genetic risk even many years after exposure to radiation. Detection of this clastogenic activity in cells is comparable with biochemical measuring of the prooxidant capacity of the cells. It was shown that CFs, being indicators of the cell oxidative stress, at the same time are indicative of the risk factor for the development of cancer, autoimmunological and inflammatory diseases. Blood plasma of Chernobyl liquidators treated at the Center of Radiation Medicine and Burns of the Armenian Ministry of Health was tested in cooperation with Prof. I. Emerit (Paris, 6 University) for clastogenic activity in blood cultures of healthy donors. The number of aberrations in the test cultures from two groups of liquidators exposed to the average dose of 0.6 and 0.2 Gy) was, 8 years after the accident, shown to be equal to 17.9% and 10.5% respectively against 5.7% in the control. Comparable results were also found in an international project that studied liquidators who had emigrated from USSR to Israel (Emerit et al., 1994).

The Comet-assay (single cell electrophoresis) has been used extensively in genetic toxicology as a sensitive, simple and rapid technique to investigate DNA damage and repair induced by various agents in a variety of mammalian cells. We applied the Comet-assay to estimate DNA damage induced by UV-C in the blood of Chernobyl accident liquidators 10 years after the accident. Irradiation doses of that group of liquidators were not higher than 0.25 Gy (Arutyunyan et al., 2000, 2001).

The Comet-assay was also used to estimate DNA damage induced by UV-C in the blood of patients with Familial Mediterranean Fever (FMF) (Arutyunyan *et al.,* 2002).

Our results showed an increased sensitivity of DNA from leukocytes of liquidators and patients with FMF to UV-C irradiation compared to leukocytes from controls. The difference between the groups is significant and we have every reason to recommend the Comet Assay for testing of mutagenic effects in groups at genetic risk. Now we apply the Comet-FISH approach to detect both the total DNA damage and specific DNA sequences.

Analysis of MCN, revealed a significant increase of chromosomal damage and nondisjunction in cells of workers exposed to mutagens from different industries in Armenia (TABLE 1).

An increase in cytogenetic parameters was the most pronounced in cancer patients undergoing antiblastic therapy (Nersesyan et al., 2001, 2002). It can be suggested that the MCN assay in exfoliated cells can be very useful as a sensitive, simple, rapid and economical endpoint for study of induced mutation in exposed subjects.

Thus, the approach outlined above has been shown to be informative for eco- and toxicogenetic monitoring programmes for Armenia.

For the investigation of groups at genetic risk we can recommend two programmes of cytogenetic monitoring:

Program-minimum - investigation of MCN and Comet-assay

Program-maximum - investigation of MCN, Comet-assay and CA (Aroutiounian *et al.*, 2001).

These programs can be expanded by considering untoward outcomes of pregnancies in the at-risk groups.

Table 1. Mean level of micronuclei in exfoliated buccal mucosa cells of workers from different industrial plants of Armenia.

Investigated group	Mean ± SE in the group	Men	Women
Control group I	0.3±0.1	0.2±0.1 (n=5)	0.4±0.2 (n=14)
Workers of metallurgical industry	1.3±0.1*	1.4±0.2 (n=30)	1.5±0.4 (n=3)
Control group II	0.5±0.1	0.5±0.2 (n=13)	0.5±0.1 (n=20)
Workers of biotechnological industry	1.2±0.2*	2.2±0.1 (n=9)	0.8±0.1 (n=26)

* - $p<0.05$

Today screening for genotoxicity in our country is virtually limited to newly synthesized drugs and the majority of experiments are performed mainly on cell cultures. Our research group together with specialists in molecular biology is responsible for investigation of genotoxic properties of cytostatics (Arutyunyan et al., 2004; Hovhannisyan *et al.*, 2005). In particular this has concentrated on new porphyrin derivatives that have been synthesised in Armenia mostly for photodynamic therapy of cancer.

Epidemiological analysis was applied in our research practice to consider severe anomalies of pregnancies e.g., spontaneous abortions and stillbirths. This approach was informative in the genetic monitoring of personnel from dozens of chemical plants and agricultural regions and provided an insight into the patterns of dominant lethal mutations in the groups under study.

The investigation of genetic load in the rural population of Armenia was carried out by interviews with 2000 woman of reproductive age in the Ararat region sub-divided into zones of varying levels of pesticides application. An increase in the abnormal pregnancy outcomes was shown for the zone with highst level of pesticide application. Most cases of hereditary pathology comprised children with mental retardation. By contrast however the cytogenetic analysis showed no changes in the levels of chromosomal aberrations in culture of lymphocytes of rural workers in contact with pesticides (Airian et al., 1990).

We also applied a reverse approach, whereby on the basis of genetic assays pathology the etiology of group pathologies can be indicated. The interview with the 293 parents of children with oligofrenia from auxiliary schools for mentally retarded children showed a high percent of heavy-drinking or drug-addict fathers compared with fathers of children with normal mental development. Moreover an increase was found in the frequency of children with oligofrenia where both parents were exposed to a polluted working environment (Airian et al., 1988).

The incidence of leukaemia in children in Armenia was analysed for the period from 1991 to 2002. During 12 years this parameter was not changed substantially. It was concluded that the incidence of leukaemia in children in Armenia was not connected with parental exposure to chemicals (Nersesyan *et al.*, 2003).

Analysis of the distribution of carcinogens and mutagens in the environment of Armenia showed that the content of heavy metals and pesticides in agriculture food products exceeds the hygienic limits. It was shown that tobacco, food and environmental pollution caused a cancer incidence higher than in developed countries (Nersesyan *et al.*, 2001b).

We investigated also the levels of environmental genotoxicants in different regions of Armenia using sensitive plant test-systems that were able to indicate mutagenicity of air, soil, subsoil, surface water and vegetation.

The highly sensitive assays using the frequency of pink mutational events (PME) in Tradescantia (clone 02) stamen hairs (Trad-SHM) and the formation of micronuclei in tetrads (Trad-MCN) gave important data on the effects of the level of pollution for a few chemical factories and ambient air in Yerevan.

The study of atmospheric pollution indicated the genotoxic effects of air pollutants from a chloroprene rubber plant that was the major contributor to air pollution in Yerevan. Ten major monitoring sites were selected which showed that not only was there a large increase in the frequencies of PME within the industrial area but that genotoxic effects were detected at sites even 1.5 km away (Arutyunyan *et al.*, 1999).

We also studied the mutagenic activity of soils from areas near to the settlements Metsamor, Aghavnatun, Armavir and Oshakan situated within 30 km of the Armenian Nuclear Power Plant (ANPP). As controls, soils from the greenhouse of the State University in Yerevan, situated outside that zone, were used. Significant correlation between the levels of ^{137}Cs and the frequency of point mutations, tetrads with MN and MN in Tradescantia was demonstrated (TABLE 2) (Aghajanian *et al.*, 2004).

Table 2. Use of the Trad-SH and Trad-MN tests for the estimation of soil genotoxicity in the vicinity of the Armenian Nuclear Power Plant.

Variant	PME, 1000± m	CME, 1000± m	Tetrads with MN, %±m	MN in tetrads, %±m	^{137}Cs Bc/kg
Metsamor	1,196±0,26*	8,03±0,67*	15,87±0,67*	26,90±0,81*	19,4
Aghavnatun	1,92±0,33*	8,13±0,68*	15,13±0,65*	24,20±0,78*	15,3
Armavir	1,44±0,28*	6,64±0,62*	14,72±0,65*	21,80±0,75*	15,7
Oshakan	0,92±0,23*	12,56±0,85*	32,0±0,85*	61,0±0,89*	65,6
Control	0,41±0,16	3,93±0,48	9,8±0,54	13,40±0,62	12,0

* - $p<0.05$

The genotoxicity of subsoil waters from nine artesian wells from different locations (the depth of wells varies from 40 to 350 m) in Armenia was investigated. It was shown that all samples of waters studied increased the frequency of PME in the Trad-SHM test. A trend was found for the frequency of PME to be decreased with the increase of artesian wells' depth. Use of the Trad-MCN test also showed an overall increase in artesian well waters' genotoxicity, compared with controls (Poghosyan *et al.*, 2005).

In other research specimens of water from 25 locations of rivers Razdan, Getar, Marmarik and Sevjur flowing through the densely populated and intensively polluted areas of Armenia were studied. The genetic monitoring undertaken over three years (2000-2002) indicated that the variations of mutagenic and clastogenic activity of the waters correlated with the non-uniform distribution of the population, different level of man-made pollution and seasonal variations (Matevosyan *et al.*, 2005).

Reproductive parameters of the male gametophyte were determined on the basis of pollen sterility in some fruit trees and vines analyzed in plants growing at distances of 3-5 km from the ANPP - near to the settlement of Metsamor. The control area was chosen more than 30 km distant from the ANPP. The results suggested a high degree of for both the trees and vines (TABLES 3 and 4) near to Metsamor that did not differ significantly from the control area. The further monitoring of pollen fertility is advisable using other plant species growing around the ANPP in order to reach a more comprehensive conclusion regarding the suitability of this assay as indicative of environment well-being (Aroutiounian *et al.*, 2004 b, 2004 c).

Table 3. Estimates of total genotoxic effect of environmental factors by the analysis of pollen fertility of fruit trees around the nuclear power plant compared with a control area.

fruit-tree variety	Pollen fertility, %	
	Metsamor	Control point
Pear, Malacha	95.44± 0.21	94.84± 0.22
Pear, Dzmernuk	79.71± 0.40	86.90± 0.34
Peach	88.69± 0.32	95.61± 0.21
Plum	96.93± 0.17	87.60± 0.33

Table 4. Vine pollen fertility around the nuclear power plant compared with a control area.

vine varity	Pollen fertility, %	
	Metsamor	Control point
Charentsi	95.36 ± 0.21	95.16± 0.21
Meghrabuyr	90.60 ± 0.29	90.80 ± 0.29
Nerkarat	98.84 ± 0.11	97.62± 0.15
Burmunk	69.60 ± 0.18	97.69 ± 0.15

We compiled a list of cultivated species growing around the ANPP and ranked them by their nuclear DNA C-values and genome size. It has been suggested that a large genome confers a selective disadvantage for plants under extreme environmental conditions (Vidic *et al.*, 2001).

The data on genome size were obtained from the Angiosperm DNA C-Values Database (release 5.0, Dec. 2004) on site:

http://www.rbgkew.org.uk/cval/homepage.html.

The genome size for 15 fruit species and 16 species of vegetables were compared. It was shown that some of them (plum, apple, pepper, onion, garlic) are more sensitive to environmental pollution than others (strawberry, apricot, peach, sweet cherry, water-melon, pumpkin, marrow, radish). The results can be used to present recommendations for future genetic monitoring and, probably, preferable planting in the environment of the ANPP (TABLES 5 and 6).

What are the current problems of genetic monitoring we need to solve in our country?

The early approach, when investigators were happy to identify any changes in cytogenetic damage or slow levels of gene mutations frequency is long gone.

Table 5. Nuclear DNA C-values of fruit species growing in the Ararat valley of Armenia.

Genus	Species	Chromosome number	Ploidy level	1C (Mbp)	1C (pg)
Fragaria	*viridis*	14	2	98	0.10
Prunus	*persica*	16	2	270	0.28
Prunus	*Armeniaca*	16	2	294	0.30
Rubus	*idaeus*	-	-	294	0.30
Prunus	*avium*	-	-	343	0.35
Vitis	*vinifera*	38	2	417	0.43
Ribes	*glutinosum*	16	2	534	0.55
Pyrus	*communis*	34	2	534	0.55
Prunus	*cerasus*	32	4	613	0.63
Ficus	*carica*	26	2	686	0.70
Punica	*granatum*	16	2	706	0.72
Cydonia	*oblonga*	34	2	711	0.73
Morus	*alba*	-	-	833	0.85
Prunus	*domestica*	48	6	907	0.93
Malus	*communis*	34	2	2205	2.25

Table 6. Nuclear DNA C-values of vegetables growing in the Ararat valley of Armenia.

Genus	Species	Chromosome number	Ploidy level	1C (Mbp)	1C (pg)
Citrullus	*vulgaris*	22	2	441	0.45
Cucurbita	*pepo*	40	-	539	0.55
Cucurbita	*pepo*	40	-	539	0.55
Raphanus	*sativus*	18	2	539	0.55
Cucumis	*sativus*	14	2	882	0.90
Cucumis	*melo*	24	2	931	0.95
Solanum	*melongena*	24	2	956	0.98
Daucus	*carota*	18	2	980	1.00
Lycopersicon	*esculentum*	24	2	1005	1.03
Spinacia	*oleracea*	12	2	1005	1.03
Apium	*nodiflorum*	22	2	1054	1.08
Beta	*vulgaris*	18	2	1225	1.25
Hibiscus	*cannabinus*	36	-	1495	1.53
Capsicum	*annuum*	24	2	3920	4.00
Allium	*sativum*	16	2	15901	16.23
Allium	*cepa*	16	2	16415	16.75

Today the emphasis has changed. It has become essential to focus on the genetical prognosis from population to individual risk, for understanding the relevance of the assay endpoints to our health.

The essential part of the research presented is supported by ISTC Grants No A-301.2 and A-773. MCN and epidemiological research was undertaken in collaboration with Dr. Armen Nersesyan.

References

Aghajanyan, E.A., Aroutiounian, R.M., Atoyants, A.L., and Pogosyan, V.S., Pyuskyulyan, K.A., 2004, Application of Trad-SH and Trad-MN tests for the estimation of soils genotoxicity in the zone of Armenian Nuclear Power Plant, *International conference «Unification and optimization of radiation monitoring on NPP location regions»*, Armenia, Yerevan: 223–225.

Airian, A.P., Pirumian, M.S., and Arutyunian, R.M., 1990, Investigation of genetic load in rural population in the Armenian SSR. *Genetika*, **26**(11): 2065–2069 (In Russian).

Airian, A.P., Shaverdyan, A.M., and Arutyunyan, R.M., 1988, Estimates of "risk factors' in parents of children, learning in auxiliary schools, *Biological Journal of Armenia*, **41** (9) : 732–735 (In Russian).

Aroutiounian, R.M., Ervandyan, S.G., and Nebish, A.A., 2004, Application of male gametophyte generation of vines for the investigation of genotoxicity of environmental factors. *International conference «Unification and optimization of radiation monitoring on NPP location regions»,* Armenia, Yerevan: 160–162.

Aroutiounian, R.M., Ervandyan, S.G., and Nebish, A.A., 2004, Estimates of total genotoxic effect of environmental factors by the analysis of male gametophyte generation of fruit trees. *International conference «Unification and optimization of radiation monitoring on NPP location regions»,* Armenia, Yerevan: 27–29.

Aroutiounian, R.M., Hovhannisyan, G.G., and Nersesyan, A.K., 2001, Monitoring of genotoxic effects in population of Armenia: Principles and results. *Mutation Research*, **483**: 59.

Arutyunyan, R., Gebhart, E., Hovhannisyan, G., Greulich, K.O., and Rapp, A., 2004, Comet-FISH using peptide nucleic acid probes detects telomeric repeats in DNA damaged by bleomycin and mitomycin C proportional to general DNA damage, *Mutagenesis*, **19**(5): 403–408.

Arutyunyan, R., Kasakyan, S., Mkrtchyan, H., Hovhannisyan, A., Manvelyan, M., and Muradyan, A., 2003, Application of DNA probes library in molecular cytogenetic analysis. *Experimental Oncology*, **25**: 307–309.

Arutyunyan, R.M., Collins, A., Hovhannisyan, G.G., Ghazanchyan, E.G., and Nersesyan, A.K., 2002, Endogenous oxidative DNA base damage in lymphocytes of patients with familial Mediterranean fever, *Central European Journal of Occupational and Environmental Medicine*, **8**(4): 283–288.

Arutyunyan, R.M., Hovhannisyan, G.G., Ghazanchiyan, E.G., and Nersesyan, A.K., 2001, DNA damage induced by UV-C in leukocytes of Chernobyl accident clean-up workers, *Central European Journal of Occupational and Environmental Medicine*, **7**(1): 15–21.

Arutyunyan, R.M., Hovhannisyan, G.G., Ghazanchiyan, E.G., Oganessian, N.M., and Nersesyan, A.K., 2000, UV-C induced DNA damage in leukocytes of Chernobyl accident clean-up workers, *Exp. Oncol.*, **22**: 219–221.

Arutyunyan, R.M., Pogosyan, V.S., Simonyan, E.H., Atoyants, A.L., and Djigardjian, E.M., 1999, In situ monitoring of the ambient air around the chloroprene rubber industrial plant using the Tradescantia-stamen-hair mutation assay, *Mut Res.,* **426**: 117–120.

Emerit, I., Levy, A., Cernjavski, L., Arutyunyan, R. et al., 1994, Transferable clastogenic activity in plasma from persons exposed as salvage personel of the Chernobyl reactor. *J. Cancer Res. Clin. Oncol.* **120**: 558–561.

Hovhannisyan, G., Rapp, A., Arutyunyan, R., Greulich, K.O., Gebhart, E., 2005, Comet-assay in combination with PNA-FISH detects mutagen-induced DNA damage and specific repeat sequences in the damaged DNA of transformed cells, *Int J Mol Med.*, **15**(3): 437–442.

Kasakyan, S., Tachdjian, G., Movsessyan, E., Dagbhashyan, S., Muradyan, L., Aboura, A., Hovannisyan, A., Muradyan, A., Mkrtchyan, H., and Arutyunyan, R., 2003, Application of FISH technique for cytogenetic analysis in chronic myeloid leukemia patients. *Experimental Oncology*, **25**: 74–76.

Matevosyan, M.B., Poghosyan, V.S., Agadjanyan, E.A., Atoyants, A.L., and Aroutiounian, R.M., 2005, Monitoring of genotoxicity of waters from rivers of Armenia by Trad-SHM and Trad-MCN tests, *The second International Conference, Modern problems of Genetics, Radiobiology, Radioecology and Evolution*, Dubna,: 79.

Nersesyan, A.K., Arutyunyan, R.M., Groger, A., and Kaiser, H.E., 2001, Spreading of carcinogens and mutagens in the environment of Armenia and possible ways of cancer prevention, *Archive of Oncology*, **9**: 29–32.

Nersesyan, A.K., Daghbasyan, S.S., Danielyan, S.H., and Arutyunyan R.M., 2003, The incidence of childhood leukemia in Armenia (1991–2002), *Archive of Oncology*, **11**: 277–279.

Nersesyan, A.K., Vardazaryan, N.S., Gevorgyan, A.L., and Arutyunyan, R.M., 2002, Micronucleus level in exfoliated buccal mucosa cells of cancer patients. *Archive of Oncology,* **10**(1): 35–36.

Nersesyan, A.K., Zalinyan, G.G., Vardazaryan, N.S., and Aroutiounian, R.M., 2001, Micronucleus level in exfoliated buccal mucosa cells, *Central European Journal of Occupational and Environmental Medicine*, **7**(1): 39–44.

Poghosyan, V.S., Agadjanyan, E.A., Atoyants, A.L., Matevosyan, M.B., and Aroutiounian, R.M., 2005, Estimation of mutagenic activity of waters from artesian wells of Armenia, *The second International Conference, Modern problems of Genetics, Radiobiology, Radioecology and Evolution*, Dubna,: 83.

Vidic, T., Greilhuber, J., Vilhar, B., 2001, Genome size is associated with differential survival of plant species, *Annals of Botany*, **87**: 719–728.

ESTIMATION OF SOMATIC GENE MUTAGENESIS IN PATIENTS WITH BENIGN TUMORS LIVING IN RADIATION CONTAMINATED REGIONS WITH DIFFERENT ^{137}CESIUM DENSITY

IRINA A. ZAMULAEVA*, LIANA S. MKRTCHYAN, LUDMILA I. KRIKUNOVA, SVETLANA G. SMIRNOVA, NINA V. ORLOVA, ALEXANDER S. SAENKO
Medical Radiological Research Center of RAMS, Korolev str.-4, Obninsk, 249036, Russian Federation

*To whom correspondence should be addressed: zamulaeva@mail.ru

Abstract. The aim of this study was to compare the level of somatic mutagenesis in women with benign tumors of the reproductive system living in radiation contaminated regions of the Russian Federation with that in unexposed healthy individuals. Frequency of peripheral blood lymphocytes bearing mutations at T-cell receptor (TCR) locus was assessed by flow cytometry in 219 patients with myoma who had been living in Novozibkovskiy, Klintsovskiy districts of Bryansk region and Uzlovaya district of Tula region for 17-19 years since the moment of the Chernobyl accident. Mean ^{137}Cs densities in these districts were 708, 322 and 171 kBq/m^2 accordingly. The control group included 42 age-matched unexposed healthy individuals. There was no significant elevation in the TCR mutant cell frequency in patients from Uzlovaya district which was contaminated with radionuclides at low level in comparison to controls. The frequency of the TCR-mutant cells was significantly higher in patients from the most contaminated district (Novozibkovskiy and Klintsovskiy) as compared to the control group by Mann-Whitney test ($p<0.05$). Median values were $4.3 \cdot 10^{-4}$, $5.9 \cdot 10^{-4}$, $3.8 \cdot 10^{-4}$ accordingly. However, only 19% of the patients had the TCR–mutant cell frequencies exceeding the 95% confidence interval in the control group ($>7,0 \cdot 10^{-4}$). The frequencies of mutant cells in other patients corresponded to those in the control group. Our results confirm that the TCR-method may be used for individual assessment of long-term health consequences after irradiation. Individuals with elevated TCR-mutant cell scores might belong to a high-risk group potentially prone to the development of neoplasm and need more thorough medical observation than the rest of population.

A.A. Cigna and M. Durante (eds.),
Radiation Risk Estimates in Normal and Emergency Situations, 137–141.

Keywords: somatic mutation, T-cell receptor, ionizing radiation, low doses, flow cytometry

1. Introduction

Radiation-induced somatic cell mutations are likely to be the principal cause of cancer risk elevation after radiation exposure. Therefore, estimation of frequency of cells with gene mutations has been suggested to be a useful tool for cancer risk assessment in irradiated individuals (Akiyama *et al.*, 1995; Kyoizumi *et al.*, 1996). One of the methods for estimation of somatic mutagenesis level is determination of mutant cell frequency at the TCR locus. Although mutations at this locus are not directly related to carcinogenesis, they are likely to reflect probability of cancer-associated mutations. To confirm this suggestion we evaluated the TCR-mutant cell frequency in a group at high risk in respect to oncological diseases.

The aim of this study was to compare the level of somatic mutagenesis in women with benign tumors of the reproductive system living in radiation contaminated regions of the Russian Federation with that in unexposed healthy individuals.

2. Materials and Methods

261 persons were investigated and divided in four groups (Table 1).

Table 1. Description of study groups.

Study groups	Number of persons	Mean ^{137}Cs density, kBq/m^2
Patients with myoma living in Novozibkovskiy district of Bryansk region	97	708
Patients with myoma living in Klintsovskiy district of Bryansk region	22	322
Patients with myoma living in Uzlovaya district of Tula region	100	171
Control group (unexposed healthy donors)	42	-

The three groups consisted of 219 patients with myoma who had been living in radiation contaminated regions of the Russian Federation for 17-19 years since the moment of the Chernobyl accident. The residents were

0-30 years old at the moment of the Chernobyl accident. Mean (±SE) age at the moment of the study was 46.0±0.4 years old. The control group included 42 unexposed healthy individuals. Mean (±SE) age in the control group was 47.5±0.3 years old at the moment of the study.

Flow cytometry was used to evaluate the frequency of peripheral blood lymphocytes bearing mutations at the TCR locus as it was described earlier in details (Saenko *et al.*, 1998). Number of variant (or TCR-mutated) lymphocytes is determined by means of enumeration of the CD4+ cells lacking the CD3 antigen on their surface. Studies, performed by the inventors of the assay S. Kyoizumi and associates from RERF (Hiroshima, Japan) (1990), have demonstrated that the absence of the CD3 on the CD4+ cells was mainly due to alterations of the TCR underlain by the mutations of the genes encoding the TCR polypeptides. Following these findings, the test has been termed the "TCR assay". Such mutations occur in mature lymphocytes after the passage through the thymus, as cells with defective TCR are eliminated in the gland by apoptosis and cannot enter the bloodstream.

3. Results and Discussion

Data on the frequency of the TCR-mutant cells in the inhabitants of radiation-contaminated regions and age-matched control donors are presented in Table 2. There was not statistically significant elevation in the TCR-mutant cell frequency in patients from Uzlovaya district in comparison to the control level ($p=0.6$). The frequency of the TCR-mutant cells was significantly higher in patients from the most contaminated districts (Novozibkovskiy and Klintsovskiy) as compared to the control group ($p \leq 0.05$).

The mutant cell frequency in the control group was normally distributed (according to Kolmogorov-Smirnov test). Upper boundary of 95% confidence interval in control group was calculated as mean+2SD ($7{,}0 \cdot 10^{-4}$). Only 19% of the patients had the TCR-mutant cell frequencies exceeding this value. The frequencies of mutant cells in other patients corresponded to those in the control group.

Number of mutations is known to increase with radiation dose at any genetic locus tested. Therefore, elevated TCR-mutant cells frequencies in an individual may imply higher probability of the occurrence of cells harboring gene mutations at loci that eventually may cover critical oncogenes and tumor suppressors. It was plausible to suggest that individuals with elevated TCR-mutant cells scores might belong to a high-risk group potentially prone to the development of neoplasm. As an argument demonstrating such reasoning, one may mention results of a study performed in Scandinavia (Hagmar *et al.*, 1994). During a long-term epidemiological investigation, cancers were found to occur more often in a cohort of individuals with elevated level of chromosomal aberrations. Similar investigations of cancer incidence in the individuals with elevated rates of gene mutations have not been published so far.

Table 2. Frequency of the TCR mutant cells in unexposed control donors and patients with myoma living in radiation contaminated regions as a result of the Chernobyl disaster.

Study groups	TCR-mutant cell frequency, x 10^{-4}		p[1]
	Range	Median	
Patients with myoma living in Novozibkovskiy district of Bryansk region	1.1 –52.9	4.3	0.05
Patients with myoma living in Klintsovskiy district of Bryansk region	3.2-12.1	5.9	<0.001
Patients with myoma living in Uzlovaya district of Tula region	1.4-23.3	4.2	0.60
Control group	1.0-10.0	3.8	

[1]Comparison with control group by Mann-Whitney test

At the same time, data on high and early onset of cancer in patients with inherited genome instability syndromes (ATM, Bloom's syndrome, etc.) who also have elevated frequency of spontaneous gene mutations are well known (Kyoizumi *et al.*,1989; Langlois *et al.*,1989). The rationale of the hypothesis was indirectly demonstrated in our work: residents of the most contaminated regions with benign tumors of reproductive system had statistically significant higher frequencies of mutant cells than individuals from the control group.

4. Conclusion

The significant elevation of the TCR-mutant frequency was observed in the certain proportion of persons with benign tumors of reproductive system who belonged to high cancer risk group. Our results confirm that the TCR-method may be used for individual assessment of long-term health consequences after the irradiation. Individuals with elevated TCR-mutant cell scores might belong to high-risk group potentially prone to the development of neoplasm and need more thorough medical observation than the rest of population.

Acknowledgements

This work was supported by the Federal agency on science and innovations in the framework of the Federal Program "Researches and elaborations in priority directions of science and technique development" for 2002-2006 (N 02.434.11.3003) and the Russian-Belarusian Program in the framework of the Union State «Liquidation of consequence of Chernobyl accident» for 2002-2005.

References

Akiyama, M., Umeki, S., Kusunoki, Y. *et al.,* 1995, Somatic-cell mutation as a possible predictor of cancer risk, *Health Phys.* **68:** 643–649.

Hagmar, L., Brogger, A., Hansteen, I.L. *et al.*, 1994, Cancer risk in humans predicted by increase levels of chromosomal aberrations in lymphocytes: Nordic study group on the health risk of chromosome damage, *Cancer Res.* **54:** 2919–2922.

Kyoizumi, S., Akiyama, M., Hirai, Y. et al., 1990, Spontaneous loss and alteration of antigen receptor expression in mature CD4+ T cells, *J. Exp. Med.* **171:** 1981–1999.

Kyoizumi, S., Nakamura, N., Takebe, H. et al., 1989, Frequency of variant erythrocytes at the glycophorin A locus in two Bloom's syndrome patients, *Mutat. Res.* **214:** 215–222.

Kyoizumi, S., Akiyama, M., Cologne, J.B. et al., 1996, Somatic cell mutations at the glycophorin A locus in erythrocytes of atomic bomb survivors: implications for radiation carcinogenesis, *Radiat. Res.* **146:** 43–52.

Langlois, R.G., Bigbee, W.L., Jensen, R.H. and German, J., 1989, Evidence for increased in vivo mutation and somatic recombination in Bloom's syndrome, *Proc. Natl. Acad. Sci. USA* **86:** 670–674.

Saenko, A.S., Zamulaeva, I.A., Smirnova, S.G. et al., 1998, Determination of somatic mutant frequencies at glycophorin A and T-cell receptor loci for biodosimetry of prolonged irradiation, *Int. J. Radiat. Biol.* **73:** 613–618.

RADIOLOGICAL HEALTH EFFECTS 20 YEARS AFTER THE CHERNOBYL ACCIDENT: DATA OF THE NATIONAL RADIATION AND EPIDEMIOLOGICAL REGISTRY

VICTOR IVANOV* ANATOLY TSYB
Medical Radiological Research Centre
Russian Academy of Medical Sciences
4 Korolev str., Obninsk, 249036 Russia
E-mail: nrer@obninsk.com

*To whom correspondence should be addressed

Abstract. On the decree of the Russian Federation Government N 948 of 22 September 1993 the National Radiation and Epidemiological Registry was established. This is a legal entity on the base of the Medical Radiological Research Center of Russian Academy of Medical Sciences. Two main tasks were imposed upon the Registry: 1) objective estimation of radiation risks of cancer and non-cancer diseases following exposure to low dose radiation (up to 0.2 Sv); and 2) producing recommendations for health-care system and medical aid for minimization of delayed consequences of the Chernobyl accident. National registry comprises personal information on 615,000 people exposed to ionizing radiation, a far larger number than in the atomic bomb survivor registry. In this paper, basic information on radiation-epidemiologic analysis related to emergency accident workers and the population of Russian territories contaminated with radionuclides is summarized.

Keywords: Chernobyl accident; radiation risk; leukemia; thyroid cancer; non-cancer diseases

1. The National Registry Today

The National Radiation and Epidemiological Registry (NRER) was set up in Obninsk on the base of the Medical Radiological Research Center of Russian Academy of Medical Sciences on the decree of the Russian Federation Government N 948 of 22 September 1993. NRER is a successor of the All-Union Distributed Registry operating in the USSR since June 1986.

A.A. Cigna and M. Durante (eds.),
Radiation Risk Estimates in Normal and Emergency Situations, 143–148.

At present, the National Registry includes personalized medical and dosimetric data for 615,000 citizens of the Russian Federation exposed to radiation as a result of the Chernobyl accident. In particular, the registry includes 190,000 emergency workers and 360,000 people living in the worst contaminated regions of Russia, namely the Bryansk, Kaluga, Tula and Oryol oblasts. This registry is a unique medical information system committed to large-scale radiation epidemiological studies and developing recommendations for public health care to minimize the consequences of radiation exposure.[1]

There are 11 regional centers involved in gathering personalized data across Russia and maintaining continuous contacts with 4000 hospitals and clinics of the country (Fig. 1).

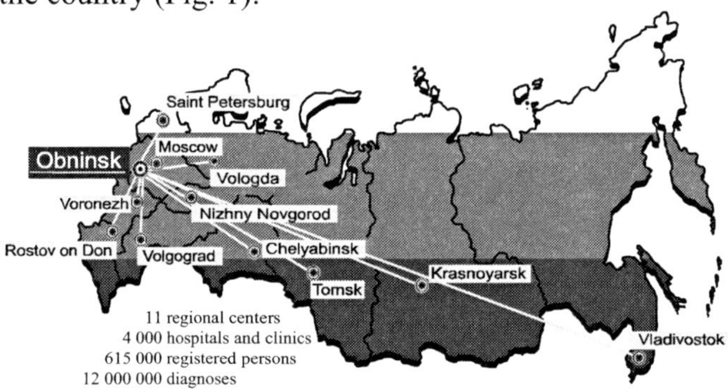

Figure 1. Regional centers of the National Registry.

One of the key issues of modern radiation epidemiology is to estimate objectively possible risks of cancers and non-cancer diseases at low radiation doses (up to 0.2 Sv). The amount of medical and dosimetric information collected by NRER and an adequate statistical power of the undertaken epidemiological studies make it possible to draw up a realistic picture of health effects of the Chernobyl accident.

1. Assessment of Leukemia Radiation Risk

Assessing the risk of radiation-induced leukemia is a special issue in radiation epidemiological studies. Among radiogenic malignant neoplasms, leukemia is associated with the highest radiation risk and the shortest latent period. An excess in the leukemia incidence above the spontaneous level can be considered as a first indication of radiation exposure for nuclear emergency workers at the Chernobyl plant, as well as for residents of the contaminated areas of Russia.

Data are summarized in Table 1. The epidemiological analysis covers a cohort of 71,870 liquidators living in the European part of Russia whose external radiation doses (mean value is 107 mGy) were known. Follow-up periods include: 1986-1996 and 1997-2003. If we compare liquidators with external radiation dose below 150 mGy with those exposed to higher doses, it can be seen that during the first ten years post-accident leukemia incidence rate was 2.2 times higher in the second group than in the first one. On the other hand, no significant differences were detected in the leukemia incidence rates for these same groups during the second follow-up period (1997-2003).

Table 1. Radiation risks of leukemias in emergency workers.

Follow-up period	1986-1996				1997-2003			
Dose groups, mGy	0-	45-	90-	150-300	0-	45-	90-	150-300
Mean doses, mGy	17	66	106	215	17	65	106	215
Number of leukemia cases	11	3	5	22	9	7	5	9
Relative risk (90% CI)	1.0 -	0.4 (0.1, 1.0)	0.4 (0.1, 1.0)	1.4 (0.8, 2.6)	1.0 -	1.1 (0.5, 2.6)	0.6 (0.2, 1.5)	0.9 (0.3, 1.8)
Comparison of two groups (90% CI)	1 -			2.2 (1.3, 3.8)	1 -			0.9 (0.5, 1.5)
Excess relative risk per 1 Gy (90% CI)	4.4 (0.0, 16.4)				-1.0 (-3.0, 3.6)			

There are two main conclusions to be drawn from the above: first, only emergency workers who received a radiation dose more than 150 mGy should be attributed to the risk group, and secondly, the risk of radiation-induced leukemia is increased at doses >150 mGy during the first ten years after the accident. A total of 55 excess leukemia cases were observed in the high risk group and can be attributed to radiation overexposure.

As to the residents of the contaminated areas, leukemia incidence rates were studied among children and adolescents of the Bryansk oblast. It was found that the excess relative risk per 1 Gy is positive in value (9.55), but is not statistically significant (-2.48; 46.55 90% CI).

2. The Problem of Thyroid Cancer

It has become clear after almost 20 years since the Chernobyl accident that one of its grave health consequences is a dramatic increase in thyroid cancer incidence among children (0-14 year old in 1986) in the contaminated areas of Russia (Fig. 2).

The dynamics of thyroid cancer incidence in the populations of the Bryansk, Kaluga, Tula and Oryol oblasts during two follow-up periods (from 1982 to 1990 and from 1991 to 2003) is shown in Fig. 3. During the second period, the incidence rate increased significantly in all age groups and both sexes. Thyroid cancer incidence rate in adults increased by 2-3 times, while for children and adolescents the increase was over 10-fold (Fig. 3). To understand the role of the radiation risk factor (thyroid exposure from incorporated iodine-131) among other factors (including the screening effect) large-scale epidemiological studies using the modern technique of cohort studies were performed within the National Registry (Fig. 4).

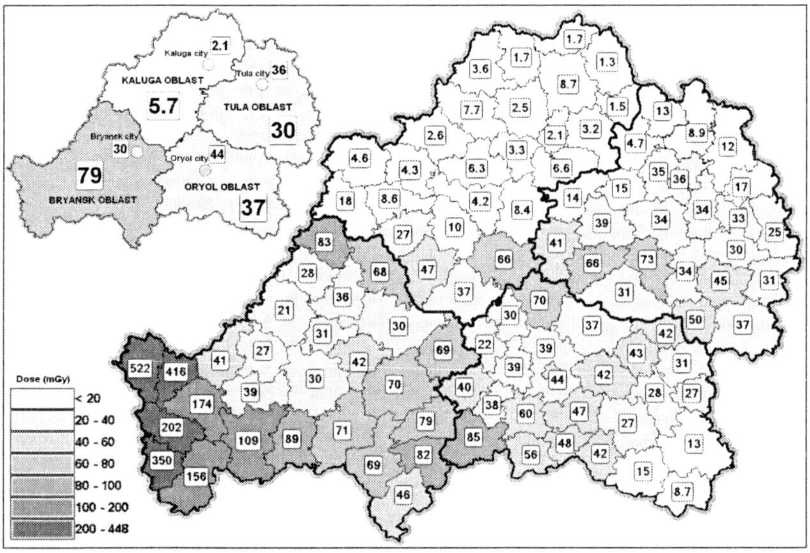

Figure 2. Region-averaged thyroid doses for children living in the Bryansk, Oryol, Kaluga and Tula oblasts.

These studies have shown that of 226 childhood thyroid cancer cases detected from 1991 to 2003 in the Bryansk oblast, 122 cases (54%) were caused by radiation exposure.

3. Radiological Risks of Non-cancer Diseases

The post-Chernobyl radiation-epidemiological studies were mainly focused on the influence of small and medium radiation doses on cancer morbidity and mortality. At the same time, the latest data of the Hiroshima and

Nagasaki Registry provide evidence of a dose-response relationship for incidence rates of non-cancer diseases (at doses higher 0.5 Sv), primarily cardiovascular morbidity.

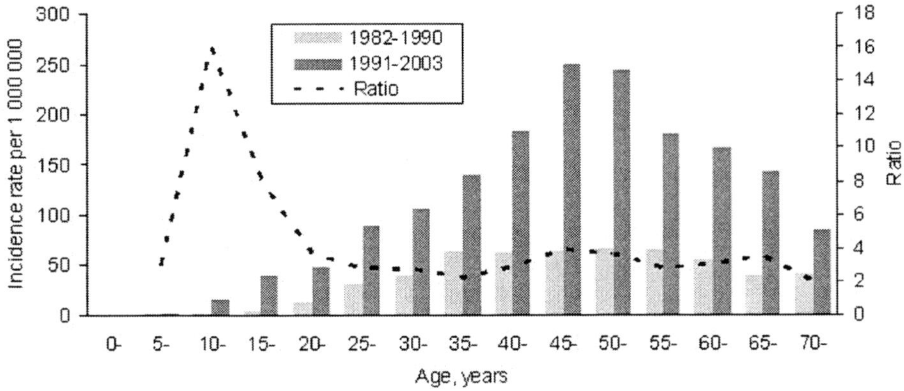

Figure 3. Thyroid cancer incidence (Bryansk, Oryol, Kaluga and Tula oblasts, women).

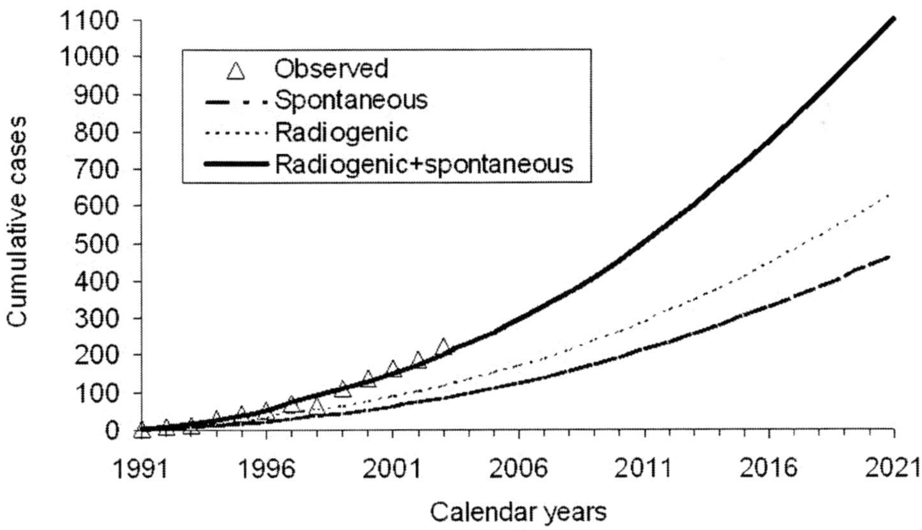

Figure 4. Prediction of thyroid cancer incidence (children at the time of the Chernobyl accident,Bryansk oblast).

4. Radiological Risks of Non-cancer Diseases

The National Radiological and Epidemiological Registry together with the Sasakava Foundation (Japan) studied non-cancer thyroid disorders in 2,457 children from the Bryansk and Kaluga oblasts whose personalized data on thyroid radiation exposure were available in May-June 1986. The mean thyroid dose from incorporated iodine-131 in this cohort was 132 mGy, and the incidence rate of diffuse goiter showed a statistically significant dose-related increase. On the other hand, no radiation risk was detected for other non-cancer thyroid disorders.The National Registry has now completed a large-scale study to assess radiation-induced cardiovascular disease risk in Chernobyl emergency workers. Special attention was given to 30,000 workers (living in the European part of Russia) who were involved in recovery activities during the first years following the accident. Statistically significant risk of cerebrovascular diseases was found for the emergency workers who received external radiation dose over 150 mGy during 6 weeks within the 30 km zone of the Chernobyl Nuclear Power Plant (Table 2).

Table 2. Relative risk of cerebrovascular diseases as a function of dose and length of stay in the 30-km zone of the ChNPP.

External radiation	Length (weeks) of stay in the 30-km zone		
dose, mGy	< 6	6-12	> 12
> 150	1.18 (1.00; 1.40)	1.02 (0.86; 1.20)	1.00 (0.84; 1.19)
50 - 150	1.03 (0.88; 1.20)	0.99 (0.83; 1.17)	0.90 (0.79; 1.04)
< 50	0.92 (0.78; 1.11)	0.97 (0.84; 1.12)	1

The data on radiation risks of non-cancer diseases obtained by the National Registry are still preliminary and need to be confirmed by radiation-epidemiological studies.

The analysis of the solid cancer incidence in the cohort of emergency workers-employees of the nuclear industry shows that the cancer incidence rate in the studied cohort does not exceed that in the respective age groups of the population of Russia as a whole. The mean value of SIR for all cancers with 95% CI is estimated to be 0.88 (0.76, 1.02). For all cancers the risk of induction of radiogenic cancers is found not to be statistically significant.

References

V. Ivanov, A. Tsyb, S. Ivanov, V. Pokrovsky, 2004, *Medical radiological consequences of the Chernobyl catastrophe in Russia: estimation of radiation risks*, Nauka, St. Petersburg.

CASUALTIES AND RADIATION DOSIMETRY OF THE ATOMIC BOMBINGS ON HIROSHIMA AND NAGASAKI

TETSUJI IMANAKA
Research Reactor Institute, Kyoto University,
Kumatori-cho, Osaka 590-0494 Japan
E-mail: imanaka@rri.kyoto-u.ac.jp

Abstract. The 60th anniversary of the Hiroshima-Nagasaki atomic bombings was commemorated in August 2005. By the end of 1945, according to the reports from both city governments, 140,000 and 70,000 deaths had occurred among inhabitants in Hiroshima and Nagasaki, respectively. In order to investigate late effects of atomic bomb radiation on humans, 94,000 survivors were selected and registered as cohort members of an epidemiological study, LSS (Life Span Study). On the other hand, comprehensive efforts have been continued to estimate radiation dose to individual survivors. The latest version of the dosimetry system, DS02 (Dosimetry System 2002) was adopted in 2003. Radiation dose at 1 m above the ground in open field at 1 km from the hypocenter was estimated to be 4.5 and 8.7 Gy in Hiroshima and Nagasaki, respectively, while at 2 km it was 0.08 and 0.14 Gy. According to the recent LSS report for the period of 1950-2000, among 86,611 survivors to whom individual dose was estimated, there have been 47,685 deaths (55%), including 10,127 from solid cancer and 296 from leukemia. The statistical analysis of the recent LSS data supports a linear-quadratic dose-response model for solid cancer, while the previous analyses indicated a linear dose-response. A linear-quadratic model is suggested for leukemia.

Keywords: Hiroshima, Nagasaki, Atomic Bomb, radiation dosimetry, DS02, LSS

1. Introduction

On August 6, 1945, the first atomic bomb, "Little Boy" exploded above Hiroshima, and three days later the second bomb, "Fatman" was dropped on

A.A. Cigna and M. Durante (eds.),
Radiation Risk Estimates in Normal and Emergency Situations, 149–156.

Nagasaki. The huge amount of energy liberated by the nuclear fission reactions instantly wiped out both cities. In summer 2005, the 60th anniversary of these tragedies was commemorated in Hiroshima and Nagasaki. During these years atomic bomb survivors have been continuing efforts to record their personal experiences and to hand them on to future generations (Hiroshima, http; Nagasaki, http).

This report presents a short overview of atomic bomb casualties in Hiroshima and Nagasaki, as well as recent efforts by a Japan-US working group to revise the radiation dosimetry system.

2. Initial Casualties

Basic physical features of atomic bombs dropped on Hiroshima and Nagasaki are summarized in Table 1 (Young & Kerr, 2005).

Table 1. Basic features of atomic bombs on Hiroshima and Nagasaki.

City	Hiroshima	Nagasaki
Name	Little Boy	Fatman
Fissile material	Enriched ^{235}U	^{239}Pu
Structure	Gun-type	Implosion-type
Bomb weight, ton	4.0	4.5
Height of burst, m	600 ± 20	503 ± 10
Yield, kton TNT	16 ± 4	21 ± 2

Of the total energy created by the nuclear explosion, about 50% contributed to blast formation. The other 35%, 5%, and 10% of energy went into thermal radiation, initial and residual nuclear radiation, respectively (Hiroshima and Nagasaki, 1979). The height of the burst and the hypocenter location were determined based on measurements of shadow angles made by heat-rays on tomb stones, buildings etc. The yield of Fatman was determined from the results of experiments using the same type of bomb. Meanwhile, the yield of Little Boy was estimated based on various data including radiation, blast and heat effects observed in Hiroshima, because Little Boy was a unique device and this type of bomb was not used in nuclear experiments.

The numbers of inhabitants at the time of the bombings were 340,000 and 220,000 in Hiroshima and Nagasaki, respectively. The hypocenter in Hiroshima was right in the middle of downtown, while in Nagasaki it was the northern part of the city. The people experienced at first a flash (PIKA in Japanese) and then felt the blast (DON). So they call the bombings "PIKA-DON". The heat, blast and subsequent fire wiped out the areas within 2 km from the hypocenter in both cities. The death rate of the people who were within 1 km from the hypocenter in Hiroshima was evaluated to be 68% on the first day and reached 92% three months later. Corresponding death rates of the people who were 1 – 2 km away in Hiroshima were 16 and 32%. In total about 57% within 2 km in Hiroshima died within three months. According to the reports from both city governments, 140,000 and 70,000 deaths occurred in Hiroshima and Nagasaki by the end of 1945 respectively (Hiroshima and Nagasaki, 1979).

3. Radiation Dosimetry

In order to investigate late effects of atomic bomb radiation on humans, the US government founded ABCC (reorganized into RERF in 1975) in Hiroshima in 1948. Through the special inquiries attached to the national census in 1950, 284,000 people were identified as atomic bomb survivors. Based on this information, ABCC registered 94,000 survivors in their epidemiological study LSS, Life Span Study. At the same time, comprehensive efforts were started to estimate individual radiation dose both by interviewing every survivor about their location and situation at the time of the bombing and by developing a methodology to estimate the radiation field in both cities.

The dosimetry system for the Hiroshima-Nagasaki bomb survivors has been revised several times, reflecting the progress of knowledge and technologies that could be used to estimate radiation dose from the atomic bombings. Modern computer technology was incorporated in the process to develop Dosimetry System 1986 (DS86, 2001). The results of DS86 were checked compared with gamma-ray data based on TL (thermo-luminescence) measurements from tile and brick samples that were exposed to atomic bomb radiation, as well as neutron data based on measurements of neutron-induced reactions such as $^{32}S(n,p)^{32}P$ and $^{59}Co(n,\gamma)^{60}Co$.

DS86 has shown a satisfactory agreement between calculation and measurement for gamma-rays, while there remained a substantial discrepancy for thermal neutron reaction producing ^{60}Co (Roesch, 1987). For the purpose of resolving neutron discrepancy in DS86 and to incorporate recent progress of computer technology, a joint Japan-US working group, including the author of this manuscript, was established in the year 2000. After two years of intensive efforts, including new calculations and new measurements of induced radioactivities such as $^{63}Cu(n,p)^{63}Ni$ and $^{35}Cl(n,\gamma)^{36}Cl$, the working group developed the latest version, Dosimetry System 2002 (DS02) (Young & Kerr, 2005).

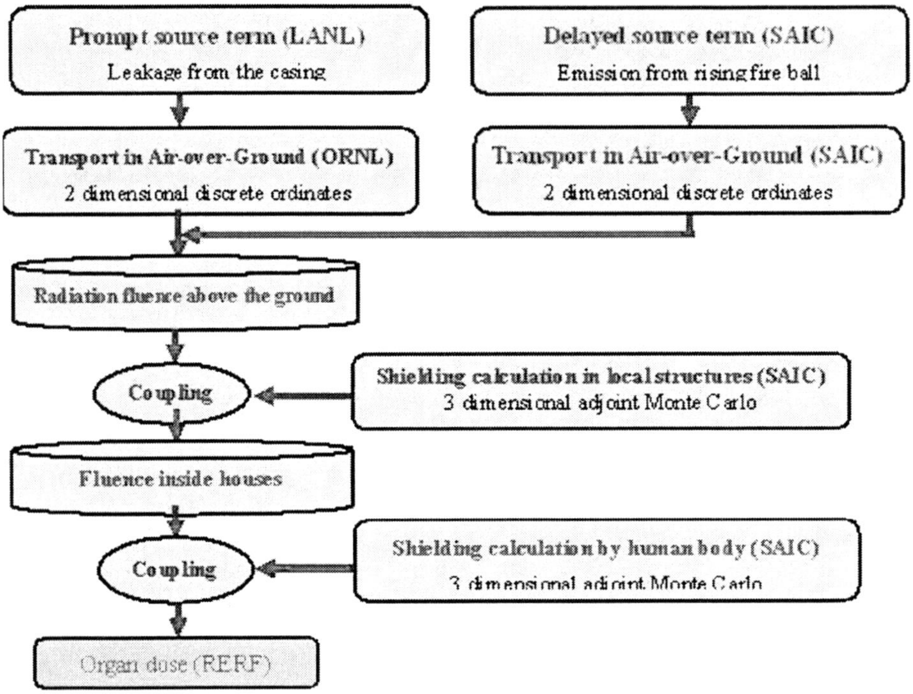

Figure 1. Computation scheme to estimate individual organ dose in DS02.

An interesting feature of DS86/DS02 is that the calculation process is divided into various modules that are replaceable if necessary. As is shown in Fig. 1, (Imanaka, 2005a) the computation process consists of the following four main parts:

- Source term calculation; prompt and delayed, neutron and gamma-ray.
- Long range radiation transport calculation in air-over-ground geometry without structures on the ground surface.
- Shielding effect calculation within local structures such as Japanese wooden houses.
- Shielding effect calculation within human body before radiation reaches target organ.

Radiation from the atomic bomb can be classified into six components depending on their origins:

- prompt neutrons and prompt primary gamma-rays, escaping from the bomb device at the moment of explosion,
- prompt secondary gamma-rays, originating from collisions of prompt neutrons with material nuclei in air and ground,
- delayed neutrons and delayed primary gamma-rays, emitted from fission products in the fireball rising after the explosion, and
- delayed secondary gamma-rays., originating from collisions of delayed neutrons with material nuclei in air and ground.

Free-in-air tissue kerma at 1 m above the ground in open field is shown in Fig. 2 as a function of distance from the hypocenter. The important distance is 1 – 2 km from the hypocenter because the people who were in this range received significant radiation and many of them survived. At 1 km from the hypocenter radiation dose was estimated to be 4.5 and 8.7 Gy in Hiroshima and Nagasaki, respectively, while at 2.0 km it was 0.08 and 0.14 Gy. As it can be seen in Fig. 2, the neutron component contributed little to the total dose at distances greater than 1 km. Prompt secondary gamma-rays and delayed gamma-rays were the major contributors to radiation dose absorbed by survivors.

There were also two kinds of residual radiation: neutron-induced radioactivities and radioactive fallout, so-called “black rain”. Cumulative external dose due to induced radioactivities at the hypocenter was evaluated to be 1.2 and 0.57 Gy in Hiroshima and Nagasaki, respectively, for the time period from 1 minute after the bombing to infinite time (Imanaka, 2005b). These values decreased rapidly with the initial time of dose accumulation as well as the distance from the hypocenter. In the case of a person who came to the hypocenter one day after the bombing and stayed there for an infinite period, cumulative dose was 0.19 and 0.055 Gy in Hiroshima and Nagasaki, respectively. At 1 km from the hypocenter, the level of induced radioactivities decreased by a factor of 300 compared with the value at the hypocenter. Radioactive contamination by black rain was observed in the Koi-Takasu area of Hiroshima about 3 km west from the hypocenter and in the Nishiyama area of Nagasaki about 3 km east from the hypocenter. Cumulative external dose at these areas was evaluated to be 1 – 3 and 20 – 40 Roentgen for the Koi-Takasu and Nishiyama area, respectively (Roesch, 1987).

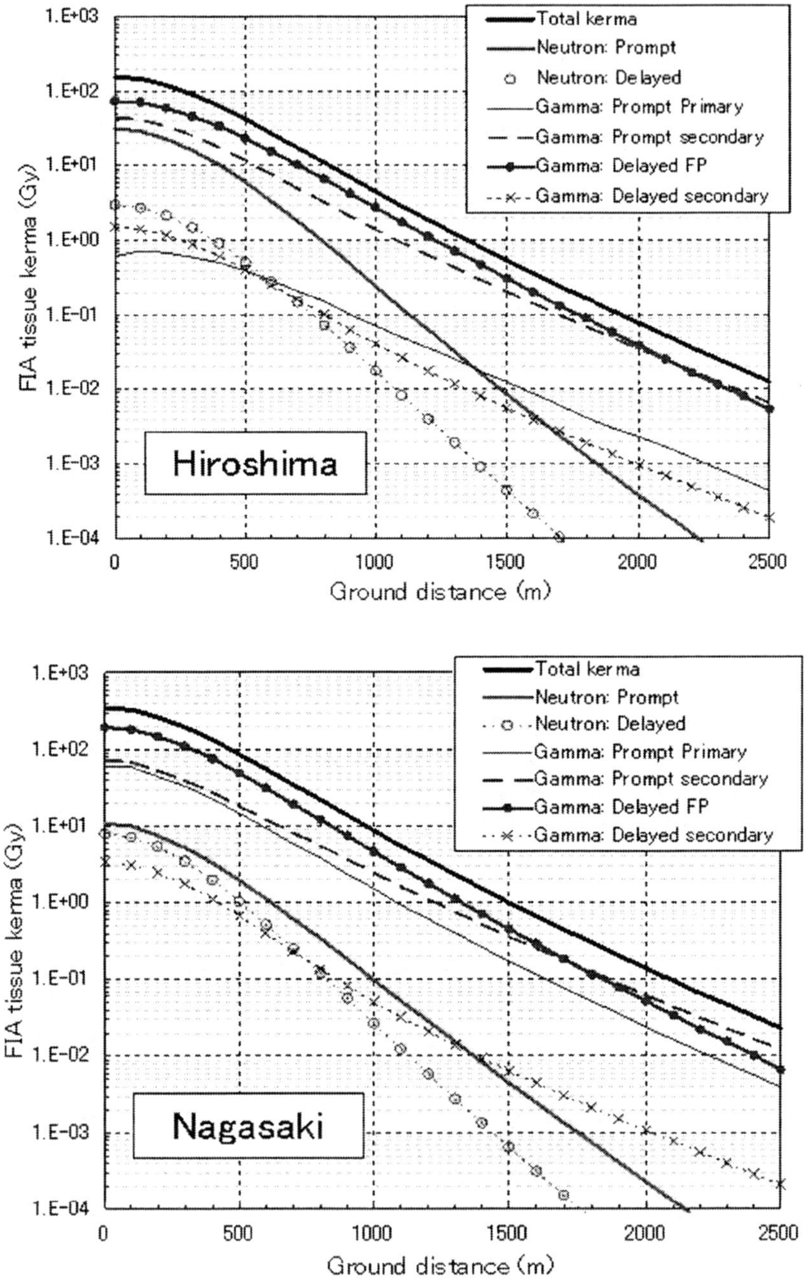

Figure 2. Free-in-air tissue kerma at 1 m above the ground as a function of distance from the hypocenter.

4. Late Effects of Atomic Bomb Radiation

Scientific merits of the LSS study from an epidemiological point of view are summarized as follows:

- The study population consists of a large number of citizens not biased by sex and generation.
- Accurate follow-up of mortality is possible through the Japanese family registration system.
- Individual dose estimates are available, ranging widely from low dose to high dose.
- The follow-up data have already been accumulated for more than 50 years.

Owing to the above features, the LSS study is still providing irreplaceable data about late effects of radiation exposure on humans. According to the latest report for the period of 1950-2000, among 86,611 survivors for whom individual dose was estimated, there have been 47,685 deaths (55%), including 10,127 from solid cancer and 296 from leukemia (Preston *et al.,* 2004), The statistical analysis of LSS data, assuming a linear dose-response model for total solid cancer, gave an excess relative risk of 0.43 (0.33-0.53; 90% CI) per Sv for exposure at age 30. Among the site-specific solid cancers, statistically significant increase with radiation dose was observed for esophagus, stomach, colon, gall bladder, lung, breast, ovary and bladder(Preston *et al.,* 2003) Meanwhile, a linear-quadratic model is suggested for leukemia. Based on the low-dose slope of the linear-quadratic model, absolute excess risk of leukemia for the age group of 0-19 was estimated to be 0.66 (0.13-1.3; 90% CI) per 10^4 person·year·Sv (Preston *et al.*, 2004).

It is also noteworthy that, other than solid cancer and leukemia, the results of recent LSS studies indicated mortality increase from non-cancer deaths among atomic bomb survivors. Statistically significant increase with radiation dose was observed for deaths from heart disease, stroke and respiratory disease. Assuming a linear dose-response, excess relative risk of total non-cancer death was evaluated to be 0.14 (0.08-0.2; 90% CI) per Sv for the period between 1968 and 1997 (Preston *et al.*, 2003).

References

DS86, 2001, *Status of the dosimetry for the Radiation Effects Research Foundation (DS86)*, Committee on Dosimetry for RERF, NRC, National Academy Press, Washington DC.

Hiroshima http, Hiroshima Peace Memorial Museum: http://www.pcf.city.hiroshima.jp/-index_e2.html.

Hiroshima and Nagasaki, 1979, *The physical, medical and social effects of the atomic bombing*. The Committee for the Compilation of Materials on Damage Caused by the Atomic Bombs in Hiroshima and Nagasaki; Iwanami, Tokyo (in Japanese).

Imanaka T., 2005a, Outline of the calculation system for DS02 and related calculations in Japan. *Proceedings of the workshop "New Radiation Dosimetry System DS02 for the Atomic Bombings in Hiroshima and Nagasaki", KURRI-KR-114*: 124–133, (in Japanese).

Imanaka T., 2005b, Neutron induced radiation dose by the atomic bombings in Hiroshima and Nagasaki based on the DS02 methodology, *KURRI-KR-114:* 150–154, (in Japanese).

Nagasaki http, Nagasaki Atomic Bomb Museum: http://www1.city.nagasaki.nagasaki.jp/na-bomb/museum/index.html

Preston D.L. et al., 2003, Studies of mortality of atomic bomb survivors. Report 13: Solid cancer and Noncancer disease mortality: 1950-1997. *Radiation Research* **160**: 381–407.

Preston D.L. et al., 2004, Effect of recent changes in atomic bomb survivor dosimetry on cancer mortality risk estimates. *Radiation Research* **162**:. 377–389.

Roesch C. (Ed.), 1987, *US-Japan Joint Reassessment of Atomic Bomb Radiation Dosimetry in Hiroshima and Nagasaki*, Final Report, RERF, Hiroshima.

Young R.W. & Kerr G.D. (Eds.), 2005, *Reassessment of the Atomic-Bomb Radiation Dosimetry for Hiroshima and Nagasaki – DS02*. Joint US-Japan Dosimetry Working Group, Radiation Effects Research Foundation, Hiroshima, Japan.

Radiation effects: the environment

RADIOECOLOGY: HISTORY AND STATE-OF-THE-ART AT THE BEGINNING OF THE 21st CENTURY

RUDOLF M. ALEXAKHIN
Russian Institute of Agricultural Radiology and Agroecology, 249032, Obninsk, Kaluga region, Kievskoe str., 109 km, RUSSIA
E-mail:alexakhin@riar.obninsk.org

Abstract. The history of radioecology is described from the end of the XIX - the beginning of the XX centuries till the present days with the indication of the major evolutionary stages of this area of knowledge. The ecological aspects of the use of nuclear power engineering for military purposes (global contamination of the environment after nuclear tests) in the 1960s-1980s are elucidated. The paper also touches upon the ecological problems of nuclear power engineering development. Some issues of the radiation protection of the environment are analysed (anthropocentric and ecocentric principles). The paper summarises the main tasks of radioecology at the present stage. The role is stressed of studying the importance of the elevating radiation background of the environment considering the ever-growing anthropogenic pressure to biota.

Keywords: radioecology, history, biosphere, radiation protection, radionuclides, migration, radiation effects, nuclear power

1. Introduction

Radioecology as a science originated at the very end of the XIX – the beginning of the XX centuries, immediately after the discovery of X-rays by W.K. Roentgen, radioactivity phenomena by A. Becquerel and discovery and identification of the first natural radioactive elements by M. Curie. The earliest experimental studies in the field of radioecology dealt with the dispersion and migration of natural radionuclides via the trophic chains in the environment and elucidation of the natural radiation background role in the evolution of biota on our planet. The term "radioecology" was introduced into the scientific vocabulary more than half a century later, in 1956, simultaneously in the USA (E. Odum) and USSR (A.M. Kuzin, A.A. Peredelsky).

A.A. Cigna and M. Durante (eds.),
Radiation Risk Estimates in Normal and Emergency Situations, 159–168.

The first investigations into the ionizing radiation effects on biota date back to the experiments of I.R. Tarkhanov with aquatic plants and animals in 1896 (Tarkhanov, 1896), and in 1923 N.V. Gajewskaja published a paper on X-ray effects on *Artemia salina* (Gajewskaja, 1923).

In 1932, a book was published on the effects of the elevated natural radiation background on plant communities inhabiting the regions of uranium ores occurrence, near Iochimstahl (present Czech Republic) "Biologie des Radiums und Uraniums". This book written by M. Curie's pupils, J. Stoklasa and J. Penkava (1932) can actually be considered one of the first monographs on radioecology dealing with the ionizing radiation effects at the ecosystem level.

A notable role in the development of radioecology as an independent science belongs to an outstanding Russian natural scientist, V.I. Vernadsky (1863-1945). The keywords in his extensive creative work were biosphere (noosphere, according to Vernadsky, is the combination of biosphere and man's intellect) and radioactivity. Actually, these two notions reflect the essence of modern radioecology (Shaw, 2005).

Among the founders of radioecology necessary is to mention N.V. Timofeev-Resovsky (1900-1981), who preferred to use the term "radiation biogeocenology" for this scientific direction. The first works of N.V. Timofeev-Resovsky go back to the late 1940s, these were implemented in the vicinity of an atomic industry facility in the USSR (Laboratory "B", now the town of Snezhinsk). In 1949, a report was written by N.V. Timofeev-Resovsky, G.I. Born and K.G. Zimmer "Calculation of ionizing radiation doses forming in a living body from the incorporation of radioisotopes. 2. Preliminary dose calculation" (classified at that time), then (in the 1990s) followed a series of declassified reports on the radionuclide migration in the environment and ionizing radiation effects on biota (Emelianov & Gavrilchenko, 2000). In the 1970s, N.V. Timofeev-Resovsky actually formulated the tasks of modern radioecology.

2. Radioecology Development in the Context of Problems of Military Application of Nuclear Power

Radioecology in the existing infrastructure of this science began to outline in the 1950s (Van der Stricht & Kirchmann, 2001). Historically predetermined was that the development of this science was dictated by the solution of tasks connected with the military application of nuclear power (Woodwell, 1965; Whicker & Schultz, 1982; Cigna, 1996).

From the late 1950s-early 1960s, radioecology saw a new stage in its development related, primarily, to the impact of nuclear weapons tests (mainly in the atmosphere), which had resulted in a global radioactive contamination of the biosphere. The man-made radionuclides dispersed in the environment (fission products, nuclides with induced activity, fissile radionuclides) were incorporated into the trophic chains of migration in the deposition-soil (water)-plant-animal-man system (Izrael, 1996). The experimental studies covered all the natural environments: atmosphere, agrosphere, terrestrial ecosystems, hydrobiocenoses. Numerous articles on dispersion of radionuclides from global fallout were published in a vast number of journals including the leading ones (Science, Nature, etc.). One of the objectives of the analysis of radionuclide accumulation in plants and animals were estimations of exposure doses to the world population from global radioactive fallout and associated health effects. The most detailed reviews on the global fallout levels were published in regular reports of the United Nations Committee on the Effects of Atomic Radiation established in 1955 and including currently 21 countries. In the family of UN organizations, UNSCEAR was one of the key institutions dealing with the acquisition and analysis of data on radioactivity levels in the natural environment from different sources of radionuclide release into the environment, and various radioecological situations (global fallout, operation of full nuclear fuel cycle facilities, radiation accidents with the release into the environment of radioactive substances, etc.) (UNSCEAR, 1996).

In 1960-1980, of fundamental importance for the development of radioecology were the studies at the largest atomic industry facilities of the leading nuclear powers in the world (mainly the USSR and USA). Technical irregularities at the enterprises with reactor and radiochemical plants for the production of nuclear weapons components caused contamination with radioactive substances of the nearby territories. On the resulting test grounds, long-time comprehensive radioecological investigations were performed to study the radionuclide migration in various natural environments and ionizing radiation effects on natural communities of plants and animals.

A retrospective comparison of programs for radioecological studies near the nuclear facilities in the USA and USSR in 1950-1985 (Hanford, Oak Ridge, Savannah River Plant laboratories in the USA and SPA "Mayak", Chelyabinsk-40, now Ozersk in the USSR) indicate virtually complete symmetric development of these programs. These radioecological studies were focused on the solution of the two main problems – study of the radionuclide migration in various natural and agricultural biogeocenoses and ionizing radiation effects on plant/animal populations and ecosystems as a whole.

Yet other regions of active radioecological studies associated with the military use of atomic energy, in addition to the areas near the nuclear facilities, have become the test sites for the development and testing of nuclear weapons (Semipalatinsk and Novaya Zemlya test sites in the USSR,

Nevada test site in the USA, Marshall Islands in the Pacific Ocean, Maralinga area in Australia, etc.). The studies at these military test grounds significantly supplemented radioecological information obtained in regions with the elevated radiation background (examples of such studies may be regions in the Komi Republic, Russia) (Titaeva & Tskev, 1983).

In the USA and West Europe countries, the major contributors to the radioecological studies in the vicinities of nuclear facilities and nuclear research centers were Sparrow A.H., Auerbach S.I., Woodwell G.M., Trabalka J.P., Platt R.B., Eyman L.D. (USA) and Bovard P., Foulquier L. (France).

For a more accurate quantitative estimation of ionizing radiation effects on natural ecosystems, the experimental works in the contaminated regions were supplemented with trials on the irradiation of large natural zones from powerful and point sources and the reactor without a biological shield (experiments in forests and meadows in the USA - University of Emory, Brookhaven National Laboratory, in the Wisconsin state; Puerto-Rico, in France – Cadarache, in the USSR – at the SPA "Mayak").

A big impetus for the development of radioecology was overcoming of the consequences of the radiation accidents accompanied by the radioactivity release into the environment. Among the first were the accident in the South Urals in September, 1957, at a radiochemical military plant "Mayak" in the USSR and not so heavy but also connected with malfunctions at a nuclear reactor for weapon grade plutonium production also in 1957, Windscale accident (Sellafield, UK). The 1957 accident in the USSR gave rise to the formation of the East Urals radioactive trail (EURT), whose area within the ^{90}Sr (the main dose forming radionuclide) isoline (2 Ci/km^2 or 74 kBq/m^2) was 23000 km^2.

The EURT has become a test site for extensive long-time studies into radionuclide migration in various natural environments and accumulation of activity by a wide range of plants and animals. A particular place at this test site is occupied by the studies on agricultural radioecology, headed by an outstanding radioecologist V.M. Klechkovsky (1900-1972). Together with R.S. Russell (1966), the author and editor of the monograph "Radioactivity and Human Diet" he may be considered the founder of agricultural radioecology. Within the EURT the various countermeasures have been tested on the return to the economic use of contaminated agricultural lands. This area itself has become the first region in the world where extensive rehabilitation works were performed on the return of these lands to the economic use (Alexakhin *et al.,* 2001).

High densities of radioactive contamination in the EURT head have caused radiation damage and death of pine and birch-pine woods, which was one of the first observations of the radiation effects at the ecosystem level in the situation of radioactive contamination of nature. The research and experimental station founded on the basis of the SPA "Mayak" was fated to play the role of the *alma mater* of radioecology in the USSR. The greatest contributors to the radioecological studies in the region of the South Urals accident in 1957-1980 were V.M. Klechkovsky, N.P. Arkhipov, A.I. Ilienko, D.A. Krivolutsky, G.N. Romanov, A.N. Sirotkin, F.A. Tikhomirov, Ye.A. Fedorov, V.A. Shevchenko*, B.S. Prister, R.M. Alexakhin, N.A. Korneyev. In 1990-2005 in the far parts of EURT actively worked I.V. Molchanova, M.Ya. Chebotina, A.V. Trapeznikov.

3. Radioecology Development and Problems of Nuclear Power Engineering

From the late 1950s the range of radioecological problems has expanded to include these of the environmental safety of nuclear power engineering. Since that moment growing in importance has become an idea that, on the one hand, the ever-increasing energy demands of the humankind were impossible without nuclear power engineering and, on the other hand, its development was fundamentally dependent on the environmental safety of this industry.

The range of radioecological problems connected with the nuclear power engineering included issues concerning releases of radioactive substances into the environment and their migration via the trophic chains, radionuclide accumulation by living organisms and transfer to humans. A special topicality has assumed a study into the management of radioactive wastes (their burial in various natural environments). In recent years these were added by the issues of transportation and reprocessing of spent nuclear fuel (Luycks and Frissel, 1996). Radioecology has become a scientific basis for the radiation monitoring of the environment, *de facto* being carried out around each even of low significance facility containing radionuclides or another source of ionizing radiation. Of paramount importance are now aspects of low-level ionizing radiation effects – a sacramental problem of radiobiology and radioecology since the moment of their origination. A program was completed with the participation of 15 organizations from 9 European countries on the estimation of ionizing radiation effects from different radionuclides on plants and animals, FASSET (framework for assessment of the environmental impact of ionizing radiation in European ecosystems) (Williams, 2004).

As is the case with the military aspects of nuclear power engineering, a noticeable milestone in the history of this science was the accident in the USSR at the Chernobyl NPP in 1986, which had provided further powerful

* See obituary in this issue.

impetus to the progress of the science on radioactivity of the biosphere. The Chernobyl accident was classified as the most severe one in the history of nuclear power engineering. Radioactive depositions covered not only the USSR territory but also some other European countries, with the "Chernobyl radionuclides" being registered outside Europe (Alexakhin *et al.,* 2001; Smith & Beresford, 2005). Thus, the total area affected by the accident with a deposition density of the main dose forming radionuclide, ^{137}Cs, of 1 Ci/km^2 (37 kBq/m^2), which is about 6-7 times higher than the background global fallout in the middle latitudes of the Northern Hemisphere, amounted to 150 000 km^2.

The influence of the Chernobyl accident on the development of radioecology has become significant to an extent that it is conventional to divide its history into the "pre- and post-Chernobyl" periods. The overcoming of health and ecological impacts, recognized as the leading ones in this accident, has given rise to the development of comprehensive and diverse projects on medical and radioecological studies and implementation of many national and international research programs. Among these, worth special noting is the IAEA Chernobyl project and a number of investigations initiated by the European Commission.

As a result of radioecological investigations in a large Chernobyl NPP region, diverse data were collected on the behavior of some biologically important radionuclides in terrestrial and aquatic biogeocenoses for a wide spectrum of environmental conditions, long-time dynamics was assessed of the radionuclide cycling in ecosystems. An independent group of reseat were the works on the study of ionizing radiation effects on natural communities of animals and plants in a wide range of dose rates and cumulative doses. As in the South Urals accident in 1957, in the Chernobyl affected area in its near (10-30 km) zone the radiation damage was more severe, whole ecosystems (for instance pine forests) were destructed.

Radioecological Chernobyl investigations brought into existence new radioecological schools in many European countries (Great Britain – B. Howard, N. Beresford; France – G. Deville-Cavellin, Italy – L. Cigna Rossi , A. Antonelli and, more recently, M. Belli, U. Sansone; Germany – G. Voigt, H. Biezold, P. Jacob; G. Prohl, Spain – G. Rauret; Norway – K. Hove, P. Strand, B. Salbu; Sweden – L. Moberg; Ireland – B. Rafferty; Belgium – G. Desmet, Ch. Vandecasteele). Among the USSR-CIS specialists, great contributors to the experimental radioecological studies in the Chernobyl affected region were: Russia – I.N. Ryabov, F.A. Tikhomirov, V.A. Shevchenko, G.N. Romanov, I.I. Kryshev, S.V. Fesenko, A.I. Taskaev, R.M. Alexakhin, N.I. Sanzharova; Ukraine – Yu. A. Kutlakhmedov, B.S. Prister, V.A. Kashparov, I.N. Gudkov, G.G. Polikarpov; Belarus – I.M. Bogdevich, S.K. Firsakova.

4. Radioecology and Environmental Protection from Ionizing Radiation

A growing societal concern over human-driven impacts on the environment (including radiation) has drastically exacerbated the problem of the biosphere protection (sustainable development and biodiversity). In this context the problem of biota protection from ionizing radiation is currently receiving increasing attention.

In 1978, an international professional organization of radioecologists was founded, the International Union of Radioecology, whose objective is coordination of a variety of research works on environmental radioactivity and organization of international conferences. Currently the UIR numbers nearly 600 members representing some 40 countries. In different years the UIR presidents were famous radioecologists (S. Myttenaere, A.A. Cigna, G. Desmet.). Since 2001, the UIR president is P. Strand and General Secretary is F. Brechignac. A special UIR (2003) declaration is being published on the principles of environmental protection against ionizing radiation at the beginning of the XXI century.

Among the most authoritative international institutions whose terms of reference include the analysis of radionuclide migration in the environment is the International Commission on Radiological Protection established in 1928. The main aim of the ICRP is radiation protection of humans, and from this point of view transport of radioactive substances in the environment via the trophic chains leading to man is considered in the aspect of formation of internal exposure dose, a crucial contributor to the overall dose burden to humans in many radioecological situations. At the same time, it is exactly the ICRP that formulated in 1977 a postulate on the principles of environmental protection (ICRP, 1977) according to which "if radiation standards protect man from ionizing radiation, then biota are also adequately protected in the same conditions", this thesis was subsequently confirmed in the ICRP recommendations on radiation protection that are now in force (ICRP, 1991).

At the same time, in recent years, the ICRP interest in problems of radiation protection of the environment has noticeably increased (separate Committee 5 has been created within the ICRP structure dealing with the problems of environmental protection against ionizing radiations). It has issued Publication 91 (ICRO, 2003), which raises a problem of changing from the mentioned above anthropocentric principle of biota protection to the ecocentric paradigm. According to the latter, the basis of the radiation protection should be that of plants and animals proper. An importance is stressed of the development of general principles of environmental protection from ionizing radiation, pollutants and agent of non-radiation nature, strengthened is the viewpoint of the importance of ecocentric approaches to the radiation protection of biota (Pentreath, 2002; Brechignac, 2003; Brechignac & Desmet, 2005).

5. State-of-the-art and Problems of Radioecology at the Beginning of the XXI Century

The major challenges of radioecology at the end of the XX – beginning of the XXI centuries are problems focused around the environmental safety and ecological advantages of a full nuclear fuel cycle. It is more and more actively stressed that along with the proper radiation safety of nuclear power engineering, increasing in importance becomes its advantage compared with power engineering on fossil fuel that the former is lacking releases of greenhouse gases. The latter becomes extremely important in the light of the solution of the problem of global warming due to the greenhouse effect.

Modern radioecology is a comprehensive scientific discipline that occupies a highly important place in the system of natural sciences and interacts with different fields of knowledge on the Earth's biosphere. It acts as the most advanced branch of ecology due to the use of highly precise techniques for determination of specific pollutants – radionuclides – in various environments and foundation on the reliable methodological basis for quantification of the acting factor – ionizing radiations (classical triangle: pollutant migration - dose from exposure to living organisms (their populations) - exposure effect). The development of nuclear power engineering, which currently continues to be considered as an alternative-free source to meet the increasing energy demands of the humankind, predetermines the development of the radioecological science, since radiation safety of the environment and human health is a necessary condition of progress in nuclear engineering.

In our opinion, among the priorities of modern radioecology are the following ones:

- Study into the ionizing radiation effects at low doses and dose rates at the levels of individuals, plant and animal populations and whole ecosystems (considering enhancement of the biosphere radiation background due to anthropogenic activity);
- Study into the combined effects of ionizing radiation, pollutants and agents of non-radiation nature (particularly at low concentrations and doses) on plant and animal communities and ecosystems as a whole;
- Development of a system of radiation protection of the environment and humans based on the integration of the anthropocentric (protected is man, protected are biota) and

ecocentric (biota protection from ionizing radiation is the focus of attention) principles;

- Introduction of the methodology of radioecology as the most advanced branch of ecology when studying the influence of different anthropogenic factors on the environment;
- Study into the applied ecological problems of the development of full nuclear fuel cycle (management of radioactive wastes, reprocessing of spent nuclear fuel, decommissioning of nuclear facilities, etc.);
- Solution of problems connected with the "radiation legacy" (rehabilitation of territories previously affected by radioactive contamination).

References

Alexakhin R.M., Buldakov L.A., Gubanov V.A., Drozhko Ye.G., Iljin L.A., Kryshev I.I., Linge I.I., Romanov G.N., Savkin M.N., Saurov M.M., Tikhomirov F.A. and Kholina Yu.B. In: Iljin L.A. and Gubanov V.A. (Eds), 2001, *Large radiation accidents: consequences and countermeasures.* Moscow, IzdAT.

Brechignac F., 2003, Protection of the environment: how to position radioprotection in an ecological risk assessment perspective. *Sci. Total Environ.* **307**: 34–35.

Brechignac F. and G. Desmet (Eds.), 2005, *Equidosimetry - Ecological Standardization and Equidosimetry for Radioecology and Environmental Ecology.* Springer, Proc. NATO Advanced Research Workshop, 2002, Kiev, Ukraine.

Cigna A.A., 1996 - Origin and aim of radioecology. In: Luykx F.F. & Frissel M.J. (Eds.) - *Radioecology and the Restoration of Radioactive-Contaminated Sites.* NATO ASI Series, 2. Environment, 13, Kluwer Acad. Publ.: 1–15.

Emelianov B.M., Gavrilchenko V.S., 2000, Laboratory "B". Sungul phenomenon. *Snezhinsk: RFYATS-VNIIFP*: 1–440 (in Russian).

Gajewskaja N., 1923, Einfluss der Roentgenstrahhen auf Artemia salina. *Verhand. Intern. Ver. Theor. und Angew. Limnol. Grund.-Ver. Z. Kiel,* **1**: 359–362.

ICRP, 1977, Recommendations of the International Commission on Radiological Protection. ICRP Publication 26, *Annals of the ICRP,* **1** (3) Pergamon Press.

ICRP, 1991, 1990 Recommendations of the International Commission on Radiological Protection. ICRP Publication 60, *Annals of the ICRP*, **21** (1–3), Pergamon Press.

ICRP, 1993, Framework for assessing the impact of ionizing radiation on non-human species. Publication 91. *Annals of ICRP*, **33** (3), Pergamon Press.

Izrael Yu. A., 1996, *Radioactive depositions after nuclear detonations and accidents.* "Progress-Pogoda", Hydrometeoizdat, St-Petersburg, (in Russian).

Luycks and Frissel (Eds), 1996, *Radioecology and the Restoration of Radioactive Contaminated Sites.* NATO ASI Series, 2, Environment, 13, Kluver Acad. Publ.

Pentreath R.J., 2002, Radiation protection of people and environment: developing a common approach. *J. Radiol. Prot.,* **22**, (2): 45–56. Polikarpov G.G., 2005, Marine radioecology in the academies of sciences of Russia and the Ukraine (1896–2004). *Problems of radioecology and related disciplines*, Ekaterinburg, **6**: 4–24 (in Russian).

Russel R.S. (Ed.), 1966, *Radioactivity and human diet.* Pergamon Press.

Shaw G., 2005, Applying radioecology in a world of multiple contaminants. *J. Environ. Radioact.* **81**: 117–130.

Smith J. and Beresford. N.A., 2005, *Chernobyl. Catastrophe and consequences.* Springer Praxis Books: 1–335.

Stoklasa J. and Penkava J., 1932, *Biologie des Radiums und Uraniums.* Verlag von Paul Parey, Berlin.

Tarkhanov I., 1896, Experiments on Roentgen X-ray effects on a living organism. *Bull. St-Petersburg Biological Laboratory,* **1** (3): 47–52 (in Russian).

Titaeva N.A. and Taskaev A.I., 1983, *Migration of heavy natural radionuclides under conditions of a humid zone.* Leningrad: "Nauka", (in Russian).

UIR, 2003, IUR. Protection of the environment in 21st century: radiation protection of the biosphere including humankind. Statement of the International Union of Radioecology. *J. Environ. Radioact.* **70**: 155–159.

UNSCEAR, 1996, *Sources and effects of ionizing radiation. United Nations Scientific Committee on the Effects of Atomic Radiation.* Report to the General Assembly, with Scientific Annex. United Nations, New York.

Van der Stricht and R. Kirchmann (Eds), 2001, *Radioecology, radioactivity and ecosystems.* Fortemps, Liège: 1–603.

Whicker F.W. and Schultz V., 1982, *Radioecology: nuclear energy and the environment.* **1** CRC Press, Boca Raton, Florida.

Woodwell G.M. (Ed.), 1965, *Ecological effects of nuclear war.* Brookhaven National Laboratory.

Williams C. (Ed.), 2004, Framework for assessment of environmental impact (FASSET) of ionizing radiation on European ecosystems. *J. Radiological Protection*, **24** (4) Special issue.

LOW DOSE RADIATION EFFECTS IN THE ENVIRONMENT: IS THE FEAR OR THE SCIENCE IRRATIONAL?

COLIN SEYMOUR* and CARMEL MOTHERSILL
Medical Physics and Applied Radiation Sciences Department, McMaster University, Hamilton, Ontario, Canada, L8S 2K1
E-mail: seymouc@mcmaster.ca

*To whom correspondence should be addressed.

Abstract. This paper discusses the recent moves to protect non-human species as distinct from Man, from the effects of ionizing radiation. The position is put, that much of the concern surrounding low dose exposures is due to irrational fear and is not based on scientific data. If anything the data show protective or negligible effects at low doses not deleterious effects. The paper looks at the consequences of the situation and explores possible reasons for the deep seated fear of radiation which has led to the political need to address what may not be a problem.

Keywords: Low dose radiation, non-human biota, radiation protection

Initially radiation protection was anthropocentric, there being two main rationales for this. The first was that man was the only thing in the environment worth protecting. The second was that if man were adequately protected, then everything else in the environment would be too.

Attitudes have now changed, to the extent that man should now be regarded as part of his environment, and if the environment fails so does man. Depletion of the ozone layer and accumulation of greenhouse gases are contributing to accelerating climate change. There is concern that loss of environment may equate with loss of man.

As a result there is now a growing lobby that feels that the environment itself should be protected. This approach is fraught with difficulty, as what is the environment that is being protected? Generally man is already managing the environment, through agriculture and industry. There is also more overt control through, for example mosquito spraying. Ethically man seems to have no problems with trying to completely eliminate bacteria and viruses that are pathogenic to man.

A.A. Cigna and M. Durante (eds.),
Radiation Risk Estimates in Normal and Emergency Situations, 169–174.

This paper will discuss whether the desire to protect the environment from radiation is a logical step in environmental protection, or whether adequate protection already exists, and concern for the environment is primarily driven by another form of expression of fear of radiation.

Science may be defined as a combination of facts and the interpretation of those facts. This may be illustrated using the painting of George and the Dragon, by Raphael (Musee du Louvre Paris 1505/6). The main character in the picture is a youthful Saint George, mounted on a stallion. He is plunging a lance into a fearsome dragon, which is pinned and writhing beneath the lance. In the background of the picture is a beautiful but pensive looking maiden.

These are the facts, which may be interpreted as follows. The brave knight is rescuing the fair maiden from the dangerous and devouring dragon. This is a perfectly acceptable interpretation, but there are others. The dragon could symbolically represent knowledge, something the maiden wishes to acquire. The knight then represents the forces of male domination, preventing women from empowering themselves through knowledge. Yet another interpretation could be Freudian, with the dragon representing the id or unconscious mind, constantly battling the conscious mind, with the maiden representing the supervisory super ego.

The point is that one set of facts (the reality of the picture) can have many different layers of meaning (the interpretation of the picture).

Facts are indisputable, interpretations are disputable. A problem in an emotive area such as radiation is differentiating between fact and opinion, with opinion often being given the same credence as fact. This is demonstrated by the underlying assumptions of the Linear No Threshold Hypothesis, which tends to be accepted as fact rather than interpretation. Radiation and environmental issues merge muddled areas of interpretation and definition.

The environment means different things to different people. Should the environment be regarded as the habitat, or the animals and plants that exist within that habitat? If we can protect the animals and plants through intervention strategies, is there any need for concern about the habitat? Or should we protect the habitat and not worry too much about individual animal populations? Is biodiversity needed?

All of these questions require answers that are a combination of facts and opinion, and introduce a sense of the unknown. This can be demonstrated by another famous painting, this time by Valesquez (Las Meninas, 1656-57, Museo del Prado, Madrid). In this painting, the foreground on the left is dominated by the back of a painter's canvas. The painter stands in front of the canvas, looking out of the painting. Grouped around him are members of the Royal family and their entourage. The question is, who or what is the subject of the painting? It is unknown, and it is this insoluble question that provides some of the allure of the painting. The unknown in radiation protection though links in to generalized human fears of death, because of the public perceptions of nuclear weapons as weapons of mass destruction. During the cold war, it was this idea of mutually assured destruction (The MAD policy) that maintained a balance of power but imprinted on the collective psyche a fear of nuclear issues. It is this residual memory that in many minds makes radiation automatically bad, and as a consequence, anyone supporting the use of nuclear power or energy a mistaken or misguided individual. Within this framework of fear and mistrust it remains difficult for individuals to assess the scientific truth of the matter. It requires considerable self analysis to recognize intrinsic bias, that is the bias that forms our own personal opinions. It is always much easier to determine extrinsic bias, that is the bias that the individual perceives as coming from others. Differences perceived between intrinsic and extrinsic bias lead to a conflict of ideas, but without objective scientific analysis this leads simply to a polarization of positions.

But radiation is often one area where science itself fails us. In low dose radiation there are many assumptions but little science. Low doses may be arbitrarily defined as low(<100mSv) or very low(<10mSv) An average X-Ray examination is 0.1mSv-20mSv. The Linear No Threshold Hypothesis is based on an imaginary extrapolation from high dose high dose rate data. The prevailing paradigm is that low level radiation effects can be predicted from high level effects. It assumes a direct relationship between cause and effect that is mathematically predictable, that is, that all radiation exposure should be treated the same way, with the use of "correction factors" commonly known as fudge factors.

This assumption tends not to be true for biological systems, where cause is not directly related to effect. These non linear biological systems can best be described using complexity or chaos theory. Non linear systems may be defined as systems which are not characterisable by linear or first order equations, but which are governed by any variety of complex, reciprocal relationships or feedback loops. Central to complexity theory is Poincare's conception of deterministic chaos. Systems have emergent properties that cannot be understood by reductionist analysis into lower order components. Within the system apparently dichotomous opposites are neither antagonistic nor fixed, but are stages in a transformational dynamic process.

Paracelsus, a fifteenth century alchemist had deduced the practical applications of this when he declared "All substances are poisons. There is none, which is not. The right dose differentiates a poison or a remedy". This remains a fundamental truth in toxicology, but tends to be ignored in radiation protection, because it would give validity to arguments of radiation hormesis. For much the same reason bystander effects, genetic instability and adaptive response arguments are not given great credence.

The uncertainty of low dose radiation effects is submerged into the administrative certainty of a Linear No Threshold response. The assumptions that modern radiobiology challenge, and that are implicit in the LNT model, are that the probability of a radiation induced mutation remains constant per unit dose irrespective of dose or dose rate. It is also assumed that after radiation induces a carcinogenic process that this process is independent and evolves similarly whatever the number of lesions present neighbouring cells and tissues.

The LNT relationship is often applied incorrectly to large numbers of people by multiplying the effects of trivial doses by large populations. An example is to "calculate" the number of deaths induced if millions of people(or animals) were exposed to a few micro-sieverts. There are many examples of this scientific scaremongering in the literature. These "guesstimates" are published in part because epidemiology cannot detect statistically significant risks at low doses even on large cohorts or populations.

The conclusions of a joint French Committee of Medicine and Science (2005) were that the LNT should not be used without precaution for assessing by extrapolation the risks associated with low doses (<10mSv). The report further suggested that empirical evidence suggested there was an overestimation of risks at low doses.

This seems to be corroborated by a recent United Nations report on the health effects of Chernobyl, (IAEA publication 2005) which suggests that psychological fear of radiation caused most of the effects. No assessment was made as to whether the fear was escalated by newspaper scare stories. On the other hand, the BEIR VII report just published (Committee on the Biological Effects of Ionising Radiation Report VII, 2005), retains the conservative view that at present, we do not understand the mechanisms of non-targeted low dose effects, nor do we understand the relationship between effects at low doses and risks of exposure to these doses and that it is therefore best to adhere to the LNT hypothesis as an operating tool allowing a framework for radiation protection. This conclusion was a grave disappointment to many working in the low dose radiobiology field, whose evidence over the last 15 years shows conclusively that low dose and high dose effects are very different and that mechanisms operating at low doses can modulate the damage induced by the biophysical energy deposition. This concern is strongly voiced by the US Department of Energy in a critical letter to the Chairman of the BEIR VII committee (Orbach 2005).

On a contemporary note, avian flu is now receiving the same kind of alarmist press coverage. There is now a suggestion that the effects of radiation on the environment should be more closely regulated. The question is why. All reasons emanating from man have to be anthropocentric – it is impossible for a fly to have the perspective of an elephant. The only logical reason is that we're happy with the amount and type of biodiversity existing on the planet, and don't wish to see it altered. This is getting very close to a fear of the unknown, of change. Logically, radiation could cause new and exotic species to bloom as it alters the environmental balance. Equally, we are all aware that the balance of the environment is constantly shifting, and that nothing can remain constant over time.

It is also logically apparent that it is the initial construction of factories and plant that destroys the local environment. By the time the plant comes into operation, much of any environmental damage has already been done.

It is obvious that man himself, and by extension his activities, is the major polluter of the environment. To concentrate exclusively on one of man's myriad activities as being the most important in terms of global pollution, because of an illogical fear of that activity, is illogical and counterproductive. It is counterproductive because it allows far greater environmental damage from less regulated industries – everyone remembers Chernobyl but nobody remembers the far larger tragedy of Bhopal.

However attacking the views of others is not constructive. It is not the purpose of this paper to say that radiation has no negative environmental effects. Rather it is an attempt to put the hazards of radiation in perspective. If the environment is to be protected, it should be protected equally from all types of environmental hazard. The only way to do this is using an additive or integrative risk model. Within this model, a threshold maximum permissible environmental burden is set for a designated land use, for example amenity or residential or industrial. It is possible to set different thresholds for different land uses. The trade off is between many small polluters or one large polluter. Instead of radiation being treated as a unique polluter, it is treated as one of many polluters. Instead of having maximum permissible discharges for radiation, for arsenic, for lead and for many other chemicals, there is a blanket maximum pollution burden, or threshold. This can be fully utilised by radiation, or arsenic, or by any other pollutant. Once the threshold is reached from any single or combination of sources, then no more discharges of any pollutant are allowable.

The difficulty with this method is in agreeing comparable endpoints in measuring toxicity, or damage to the environment. The advantage of this method is that it allows a permissible level of environmental stress only, and can be further refined by allowing for synergistic interactions as well as additive ones.

To treat radiation as an environmental hazard that can be viewed in isolation is giving in to the unconscious dragons of the mind. Saint George, representing the conscious mind, protected by his armour of logic and rational thought, must do battle with the dragon. His objective though is not to kill the dragon, but to wrestle it from the unconscious to the conscious, where it can be critically examined. It is our belief that a critical examination of radiation will reveal that legislation for environmental radiation protection should not be isolated, but should be part of a systemic environmental legislative programme.

References

Aurengo A., Averbeck D., Bonnin A., Le Guen B., Masse R., Monier R., Tubiana M., Valleron A.-J., and de Vathaire F., 2005, Dose-effect relationships and the estimation of the carcinogenic effects of low doses of radiation, *Joint Report of the Academie des Sciences (Paris) and the Academie Nationale de Medicine*.

BEIR, 2005, *Report VII of the Committee on the Biological Effects of Ionising Radiation.*

IAEA, 2005, Chernobyl's Legacy: Health, Environmental and Socio-economic Impacts, The Chernobyl Forum, *IAEA publications,* IAEA/P1/A87/05–28601.

Orbach, R.L., 2005, Letter to R. Cicerone, Chairman BEIR VII committee.

RADIOCAPACITY: CHARACTERISTIC OF STABILITY AND RELIABILITY OF BIOTA IN ECOSYSTEMS

YURIY KUTLAKHMEDOV*
Institute of cell biology and genetic engineering NASU, Kiev, Ukraine
E-mail: ecoetic@mail.vtv.kiev.ua
VLADIMIR KOROGODIN[1]
Joint institute for nuclear research, Dubna, Russia
VIKTOR RODINA
Institute of geography NASU, Kiev, Ukraine
SVETLANA PCHELOVSKYA
Institute of cell biology and genetic engineering NASU, Kiev, Ukraine

*To whom correspondence should be addressed

Abstract. In models of radiocapacity for miscellaneous types of ecosystems (slopes, water ecosystems etc.) rather small levels of primary contamination of ecosystems are established, which, due to the laws of reallocating of radionuclides, may result in concentration in biota of critical levels of the contents of radionuclides and formation of noticeable radiation doses. From a number of estimations, dose rates in biota of 0.4 Gy/year for animals and 4 Gy/year for plants can notably oppress growth and condition of biota in ecosystems. If we accept ecological risk arising at these doses, it is possible to define for a unit of biomass, concentrations of radionuclides which are capable of giving similar radiation doses. This give a value for ^{137}Cs of 100 kBq/kg for animals and 1000 kBq/kg for plants, as an example. Based on a hypothesis about linear increase of ecological risk for biota from 0 up to 1, with increase in the level of radionuclide contamination of ecosystems, the general ecological risk from miscellaneous radionuclides which accumulate in a real ecosystem can be estimated. Thus, for each type of ecosystem (aqueous or terrestrial), depending on dynamics of reallocating of radionuclides, with the help of models of radiocapacity, it is possible to establish marginal primary levels of radionuclide contamination of

[1] Dr. Korogodin past away shortly after the meeting.

A.A. Cigna and M. Durante (eds.),
Radiation Risk Estimates in Normal and Emergency Situations, 175–185.

ecosystems. It is shown that these primary levels of radionuclide contamination can be very rigid. We for the first time demonstrate a method for ecological standardisation of miscellaneous ecotoxicants-pollutants on the basis of the theory of both models of ecological capacity and radiocapacity of a miscellaneous type of ecosystem. The analysis of dynamics of distribution and reallocating of tracer (^{137}Cs) in model ecosystems - aquacultures of plants - is conducted. This analysis has allowed us to establish that in response to stressful factors, the biota of ecosystems is able to change rapidly the TF (Transfer Factor) and consequently to cause fast reallocation of tracer and its removal from biota in the environment. As stressful factors we have used in experiments -acute gamma-irradiation of seeds and deposition of salts of heavy metals ($CdCl_2$) in different doses. In laboratory experiments the dose relation is established as the rate of change of parameters of radiocapacity at operating doses of irradiation and heavy metals. The given approach and method allows one to predict an ecological setting of permissible resets and outlets of pollutants in environment using unified idealised standards. The given method obeys the fundamentals of equidosimetry showing that stressful factors affect biota in the same way as influencing the parameters of radiocapacity. Thus, it is feasible to sum the total effect of miscellaneous factors on biota in terms of radiation doses.

Keywords: Ecosystem, radiocapacity, radiocapacity factor, tracer, stress-effect, mathematical models, box models

1. Introduction

In order to characterise the influence of various factors on biota, more than 30 different indices are used. The main ones are: biodiversity, biomass, population and reproduction rate (Grodsinsky *et al.,* 1991). Such indices are known to have a delayed response to negative environmental factors and can only be measured when the condition of biota irretrievably deteriorates. There is a need to find an index of well being of biota that could precede the response of biological growth indices and thus allow estimation of the condition of biota promptly. The investigation of behaviour of radionuclide tracer ^{137}Cs and parameters of radiocapacity using this tracer has allowed us to propose an estimation of the condition of the biota in ecosystem using the responses and changes of radiocapacity indices (Kutlakhmedov, 1998; Agre & Korogodin, 1960). The experiments have been carried out on plant aquacultures as simplified models of plant ecosystem. It is known that the condition of biota can be characterized by its ability to take up nutrients, in particular potassium that participates in all physiological reactions. Thus, the dynamics of absorption of artificially introduced tracer ^{137}Cs, that is an analogue of potassium, by plants can reflect well-being of biota within an ecosystem.

After the Chernobyl disaster such a tracer is an unavoidable element of life in almost all ecosystems. The investigations have shown that distribution and re-distribution of this tracer in aquatic and terrestrial ecosystems clearly changes with all significant external influences (climate, floods, counter-measures etc.) and various pollutants (thermal, radiation doses, chemical pollutants etc.) (Kutlakhmedv *et al.*, 1997; 1998). It has also been shown that no serious influence on the ecosystem can occur without an effect on this tracer distribution and radiocapacity parameters.

1.1 THEORETICAL ANALYSIS OF RADIOCAPACITY USING THE EXAMPLE OF A SIMPLE MODEL ECOSYSTEM

Let us examine the problem of radiocapacity using the example of a two-chamber model ecosystem that includes the environment (water) and biota. This consists of a two-compartments model of the environment: the Environment compartment (water, soil, etc) and the Biota compartment (terrestrial and aquatic plants, forest etc.). Let us examine a box model of a lake ecosystem:

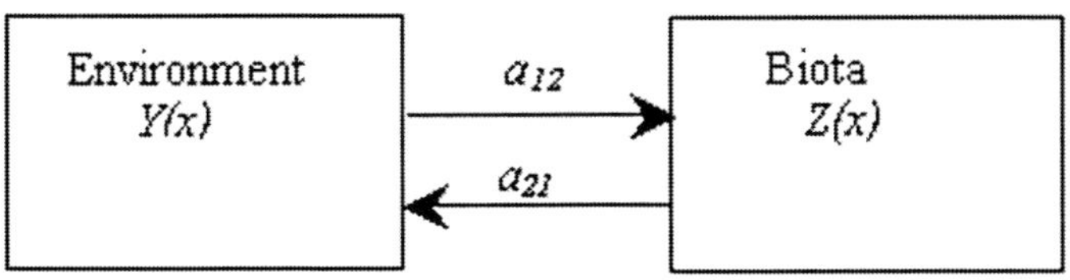

Figure 1. Block-scheme of simple box-model ecosystem.

Hence we have two compartments (Fig. 1) that contain Y(x) and Z(x) of radionuclides with time parameter x; a_{12} is the rate of absorption of tracer nuclides (and proportionally the rate of absorption of nutrients, for example, potassium); a_{21} is the rate of their return to the environment (water).
Box model can be described using 2 equations.

With protracted observation these equations can be used for the estimation of radiocapacity factors both for water and for biota in the following way:

$$F_{á} \cong \frac{a_{12}}{a_{21}+a_{12}}, \quad F_{â} \cong \frac{a_{21}}{a_{12}+a_{21}} \tag{1}$$

If the lake's ecosystem consists of 2 compartments, biota and water, then the formulation of radiocapacity for biota and water is simpler: in terms of model of lake ecosystem's radiocapacity, the radiocapacity factor can be described by the following expression:

$$F_{á} = \frac{PK}{1+PK}, \quad F_{â} = \frac{1}{1+PK} \tag{2}$$

The comparison of 2 and 1 results in the following:

$$\frac{a_{12}}{a_{21}} = PK = \frac{F_{á}}{F_{â}} = \frac{1-F_{â}}{F_{â}}. \tag{3}$$

Thus absorption/return ratio for tracer and mineral nutrient potassium is proportional to the biomass of biota and the coefficient of accumulation in the system "water-biota". The higher the biomass and the coefficient of tracer accumulation by biota, the better the condition of biota, and the higher the absorption/return ratio for the tracer. The same applies to other nutrients transferred from water into biomass. In this case there is a strong link between the tracer-dependent index of radiocapacity and the biological indices such as the rate of absorption and return of tracers and nutrients. The decrease of the K–accumulation coefficient (water-biota) under the influence of stress-factors (radiation, heavy metals etc) is enough to change the parameters of radiocapacity. If the influence of pollutants leads to the decrease of biological indices (as biomass, rate of its growth) then there will be a stronger change of indices and parameters of radiocapacity.

2. Experimental Test of the Possibility of Using the Radiocapacity Factor as the Key Parameter for Estimation of the Response of Biota to Various Pollutants

2.1 MATERIALS AND METHODS

In order to investigate the possibility of using an index of a plant ecosystem radiocapacity for characterising its condition and for prognosis of its alteration under the influence of external factors the following set of experiments was performed. Aquaculture of maize was chosen as a simplified model of a plant ecosystem. 3-day old germs (grown at 23^0C) were exposed to gamma-radiation at the cobalt irradiator "Issledovatel" The absorbed dose was 15 Gy. After exposure the germs were planted onto 0.5 L glass cans filled with water. Water from all experimental variants contained the tracer, radioactive ^{137}Cs. Some variants contained also $CdCl_2$. Control plants, which were not irradiated or exposed to CdCl2, were also planted onto water with tracer. During the experiment (14 days), water samples were taken regularly for estimation of activity by measurement of tracer concentration. The length of the main root of the plants was also measured.

2.2 RESULTS AND DISCUSSION

Time dependence has been obtained for indexes of radiocapacity and for growth processes for a model system.

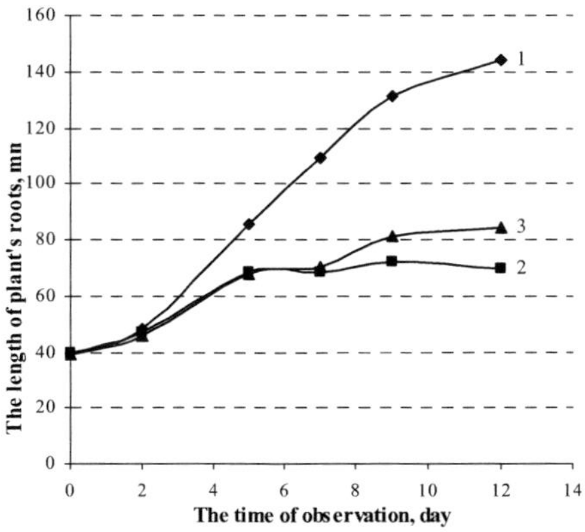

Figure 2. The influence of single acute exposure of maize germs to gamma-radiation with the absorbed dose 15 Gy and introduction of $CdCl_2$ with the concentration 50 μmol/L on the dynamics of aquaculture root growth: 1- blank; 2-15 Gy exposure 3- introduction of cadmium salt 50 μmol/L.

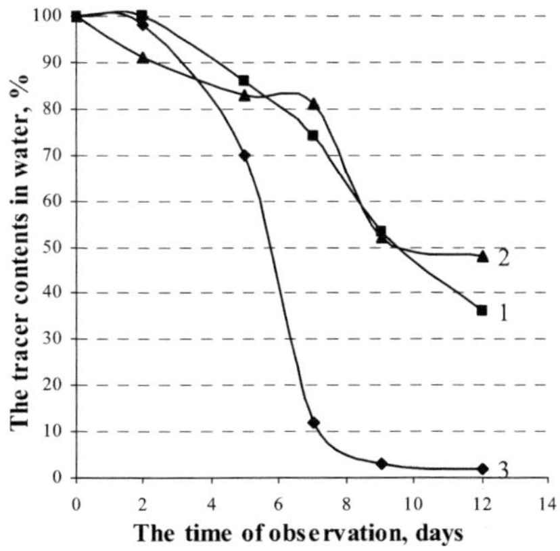

Figure 3. Dynamics of tracer content in water: 1- introduction of 50 μmol/L L $CdCl_2$; 2-acute 15 Gy exposure 3-blank.

The data obtained (Fig. 2) show that single-time acute gamma-exposure of maize germs to a 15 Gy dose coupled with introduction of a low-concentration of $CdCl_2$ into the environment suppress root growth. Growth indexes for plants that were exposed to $CdCl_2$ were 40% lower than for control plants.

Because of this, a decrease in the absorption capacity of plants (tracer-dependent radiocapacity index) was also expected. The dynamics of water radocapacity factor as a component in this simplified system is shown in

Fig. 3. This demonstrates that control plants absorb almost 100% of the tracer towards the end of experiment. Curve I that shows the content of ^{137}Cs in water (%) approaches zero on the 5th day of the experiment whereas experimental plants absorbed up to 55% of tracer.

It can be seen that the influence of 2 different factors such as gamma-radiation and a toxic heavy metal have approximately equal biological effects on root growth suppression and a lowering of plant radiocapacity. The suppression of radiocapacity characteristics of biota has clearly been observed.

To check the robustness of the relationship between radiocapacity factor and the influence of negative factors, the following set of experiments using radiation and toxic compounds was performed. Aquacultured maize germs were exposed to acute gamma-radiation with absorption doses: 5.10, 20 and 40 Gy. The plants of another 4 variants were exposed to $CdCl_2$, which was added to growth media (water) at concentration of 22, 44, 78 and 100 μmol/L. All plants were simultaneously planted onto water containing tracer ^{137}Cs. Water samples were collected the next day for the estimation of remaining ^{137}Cs activity. The dependencies of water radiocapacity factor on Cd concentration and absorbed dose were obtained. (Fig. 4 and 5).

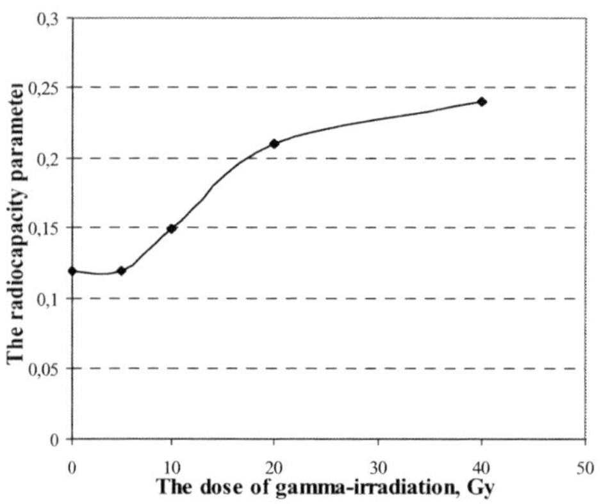

Figure 4. Dependence of the dynamics of water radiocapacity changes on introduced concentration of cadmium.

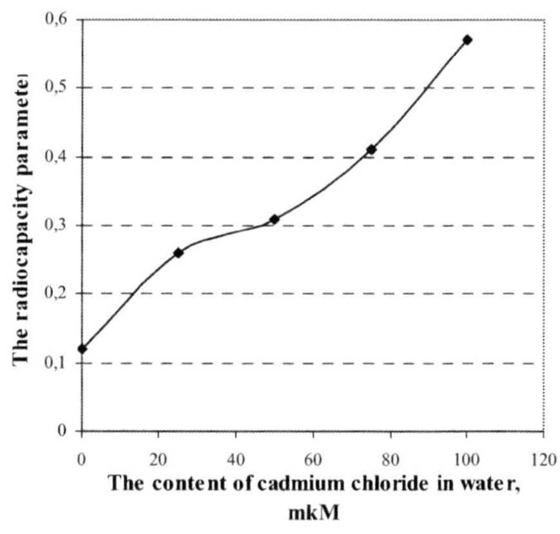

Figure 5. Dependence of the dynamics of water radiocapacity changes on dose absorbed by plants.

Water radiocapacity increases along with the absorbed radiation dose or Cd concentration (Balan & Kutlakhmedov, 2003). This increase is influenced by selected factors which means that the radiocapacity of biota decreases (Kutlakhmedov *et al.,* 2003). The well-being and vitality of the ecosystem, in general correlates with high radiocapacity and vice versa. On the other hand a high radiocapacity and stable high radiocapacity factor indicate viability of the ecosystem. Radiocapacity and radiocapacity factor control, can serve as an objective indicator and method of estimation of well-being of any ecosystem (aquatic, terrestrial etc).

3. The Estimation of Synergism of Various Factors Exerting Influence on Biota, by use of a Tracer-dependent Radiocapacity Parameter

The problem of combined impact of environmental factors on the ecosystem has already been discussed in literature (Petin *et al.*, 1997a; 1997b) Let us analyse the possible impact of different factors (e.g. radiation and a chemical factor such as the introduction of cadmium) on the radiocapacity parameter of a simplified ecosystem. This will give an evaluation of the limit of quantitative estimation of synergism or antagonism of different factors affecting the biota of an ecosystem. Let us suppose that in a control trial the biota absorb and return the nuclide with rates a_{12} and a_{21} respectively. Their ratio can be defined as.

$$Z = \frac{a_{12}}{a_{21}}$$

Synergism coefficient has been defined as

$$P = Z_0 \frac{Z_{Cd+ir}}{Z_{Cd} Z_{ir}}$$

where Z_0 is a ratio of radiocapacity factors of control variant. Z_{Cd+ir} is a ratio under the combined impact of radiation and heavy metal; Z_{Cd} and Z_{ir} are independent ratios for the impact of each factor. When p=1, there is no synergism in the impact of different factors on radiocapacity parameters. When p<1, it can suggest a significant synergism and in this case 2 factors strengthen each other, when p>1 it indicates antagonism, i.e. the first factor lowers the negative impact of the second one and viceversa. Thus we have created the scheme and introduced the parameter of evaluation of synergism of different factors by coefficient p.

Using the results of experiments where aquatic cultures of plants underwent acute gamma-radiation exposure to 20 Gy separately and in combination with 50 μmol/L $CdCl_2$ (added stepwise) we have estimated synergism of these two factors via radiocapacity indices. Time intervals of $CdCl_2$ introduction (2 fractions 25 μmol/L each) were 6, 10 and 24 hours. Obtained ratios of radiocapacity factors Z enabled us to define the means of synergism coefficients P (using eq. 3) and to build their time dependence (Fig. 5).

It can be seen that radiation and chemical toxins interact during their impact on a plant ecosystem. The interaction in all combinations of radiation and $CdCl_2$ is non-additive and relates to a rehabilitation process.

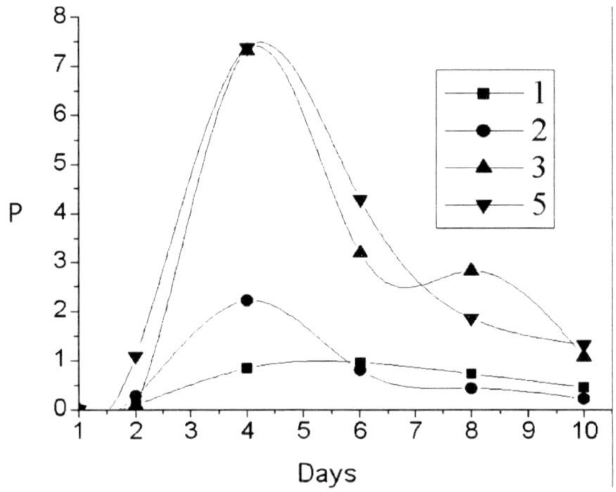

Figure 6. The dependence of interaction index (P) of two factors (gamma-radiation and cadmium salt introduction in acute and fractionated modes: *1* – irradiation on dose 20 Gy in combination with $CdCl_2$ adding with the time between fractions 6 h; *2* – irradiation with $CdCl_2$ adding with the time between fractions 10 h; *3* – irradiation with $CdCl_2$ adding with the time between fractions 24 h; *4* irradiation with $CdCl_2$ adding without fractionating.

3.1 APPLICATION OF GIS-TECHNOLOGY TO THE ANALYSIS OF RADIOCAPACITY OF A REAL LANDSCAPE ECOSYSTEM

Using analytical GIS allows to estimate and predict the dynamics of pollutant re-distribution in real landscape by using radiocapacity models. Using the maps of landscape and radiocapacity indices we have predicted the re-distribution of the ^{137}Cs pollutant over the real landscape and the concentration of radionuclides in 30 years in the declines of this landscape (Fig. 8)

Thus we have demonstrated the possible application of radiocapacity models for evaluation and prediction of dynamics of pollutant redistribution using a specially developed variant of the analytical model.

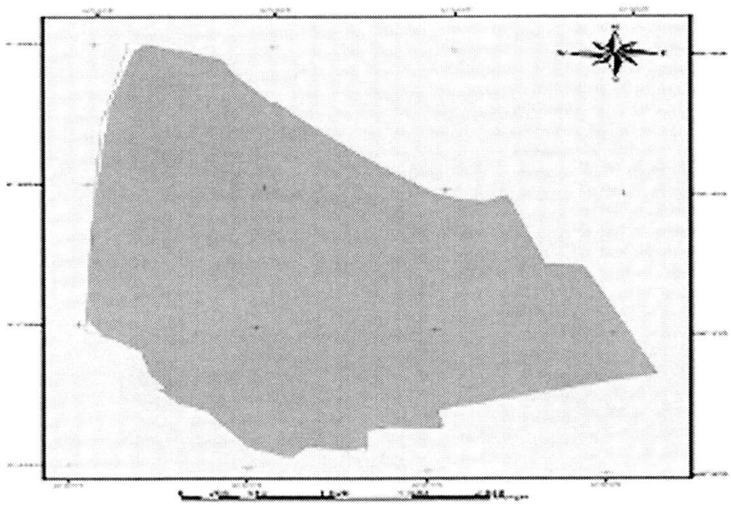

Figure 7. The initial uniform distribution of pollutants over the real landscape – testing area "Lesniki" in the suburbs of Kiev.

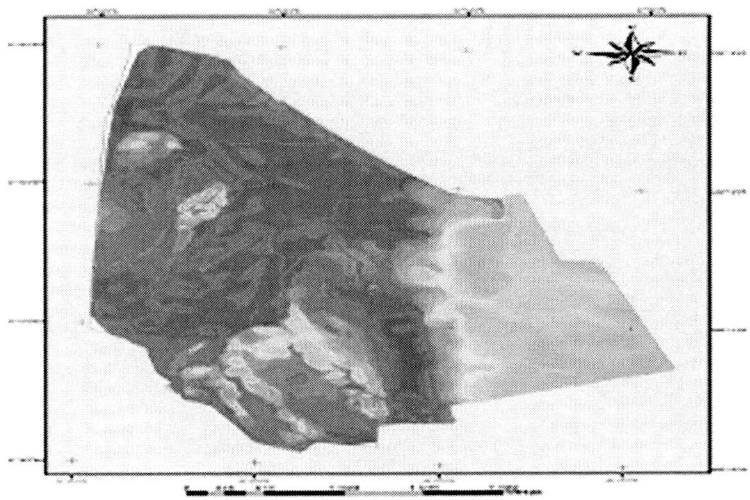

Figure 8. Prediction map of radionuclide redistribution in 30 years after the initial uniform pollution.

4. Conclusions

1. Our experimental and theoretical investigations have shown that tracer (^{137}Cs)-dependent index of radiocapacity of model ecosystem responds adequately to alteration of biota's conditions. In experiments carried out using a simplified model of a plant ecosystem the impact of radiation and a heavy metal induces a significant decrease of radiocapacity of biota. This phenomenon indicates changes of condition and well-being of biota that are reflected by re-distribution of the tracer as a test-index.
2. It is established that radiation and chemical factors (heavy metals) impact the growth rate and condition of biota and this impact is adequately reflected by means of radiocapacity factors. Redistribution of the tracer within the system illustrates internal principles of the condition and behaviour of biota in different ecosystems.
3. The mathematical model for estimation of synergism of several harmful factors using tracer-dependent radiocapacity indices has been developed. It can be shown that along with the processes of growth and repair of the biota, significant changes of synergism occur. These include additivity and antagonism of different factors.
4. Repair/recovery processes in biota are shown by alterations both of biological growth indices and improvement of radiocapacity indices. This means that rehabilitation measures can increase radiocapacity indices of biota.
5. It has been shown that tracer (^{137}Cs)-dependent index can be used for estimation of stability of biota following human impact on the ecosystem.
6. The possibility of application of radiocapacity models for evaluation and prediction of pollutant redistribution dynamics using specially developed variant of analytical is demonstrated.

References

Agre V., Korogodin V., 1960, Distribution of radioactive pollution in slow-exchangeable water sources. *Med.radiology*, **1**: 67–73. (in Russian).

Balan P., Kutlakhmedov Yu., 2003, Estimation of the influence of rehabilitation processes on radiocapacity index of plant aquaculture. *Anthropogenously changed environment of Ukraine: health and ecological risks*. Papers Int. Conference, Kiev, Chernobulinaform: 477–484. (in Ukrainian).

Grodsinsky D., Kolomiez O., Kutlakhmedov Yu., 1991, *Antropogenic radionuclide anomaly and plants*. Kiev, Lubid: 1–160 (in Russian).

Kutlakhmedov Yu., 1998, Medical and biological consequences of Chernobyl accident **Vol.1** *The long-term radioecological problems of Chernobyl accident and the countermeasures*. Kiev: Medecol: 1–172 (in Russian).

Kutlakhmedov Yu., Polikarpov G., Korogodin V., 1998, Principles and methods of estimation of ecosystems' radiocapacity. *Heuristics of radiobiology*, Kiev, Naukova dumka: 109–115 (in Russian).

Kutlakhmedov Yu., Polikarpov G., Korogodin V., 1997, *Methodology and methods of investigation of radionuclides and other technogenic pollutants in terrestrial and aquatic ecosystems*. Kiev, Vucha schola : 1–48 (in Russian).

Kutlakhmedov Yu., Korogodin V., Koltover V., 2003, *Basics of radioecology*.-Kiev, Vucsha schcola: 1–319.

Petin V., Ryabchenko N., Surinov B., 1997a, Concepts of synergism in radiobiology. *Radiation biology- Radioecology*, **37**(4): 482–487.

Petin V., Dergacheva I., Romanenko A., Ryadova S., 1997b, New concept of optimization and prognosis of synergism effects under combined impact of physical and chemical environmental factors. *Russian Chemical Journal,* **41**(3): 96–104.

TRITIUM IN WATER SYSTEMS OF URAL REGION

MARGARITA CHEBOTINA*, OLEG NICOLIN
Institute of Plant and Animal Ecology, Russian Academy Science, Ural Division, 8 Marta 202,Ekaterinburg, Russia
E-Mail: chebotina@ipae.uran.ru

*To whom correspondence should be addressed

Abstract. The paper gives the data based on the many-year observations over tritium contents in the water ecosystems of Ural region. Almost all tritium in Ural region, including the control plots, was of anthropogenic origin. Nuclear enterprises are the main sources for the environmental contamination which tritium. In the areas where such enterprises are situated technogenic complexes for the retention and localisation of tritium should be developed and constructed.

Keywords: tritium, water, Beloyarskaya Atomic Power Station, the "Mayak" enterprise, river ecosystem

1. Introduction

Tritium, a radioactive isotope of hydrogen, is a widely spread pollutant of natural waters. Its natural background level is about 1 Bq/L, the technogenic background level is 5 Bq/L. Nuclear energy enterprises release tritium into the environment is additional quantities resulting in the local contamination of the water ecosystems around them.

In Institute of Plant & Animal Ecology carries out systematic studies of the contribution of the Beloyarskaya Atomic Power Station (BAPS, in Sverdlovsk region) and the "Mayak" enterprise (in Chelyabinsk region) to the contamination of the water ecosystems with tritium in the Urals. Our method to determine tritium contents in the water was the following. Water samples (0.5 L) were filtered for cleaning and kept in closed glass vessels until the end of the analysis. During to analysis the samples were purified

A.A. Cigna and M. Durante (eds.),
Radiation Risk Estimates in Normal and Emergency Situations, 187–196.

from soluble salts. For this purpose the filtered water was placed in thermoresistant glass retorts and distilled dry in a reverse refrigerator. Tritium content was estimated by the method of a single-stage electrolyte enrichment on a special installation. The count was kept on an American installation “Delta-300”. The reliability of our method was proved by comparison of our method was proved by comparison of the methods of quantitative & qualitative tritium estimations used in the Institute of Plant & Animal Ecology and VSEGINGEO, Ministry of natural resources; the methods proved to have much similarity.

2. Contents

Since 1980 we estimated tritium contents in the areas adjacent to the BAPS: in the cooling reservoir, the Olkhovskoye bog-river ecosystems, snow and rain precipitation’s, drinking water, some soil types.

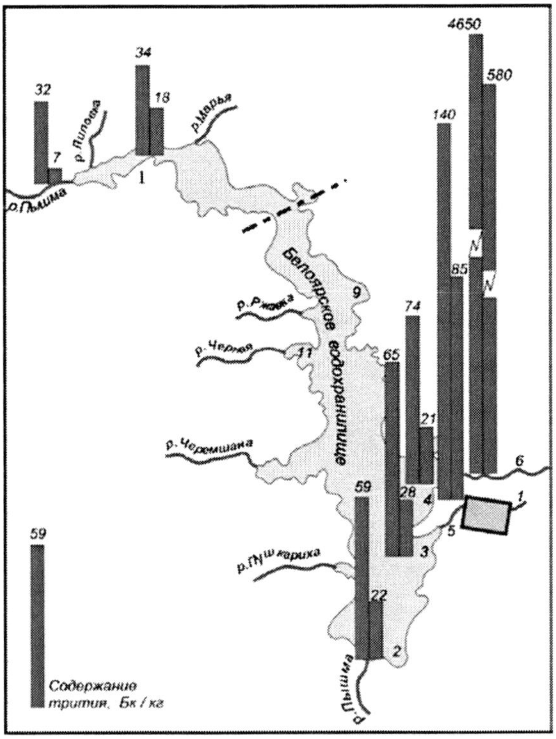

Figure 1. Permanent observation points in the water area of Beloyarskoe reservoir and average tritium concentrations (Bq/L) during the joint operation of units 2 and 3 (left column) and since unit 2 was out of operation (right column). 1: upper reach; 2: damn; 3: Teply Bay; 4: Biophysycal station area; 5: discharge channel; 6: upland ditch.

In the cooling reservoir tritium monitoring involved the following permanent points of observation: 1 - an area in the upper reaches, 15 km upstream; 2 – dam area, outlet to the Pysma river; 3 - Teply Bay, entry of the warm water from the BAPS cooling system; 4 – Biophysical Station adjacent to the BAPS; 5 – discharge channel by which the wastes from the BAPS and the neighbouring SFINIKIET enterprise enter the Beloyarskoe reserve; 6 – upland ditch draining the area around the BAPS (Fig. 1).

The area in the upper reaches 15 km upstream from the BAPS was chosen as the control because the low-radioactive run-off from the BAPS followed mainly downstream towards the dam area and could not directly enter the upper reaches. During the investigation period tritium concentrations in the upper reaches varied from several units to 60-70 Bq/L (Fig. 2). In 1989-2001 tritium concentrations were averagely twice lower than in 1980-1989 (34 and 16 Bq/L, correspondingly; significant differences at $p < 0,01$) due to the fact that since the end of 1989 one of two energy units at the BAPS (units 2 and 3) was out of operation (unit 2). As a result, tritium concentrations in the water notably reduced.

The dam area characterized the water of the reservoir at its outlet into the Pyshma River. Figure 3 shows that during the joint operation of units 2 and 3 tritium concentrations in this part ranged from 40 to 80 Bq/L

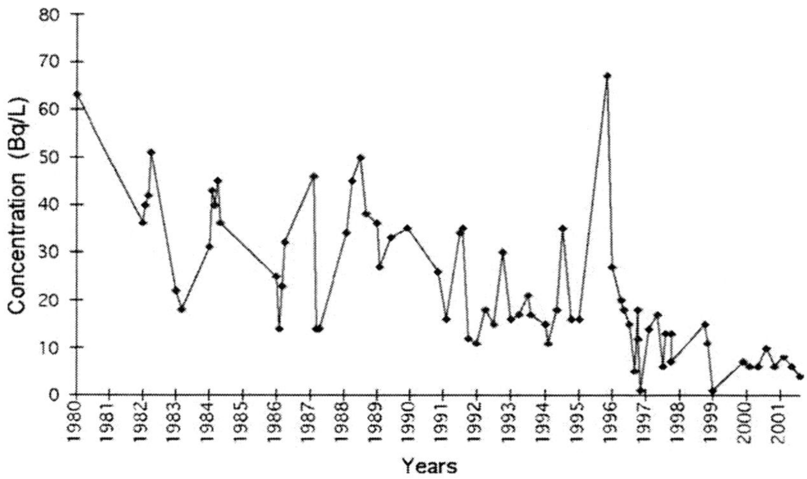

Figure 2. Tritium concentrations in the upper reaches of the Beloyarskoe reservoir (Bq/L).

Against the background of the general reduction of tritium concentrations 2 peaks were recorded in February 1990 and February 1998 (the peak values were not included in calculations of the average values). The peaks evidenced of an increased tritium discharge caused by disturbances at the BAPS. The comparison of tritium contents in the upper reaches and in the dam area revealed the contribution of the BAPS to the contamination of the water. In 1980-1989 the average value in the dam area was significantly higher ($p < 0,01$) than in the upper reaches (58 and 34 Bq/L correspondingly). Since unit 2 was made out operation in the difference between the background of a general decrease of the isotope concentrations.

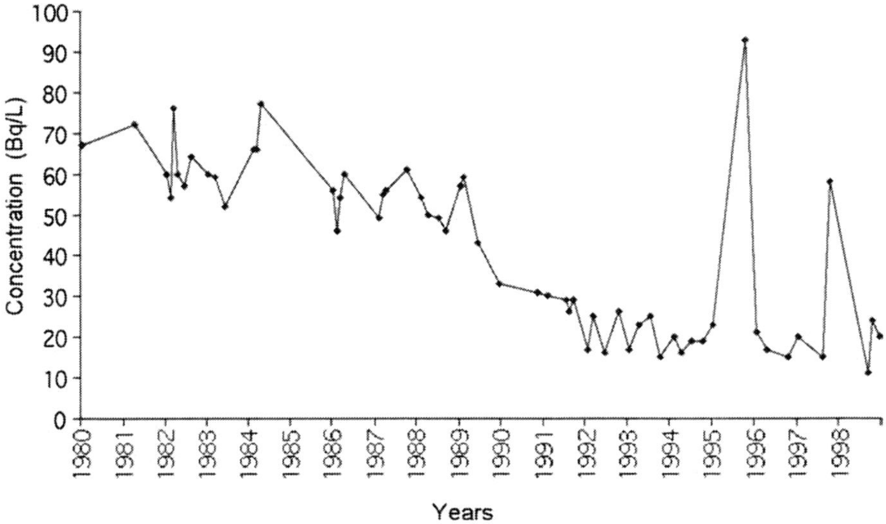

Figure 3. Tritium concentration in the dam area (Bq/L).

The inspection of the Teply Bay area was very important, as fish is reared there. The bay is isolated from the reservoir with a dam and receives the warm water from the BAPS cooling system. In its turn the cooling system receives by the water-capture channel the water from the upper part of the reservoir where low-radioactive run-off by discharge channel and the upland ditch enter the reservoir. During the joint operation of units 2 and 3 tritium concentrations in the Teply Bay varied from 45 to 90 Bq/L. With unit 2 out of operation concentrations reduced about trice (up to 22 Bq/L). However, the peaks of tritium recorded in the upper part were also registered in the Teply Bay zone, the highest ones after unit 2 was made in operative.

Near the Biophysical Station there is the upland ditch, which drains the territory around the BAPS, and brings higher quantities of tritium into the reservoir. Therefore, sharp changes in the radionuclide concentrations were registered in this area. Peak discharges of tritium in January 1988. October 1989, December and February 1998, 1999 resulted in higher contents of the radionuclide in the Teply Bay water (up to 140-1000 Bq/L). High concentrations in this area were also recorded in 1980-1981 (90-95 Bq/L) when unit 1 and unit 2 were in joint operation. Between 1982 and 1989, before unit 2 was made inoperative, the average value was 66 Bq/L, after that it was 17 Bq/L.

Our research showed that tritium entered the cooling reservoir mainly by the discharge channel and the upland ditch. The first one received wastes discharge from the BASP and the SFINIKIET enterprise, the other one drained the area around the BASP. Sharp fluctuations in the radionuclide contents evidenced of its uneven entry into the reservoir.

Another water ecosystem suffering from the BASP was the Olkhovskoe bog and the Olkhovka River about 5 km southeast of the BASP. The wastes passed through the bog and the river entered the Pysma and further the Tura-Tobol-Irtysh river system.

The investigations of 1981-1982 (when units 1 and 2 were in operation) showed that tritium contents in the whole bog-river ecosystem exceeded the technogenic background level several orders. The isotope contents in the discharge channel, in the bog and in the Olkhovka river rising from the bog, varied in a wide range depending on the water sampling time – an evidence of periodical discharge of tritium with the BAPS waste water. The average data are given in Fig. 4. It is seen that tritium concentrations were average twice lower in the lower part of the bog than in its upper part. Although the differences were statistically insignificant, the tendency towards the decrease of tritium in the lover part suggested that the wastewater be diluted with the water from underground springs. The suggestion was proved by a geophysical research: the bog was found to be located in a tectonically disturbed zone with pressure water supply from the tectonic fracture (Outkin *et al.*, 2004). The wastewater was diluted in the upper part of the bog.

Figure 5 gives the average data on tritium contents in the Olkhovskoye bog during 1980-1990. In spite of the great variability one can see that during the joint operation of units 2 and 3 (1980-1989) the contents of tritium were higher (3380±480 Bq/L, n= 62 than after unit 2 was made inoperative (since the end of 1989) and only unit 3 was in operation (1034±156 Bq/L, n=46). The difference was significant at $p < 0{,}001$. According to calculations (Chebotina *et al.,* 1994) the average annual discharge of the radionuclide from the DAPS into the bog became (300-400)$\cdot 10^{10}$ Bq/year after unit 2 was rendered out of operation, while in 1990 the figure was (110-130)$\cdot 10^{10}$ Bq/year.

To characterise trace the tritium entry from the bog into the river system downstream we chose a location at the outlet from the bog – a point of constant monitoring at this point (Fig. 6). One can see that the values varied within the range of several orders (from 10 to 10000 Bq/L). In spite of the variability of the data the decrease of tritium concentrations in time is distinct. After unit 2 was removed from service in 1989 the contents of tritium decreased circa 10 times (up to 250±50 Bq/L at n=8) compared to the earlier period (2600±260 Bq/yeas at n=33). The statistical treatment of the results showed that the differences were highly significant at p< 0,01.

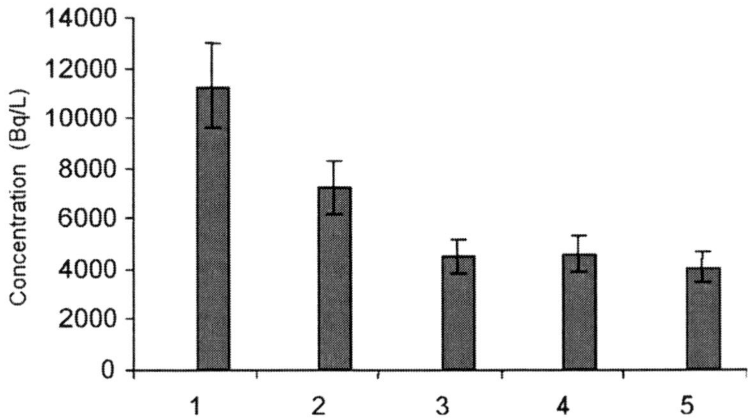

Figure 4. Tritium concentrations (Bq/L) in the bog-river ecosystem in 1981-1982: 1 -Disharge channel; 2 - Upper reaches; 3 - Middle part; 4 - Olkhovka source; 5 - Olkhovka mouth.

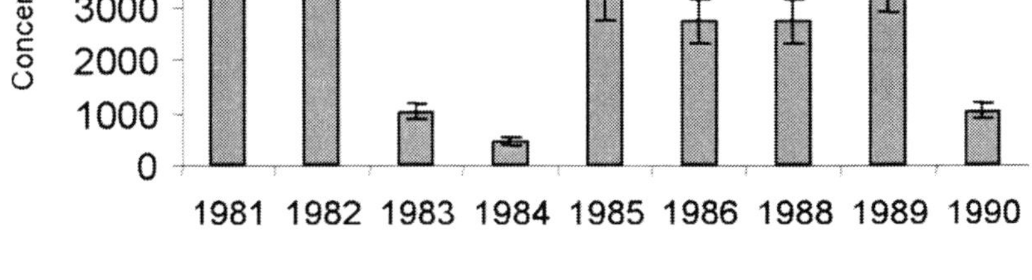

Figure 5. Tritium concentrations in the Olkhovskoe bog in various years.

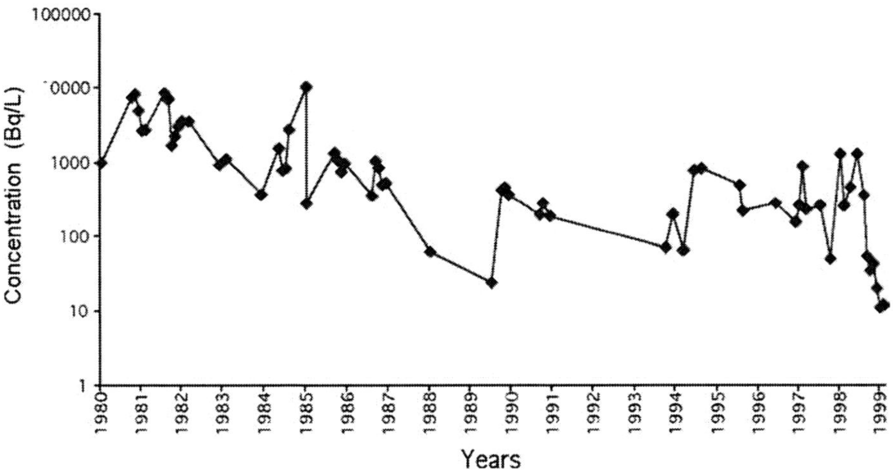

Figure 6. Tritium concentrations at the outlet from Olkhovskoe bog in various years (Bq/L).

As the saturation of the soils with radionuclides was the highest in the upper and middle parts of the bog, a reconstruction's of the waste discharge system was made: debalance waters were withdrawn into the lower part of the bog by special pipeline, while the sewage system run-off remained entering the upper part. After the reconstruction the volume of the water in the bog sharply decreased, the water flows by a channel falling into the Olkhovka River. Presently we observe a further decrease in tritium concentrations in the bog-river ecosystem (Table 1).

In 2003 a map of the bog with tritium concentrations in it was made, the coordinates and locations of the observation points were estimated with the help of the GPS sputnik system. Most of the bog was found to be cleaner of tritium: along the periphery and deep in the basic part of the bog the concentrations exceeded the technogenic background level only several times. However, in some places, basically in the area of the old and the new channels, the radionuclide was found in significant quantities – up to 800-1300 Bq/L.

Thus, making two units (1 and 2) out of operation and the reconstruction of the discharge system improved the situation in the Olkhovskoye bog and will improve the environment in the region.

Table 1. Tritium concentration in the bog-river ecosystem after the reconstruction of the discharge system, Bq/L.

Sampling time	Discharge into the Olkhovskoe bog		Discharge into the Olkhovka
	Upper part	Lower part	
April 2000	8 ± 0,7	49 ± 0,5	6 ± 0,4
May 2000	14 ± 0,7	83 ± 0,4	78 ± 0,4
June 2000	841 ± 0,4	10 ± 0,3	181± 0,3
August 2000	2120 ± 298	116 ± 28	1027 ± 108
March 2004	48 ± 1	12 ± 0,5	34 ± 1
June 2004	93 ± 3	16 ± 1	48 ± 1
July 2004	375 ± 6	27 ± 0,4	292 ± 8
August 2004	1825 ± 2	163 ± 1	1040 ± 1

The Pyshma is the largest river in the BAPS environs. It receives the waters from the Beloyarskoe reservoir and the low-radioactive run-off downstream, which had passed by the Olkhovskoye bog and the Olkhovka. Along the Pyshma there are settlements and holiday homes, its water is used for various economic purposes. To estimate the contribution of the Olkhovskoye bog to the contamination of the Pysma tritium monitoring was performed upper (Jalunino) and lover (Kumovskaya Mill) the confluence with the Olkhovka during the joint operation of units 1, 2 and 3 (1981), during the operation of unit 2 and 3 (1982-1983) and during the operation of unit 3 (after 1989).

Table 2 gives the average tritium concentrations in the Pyshma during these periods. Between 1981 and 1983 the figure was circa 47 Bq/L upstream the confluence with the Olkhovka and much higher downstream. After unit 2 was put out of operation (1989) tritium contents decreased in both area. On the whole, at present tritium-containing wastes entering the Pyshma from the Olkhovskoye bog and the Olkhovka, are diluted with the water of the Olkhovka and almost do not affect the Pyshma. However, against the background of the generally stable situation tritium peaks were recorded in some periods. Thus, on the 15th-20th of June 2000 a peak of 170 Bq/L was registered in Mezenskoe (upstream the confluence). This case also marked in other areas upstream the Pyshma. The source of tritium was not the Olkhovskoye bog but the Beloyarskoe reservoir, which received tritium by the storm run-off channel and the upland ditch. Evidently uncontrolled emergency situation at the BAPS or SFINIKIET and are temporary.

Table 2. Average tritium concentration in the Pyshma water upstream and downstream the confluence with the Olkhovka, Bq/L.

Observation period	The Pyshma upstream the confluence with the Olkhovka	The Pyshma downstream the confluence with the Olkhovka
1981	47 ± 14	618 ± 325
1982-1983	47 ± 2	149 ± 31
After 1989	10 ± 1	13 ± 1

To trace tritium migration downstream the Pyshma we periodically investigated the water at a distance of about 120 km from the confluence high the Olkhovka. Permanent points of observation were parts of the river near 10 settlements located along the Pyshma: Kumovskaya Mill, Malinovka, Petushki, Belokamenny, Khimleskhoz, Glyadeny, Znamensky, Kurii, Filatovskoe, Kamyshlov, "Mayak".

On the whole, during the investigation period tritium concentrations varied from 10 to 109 Bq/L. In 1983-1984 the value was the highest between Kumovskaya Mill and Kamyshlov and decreased 1,5-2 times downstream due to the dilution. After unit 2 was made inoperative tritium concentration there was in general lower than during the earlier period and continued decreasing in course of time (1990: 42±1 Bq/L; 1991: 27 ±2 Bq/L; 1999-2001: 14±1 Bq/L).

We investigated the Pyshma several times in its middle and low streams. In the period of lack of wastes discharge tritium concentrations did not exceed the technogenic background level more than twice (up to 13 Bq/L). Episodically peaks registered in some period's evidence that the river requires constant monitoring.

Another important source of tritium in the water ecosystem is the "Mayak" enterprise (Demin, Telushkina, 1987; Demin, 2001). In 2001-2003 we investigated 34 points at a distance of up to 90 km around it. It is seen from Fig. 7 that tritium concentrations in the pond decreased with the distance from the "Mayak". The highest value was registered in the lakes Tatysh (107 Bq/L) and Ulagach (113 Bq/L) located 6-7 km south of the "Mayak". In the lakes located at a distance of 20-25 km the value decreased sharper than in the lakes located farther from the enterprise. At the distance of over 50-km the value was averagely 10 Bq/L, which twice exceeded the technogenic background level (Lakes Tishki, Shablish, Chervyanoe, Shchuchie, and Tygish).

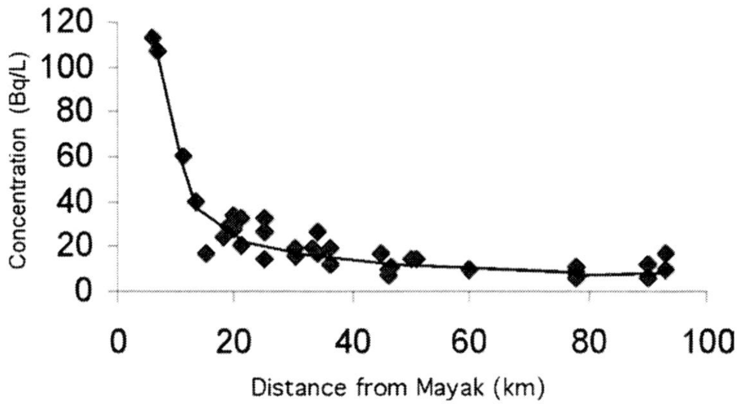

Figure 7. Tritium concentrations in the lake water around the enterprise "Mayak".

3. Conclusion

On the whole, the investigation revealed that in a great part of the Ural region, especially in the area adjacent to the BAPS and the "Mayak" tritium contents in the waters were higher than in the control technogenic background. Considering the fact that the global background level for tritium is 1 Bq/L we conclude that almost all tritium in the investigated region, including the control area, was of anthropic origin. Nuclear enterprises are the main sources for the environmental contamination which tritium. In the areas where such enterprises are situated technogenic complexes for the retention & localisation of tritium should be developed & constructed.

Acknowledgement

The investigation was supported by RFFI (grant 03-05-64295)

References

Chebotina, M.Ja., Kuligin, A.P., Rech T.A. *et al.*, 1994, The entry of tritium from the BASP to water ecosystems, *BASP Exploitation Safety*: 187–191.

Demin, S.N., Telushkina, E.Z., 1987, Radioactive hygienic estimation of the environmental pollution whish tritium and dose effects on the population in a region of a radiochemical enterprise, *Bull. Radiochemical Medicine*, **1**: 23–28.

Demin, S.N., 2001, The problem of tritium-hygienic aspects, *Tritium is dangerous*, Cheljabinsk: 13–21.

Outkin, V.I., Chebotina, M.Ja., Evstigneev, A.V., Lyubashevsky, N.M., 2004, *Peculiarities in the radioactive situation in the Urals,* RAS Ural Div., Ekaterinburg: 1–150.

THE PROBLEM OF MULTIPLE STRESSORS INCLUDING LOW DOSES OF RADIATION IN THE ENVIRONMENT

CARMEL MOTHERSILL* and COLIN SEYMOUR
McMaster University, Hamilton, Ontario, L8S 2C1, Canada
Email: mothers@mcmaster.ca

*To whom correspondence should be addressed

Abstract. This paper addresses the issue of multiple stressors in the environment all acting by common mechanisms, to produce a variety of non-targeted effects.

Low dose radiation and other pollutants all appear to induce these effects but what outcome occurs depends on genetic and environmental factors not on dose. Possible cellular outcomes include cell death, terminal differentiation, mutation [lethal or carcinogenic]. The issue is defining outcome susceptibility and translating to ecosystem burden. Possibly relevant low dose effects include: genomic instability, bystander effects, low dose hypersensitivity and adaptive or inducible responses. All are expressed at very low doses, are probably related and appear to have an epigenetic basis. What we need is a simple assay, which detects low dose effects, is preferably non lethal to the test organism (i.e., collecting not essential), which works for a wide range of species and where field sampling possible. In the paper we review potentially useful low dose effects, the bystander effect in particular, and discuss their suitability as bio-indicators of environmental risk.

Keywords: stressors, environmental effects, low doses, bio-indicators

1. The Bystander Effect

The bystander effect is a now an accepted consequence of low dose irradiation using high or low LET sources (Morgan, 2003, Mothersill and Seymour, 2004, Lorimore and Wright, 2003, Little and Morgan, 2003). A vast amount of data is discussed in the referenced reviews and many other excellent papers. Bystander effects can be detected using a simple media transfer assay where medium is harvested from irradiated cells some time

A.A. Cigna and M. Durante (eds.),
Radiation Risk Estimates in Normal and Emergency Situations, 197–208.

after exposure, filtered and transferred to cells which are not ever exposed to direct irradiation. Responses similar to those seen in irradiated cells can be found in the recipients of irradiated medium. Figure 1 is a schematic showing the bystander phenomenon.

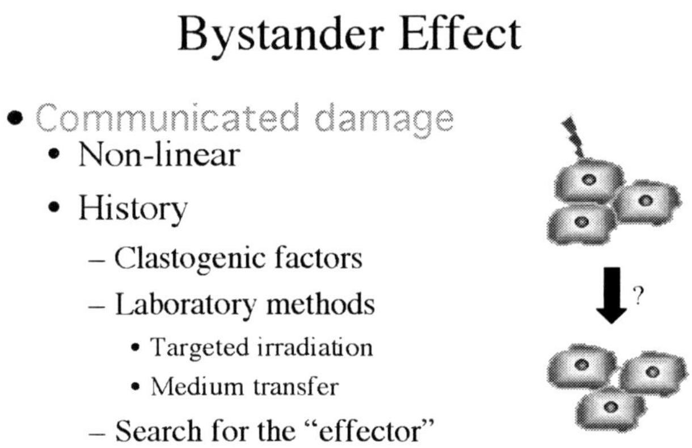

Figure 1. The bystander effect.

Bystander effects can also be detected using microbeams, where the radiation dose is targeted to particular cells in a population and effects are detected in non-targeted cells. Alpha particles have also been used. Here very low fluences are used such that all the cells do not receive a hit. Statistically, the frequency of cells showing effects is greater than the number of hit cells. There is controversy over whether gap junction intracellular communication [GJIC) is a mechanism involved in transfer of signals from irradiated to unirradiated cells. In microbeam and alpha particle experiments there is evidence of GJIC involvement but it is hard to see how they could be involved in the media transfer effect. It is likely that both GJIC and diffusible factors are possible mechanisms. In terms of what responses have been seen – virtually all radiation endpoints have been detected - apoptosis, mitotic death, terminal differentiation, mutation, micronucleus formation, chromosome aberration, transformation, and induction of stress proteins, early response genes and oncogenes.

The phenomenology of the bystander effect has been well worked out (Morgan, 2003, Mothersill and Seymour, 2004, Lorimore and Wright, 2003, Little and Morgan, 2003) and it is clear that whatever means are used to detect it, it has an induction threshold in the 1-5mGy range (Schettino *et al.,* 2005, Seymour and Mothersill, 2000, Liu *et al.,* submitted), is reduced or absent as a death inducing signal in tumour cell lines (Mothersill *et al.,* 2002) and is expressed as an apoptosis inducing signal in many repair deficient cell lines (Mothersill *et al.,* 2004, Nagazawa *et al.*, 2003, Little *et al.*, 2003).

2. What is the Signal?

Nature of the signal is unknown at present although the experimental evidence suggests either it is very small or that some sort of perpetuation by reactive oxygen or nitrogen species generation is involved. It is destroyed by repeated freeze thaw cycles and by heating. It is also not produced in cultures held at 4°C during and after irradiation suggesting the need for protein synthesis in the production of the diffusible factor. Evidence from molecular filtration experiments however has failed to find a filter small enough to exclude it from culture medium and in much of the older literature very small size elements are suspected of being involved (Kahn and Emerit, 1985, Emerit *et al.,* 1995).

Much more is known about the consequences of receipt of the hypothetical signal by unirradiated cells. Work by past and present members of our group has shown that the apoptotic pathway seems to involve a rapid influx of calcium from the extracellular fluid. This calcium is pumped out within 2 minutes but triggers downstream events in the apoptotic cascade (Lyng *et al.,* 2000, 2002a, 2002b, Maguire *et al.*, 2005). A key long-term consequence of receipt of bystander medium is the induction in recipient cells of persistent oxidative stress shown by persistently high levels of reactive oxygen species (ROS) in the cells (Lyng *et al.*, 2002b, Limoli *et al.,* 1998). This can be shown (Fig. 2) to persist over several generations of cell division and indeed is not diminished over 7 serial passages in cultured cells. Since persistent oxidative stress is known to be associated with the perpetuation of genomic instability (Clutton *et al.,* 1996), it has been proposed that radiation-induced non-clonal genomic instability is actually driven by ROS generated by bystander effects.

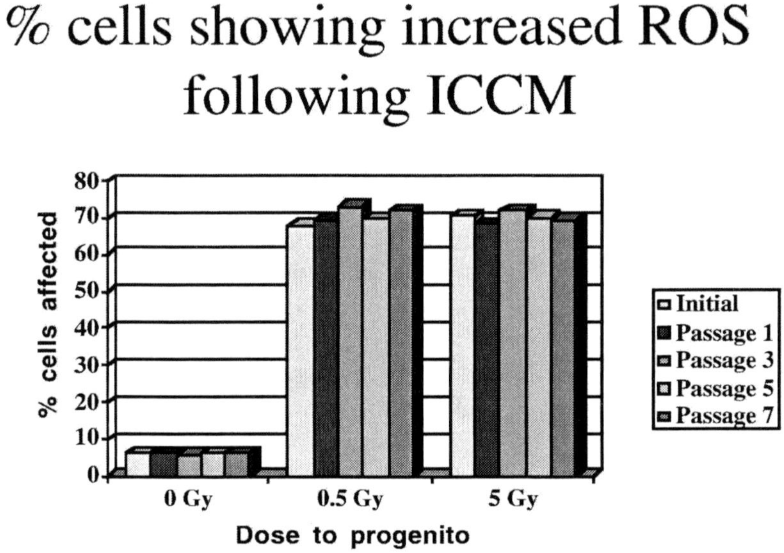

Figure 2. Induction in recipient cells of persistent oxidative stress shown by persistently high levels of reactive oxygen species (ROS).

3. Towards a Relevant Bioassay

In order to examine whether bystander effects or consequent induction of genomic instability are relevant to radiation protection or adaptive evolution, it is necessary to show they occur in tissues and in vivo and that they are evolutionarily conserved. Recent work by our group and collaborators within the EU radiation protection programme, has demonstrated all of the above (Mothersill *et al*., 2001, Mothersill *et al.,* 2001, Mothersill *et al.,* 2005). The effect has been measured in a range of tissues in vertebrates and invertebrates and have been recorded in mice irradiated in vivo although to our knowledge, it has not been looked for in plants. There is a history going back to 1954 (reviewed in Mothersill and Seymour 2001) showing clastogenic factors in blood from exposed rodents and in humans exposed through accident, warfare or for medical reasons, to radiation. These are documented to persist for at least 30 years in humans. They probably involve the same basic process as the bystander effect.

4. Explant Culture Technique

A typical result from an explant culture experiment is shown in Figure 3. This was aimed at comparing the delayed effects of low level radiation exposure on growth and differentiation of tissue harvested from an exposed animal and cultured in vitro.

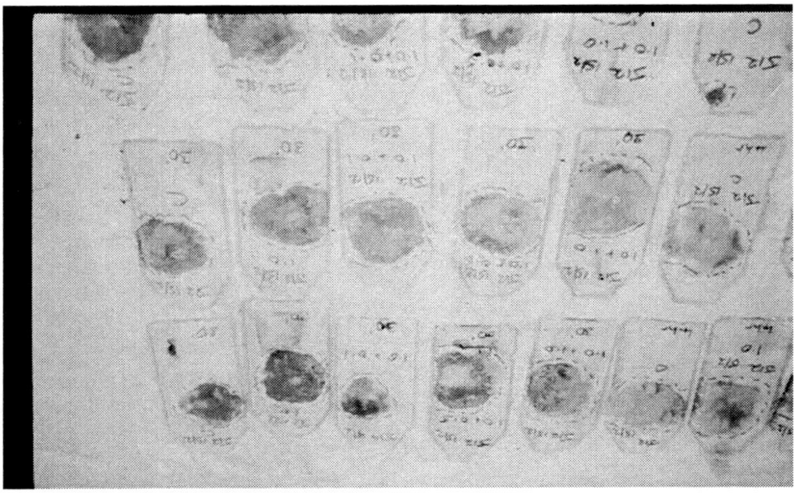

Figure 3. Typical result from an explant culture experiment aimed at comparing the delayed effects of low level radiation exposure on growth and differentiation of tissue cultured in vitro.

In Figure 4 a schematic diagram shows how the bystander assay can be done for this tissue. Basically a reporter cell which is known to produce a significant bystander effect and which grows and forms colonies in vitro, is used to quantify the amount of toxicity induced by medium harvested from in vivo exposed tissues, which are later harvested and cultured as explants. The medium from the cultured explant is a vehicle for the secreted factor. While this does not prove that the signals produced in vivo actually have an effect on the tissue in vivo, it does prove signal generation occurs following exposure of in vivo irradiated animals.

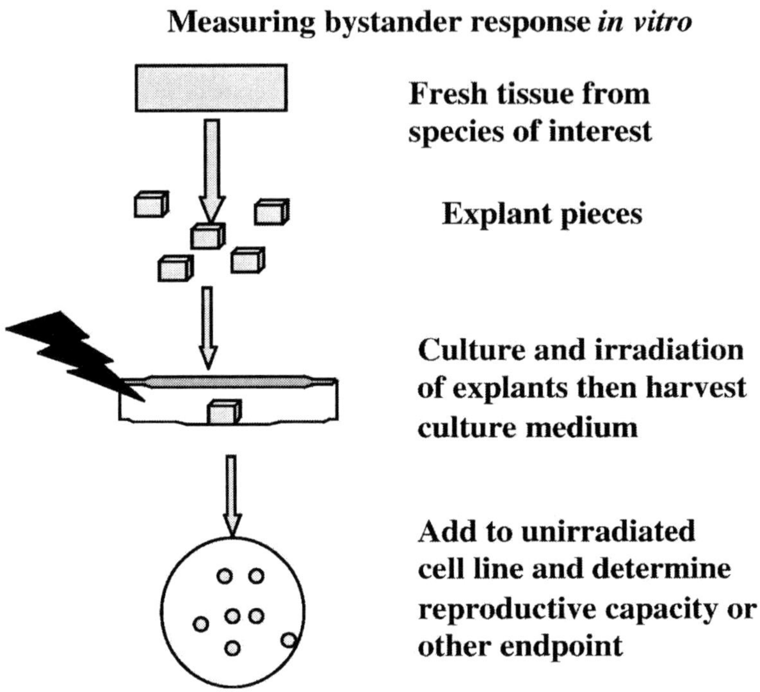

Figure 4. Schematic diagram of the bystander assay.

We have extended this work further to show that mice from different genetic backgrounds with respect to radiosensitivity and radiation induced apoptosis, produce signals leading to different types of calcium pulse and different downstream consequences including different levels of cell death (Mothersill *et al.,* 2005). A further demonstration of relevance in vivo is shown by Fig. 5. This gives levels of toxicity seen when reporter cells were exposed to serum samples harvested from patients who received radiotherapy. Before, during and after data are shown. Two things are apparent, first, there are differences before and after therapy, and second, there is considerable inter individual variation in response. This was not seen in the normal control where blood from volunteers not receiving therapy was harvested after the same time intervals.

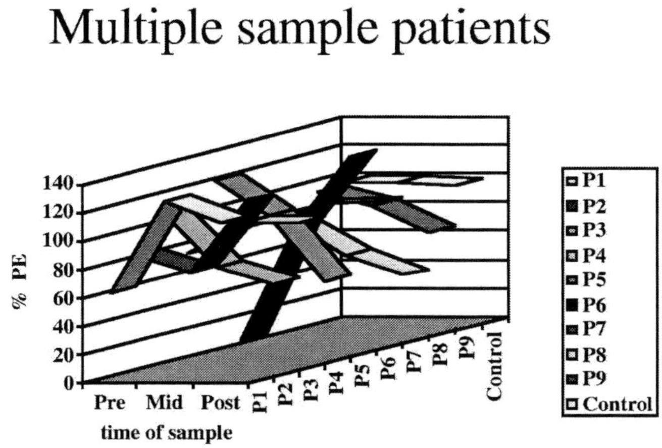

Figure 5. Levels of toxicity when reporter cells were exposed to serum samples harvested from patients who received radiotherapy.

5. Application in Environmental Protection

One of the most important factors in environmental protection is that stressors or pollutants are rarely found alone. Usually radiation is accompanied by the presence of the chemical component of a radionuclide. This means that the chemistry of the radioisotope and the physics of the radiation both contribute to the overall effect. Also, most cases of radioisotope exposure are in a background of industrial or agricultural contamination with other organic or inorganic compounds. Almost always, chemicals are regulated as individual agents to levels established as safe. Multiple stressors, if considered at all are treated by assuming additivity. Radiation is regulated as a carcinogen using dose limits based on extrapolated projections of human cancer incidence based on the Japanese Atomic bomb data set. There are many problems with this approach of which most people are aware but in the absence of anything else, the decision is generally to adopt the *Status Quo*. For the environment, this is not necessarily a good option. It takes no account of social and hierarchical aspects of ecosystems, which can greatly influence survival of individual components. It takes no account of the complexity of biological interactions or of our lack of knowledge about relative radiation –or chemical sensitivity in different species, and what the consequences of perturbations to one species in an ecosystem might be on others. Equally, the question of synergistic interactions between metal, organics and radiation is not considered. Clearly what is needed is a biological endpoint which can indicate the existence of a stress response in the population and can monitor changes. The issue is not necessarily that all the effects detected may be deleterious. Quite the contrary, there is a huge literature on adaptive responses and other beneficial effects following low level exposure to radiation and / or chemicals (reviewed in Calabrese *et al.,* 2005, Schollnberger *et al.,* 2002). Some of this deals with interactions, where exposure to a very low dose of radiation can for example, make the organism more resistant to a subsequent high dose of a heavy metal. Such findings are very welcome to industry faced with expensive pollution control costs. Again a bioassay, which is effect rather than dose based, is badly needed for monitoring purposes. Choice of endpoint and the relationship that endpoint has to "harm" and "risk" is a challenging issue. Harm at one level may not translate to harm at another. Similarly, a "beneficial" effect such as cell sparing by an adaptive response mechanism may not translate into a good effect for the whole organism. Possibly, monitoring of perturbation or change may work. This would indicate stress or more accurately, "response" to the totality of changed environmental conditions and would have folded in, the issues related to hierarchy and complexity.

To this end, our laboratory has been examining the possibility of measuring ecosystem change by monitoring bystander effects as an indicator of stress in key species in the environment. The group is focusing on rainbow trout as a key sensitive species. Advantages are that the tissues can be cultured as explants relatively easily. Also it is possible to establish cultures from fin clips or to obtain blood, which removes the need to kill the animals to do the test.

Preliminary data from fish exposed in vivo to estrogenic compound then cultured (dorsal fin) and exposed to 0.5 Gy suggests that estrogenic treatment sensitised fish cells to subsequent irradiation. Growth of cells was reduced, many explants did not grow at all.

Obviously considerably more work needs to be done to establish the usefulness of this assay for monitoring purposes. The complexity of the response coupled with other data in the lab suggesting that different tissues within a fish respond with different inducing signals, may mean that at low doses, it may not be possible to predict consequences of exposure in any meaningful way.

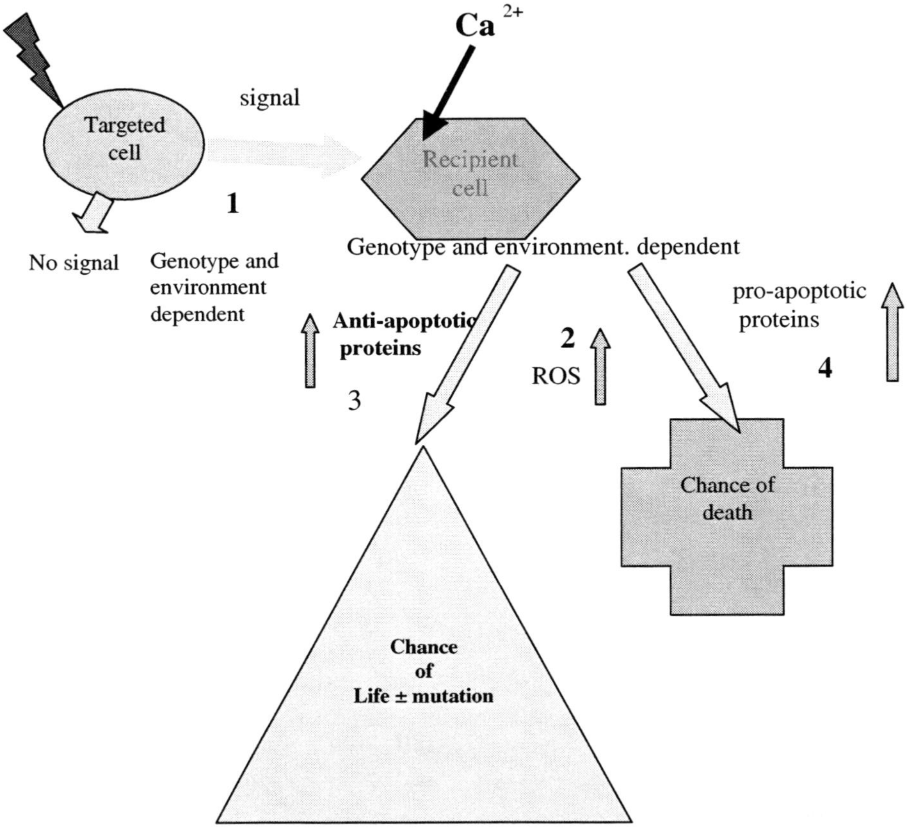

Figure 6. Possible model for expression of bystander effects in biota with intervention points for protective strategies.

Effects can be demonstrated but the meaning of these effects may never be clear. These authors have coined the term "zone of uncertainty" to encompass the region where outcome can only be assessed relative to the context in which the dose is delivered. Outcome possibilities in the zone of uncertainty include A dose related induction of harmful effect, adaptation/induced response, negation/repair of the damage, hormesis (a beneficial effect characteristic of low dose exposure to many chemicals) or no effect.

6. Implications for Risk to Biota Including Man?

Whatever the relevance of the zone of uncertainty concept, it is clear that where less than one ionisation occurs per cell, the bystander effect may amplify the dose. Therefore deviations from LNT dose response might be expected. Which way the curve goes will depend on determinants of cellular response not on dose. A NOEL/LOEL style protection framework seems more likely to work. Taking bystander effects as an example, the uncertainty arises from the issue of which response predominates – an apoptotic or an anti-apoptotic (pro-survival) one?

Which effect predominates depends on factors *independent of dose* (genetic and environmental). Death responses or life responses are major choices, but the *consequences* of these choices need to be assessed at several levels of organisation. Figure 6 is a conceptual diagram aimed at clarifying how we see these options. Having said this, radiation dose in terms of the amount of damage caused in the system is relevant to the *determination of consequences*

The challenge in this field is how to extrapolate from effect to harm, from harm to risk, from individual risk to population risk and from population risk to ecosystem risk.

7. Summary

Response associated biomarkers are required to detect low dose effects of multiple stressors in the environment. The response of cells to bystander signals was investigated for possible usefulness as a response associated biomarker. Data for mice and fish exposed in vivo suggest that bystander signals are produced in vivo and vary according to tissue tested and treatment applied.

References

Calabrese E.J., 2005, Paradigm lost, paradigm found: the re-emergence of hormesis as a fundamental dose response model in the toxicological sciences. *Environ Pollut.* **138**: 379–411.

Clutton, S.M., Townsend, K.M., Walker, C., Ansell J.D. and Wright, E.G., 1996, Radiation-induced genomic instability and persisting oxidative stress in primary bone marrow cultures. *Carcinogenesis*, **17**:1633–1639.

Emerit, I., Arutyunyan, R., Oganesian, N., Levy, A., Cerniavsky, L., Sarkisian, T., Pogossian, A. and Asrian, K., 1995, Radiation-induced clastogenic factors; Anticlastogenic effect of Ginkgo biloba extract: *Free Radic. Biol. Med.* **18:** 985–991.

Kahn, S.H. and Emerit, I., 1985, Lipid peroxidation products and clastogenic factors in culture media of human leukocytes exposed to the tumour promoter phorbol-myristate-acetate: *Free Radic. Biol. Med.* **1:** 443–449.

Limoli, C.L., Hartmann, A., Shephard, L., Yang, C.R., Boothman, D.A., Bartholomew, J.A., and Morgan, W.F., 1998, Apoptosis, reproductive failure, and oxidative stress in Chinese hamster ovary cells with compromised genomic integrity. *Cancer Res.* **58** :3712–3718.

Little J.B. and Morgan W.F., (Guest editors), 2003, *Special issue of Oncogene.*, **13** (22: 6977.

Little, J.B., Nagasawa, H., Li, G.C. and Chen, D.J., 2003. Involvement of the nonhomologous end joining DNA repair pathway in the bystander effect for chromosomal aberrations. *Radiat Res.,* **159:** 262–267.

Liu, Z., Mothersill, C.E., McNeill F.E., Lyng F.M., Byun, S.H., Seymour C.B., Prestwich W.V., 2005, A dose threshold for bystander effect, *Radiat. Res*. (submitted)

Lorimore, S.A. and Wright E.G., 2003, Radiation-induced genomic instability and bystander effects: related inflammatory-type responses to radiation-induced stress and injury? A review. *Int. J. Radiat. Biol.* **79**: 15–25.

Lyng F.M., Seymour C.B., Mothersill C., 2000, Production of a signal by irradiated cells which leads to a response in unirradiated cells characteristic of initiation of apoptosis. *Br J Cancer* **83**: 1223–1230.

Lyng F.M., Seymour C.B., Mothersill C., 2002a, Initiation of apoptosis in cells exposed to medium from the progeny of irradiated cells: a possible mechanism for bystander-induced genomic instability? *Radiat Res*. **157**: 365–370.

Lyng, F.M., Seymour, C.B., Mothersill, C., 2002b, Early events in the apoptotic cascade initiated in cells treated with medium from the progeny of irradiated cells. *Radiat. Prot. Dosimetry,* **99:** 169–172,

Maguire P., Mothersill C., Seymour C., Lyng F.M., 2005, Medium from irradiated cells induces dose-dependent mitochondrial changes and BCL2 responses in unirradiated human keratinocytes. *Radiat Res*. **163**: 384–390.

Morgan W.F., 2003, Non-targeted and delayed effects of exposure to ionizing radiation: I. Radiation-induced genomic instability and bystander effects in vitro.*Radiat. Res.* **159**: 567–580.

Mothersill C., Seymour C.B., 2001, Radiation-induced bystander effects: past history and future directions. *Radiat Res.* **155**: 759–767.

Mothersill C., Lyng F., Mulford A., Seymour C., Cottell D., Lyons M., Austin B., 2001, Effect of low doses of ionizing radiation on cells cultured from the hematopoietic tissue of the Dublin Bay prawn, *Nephrops norvegicus*. *Radiat Res.* **156**: 241–250.

Mothersill C., Rea D., Wright E.G., Lorimore S.A., Murphy D., Seymour C.B., O'Malley K., 2001, Individual variation in the production of a 'bystander signal' following irradiation of primary cultures of normal human urothelium. *Carcinogenesis*. **22**: 1465–1471.

Mothersill C., Seymour C.B., Joiner M.C., 2002, Relationship between radiation-induced low-dose hypersensitivity and the bystander effect. *Radiat Res.***157:** 526–532.

Mothersill, C., Seymour R.J. and Seymour C.B. (2004), Bystander effects in repair-deficient cell lines, *Radiat. Res.,* **161**, 256–263,

Mothersill C. and Seymour C.B., 2004, Radiation-induced bystander effects – implications for cancer, *Nature Rev. Cancer*, **4**: 158–164.

Mothersill C., Lyng F., Seymour C., Maguire P., Lorimore S., Wright E., 2005, Genetic factors influencing bystander signaling in murine bladder epithelium after low-dose irradiation in vivo. *Radiat Res*. **163**: 391–399.

Nagasawa, H., Huo L. and Little, J.B., 2003, Increased bystander mutagenic effect in DNA double-strand break repair-deficient mammalian cells. *Int. J. Radiat. Biol.* **79:** 35–41,

Schettino G., Folkard M., Michael B.D., Prise K.M., 2005, Low-dose binary behavior of bystander cell killing after microbeam irradiation of a single cell with focused c(k) x rays. *Radiat Res*. **163**: 332–336.

Seymour, C.B. and Mothersill, C., 2000, Relative contribution of bystander and targeted cell killing to the low-dose region of the radiation dose–response curve: *Radiat. Res.* **153**: 508–511.

Schollnberger H., Mitchel R.E., Azzam E.I., Crawford-Brown D.J., Hofmann W., 2002, Explanation of protective effects of low doses of gamma-radiation with a mechanistic radiobiological model. *Int J Radiat Biol.* **78**: 1159–1173.

Radiation effects: humans

DIRECTED EVOLUTION OF MANKIND AND BIOSPHERE

IGOR POPOV
Institute for the Lake and River Fisheries (GosNIORKh)
Laboratory for salmon population monitoring
Russia 199053 Saint-Petersburg, Makarova n., 26
E-mail: igorioshapopov@mail.ru

Abstract. Directed evolution means the following: living organisms have a predisposition to vary in certain directions, and this very predisposition determines trends of evolution first of all; as crystals grow, taking a certain form, so phylogenetic trends evolve following their internal laws irrespectively of adaptation and natural selection. The evolution of mankind and biosphere will be analysed in this context.

Keywords: laws of evolution, directed evolution, human species, biosphere, ecological history, fisheries

1. Introduction

Aiming to characterise the mankind and biosphere evolution tendencies I referred to the history of evolutionary biology looking for the most general regularities of species evolution, and to the environmental history, i.e. the history of human impact on wildlife (by the example of the history of fisheries).

2. Looking for Evolution Laws

The attempts to reveal the laws of evolution, which would be similar to the laws of chemistry and physics, were undertaken mainly within a framework of the directed evolution concepts (orthogenesis). The advocates of these concepts claimed the following: living organisms have a predisposition to vary in certain directions, and this very predisposition determines trends of evolution first of all; as crystals grow taking a certain form, so phylogenetic trends evolve following internal laws. This hypothesis was originated in the second half of the 19th century. Afterwards, several models of directed evolution concept were regularly formulated (table 1. see Popov 2002, 2005 for details). They were usually strongly challenging the Darwin's theory. Nevertheless now the idea of directed evolution is penetrating «through the

A.A. Cigna and M. Durante (eds.),
Radiation Risk Estimates in Normal and Emergency Situations, 211–218.

back entrance» even those studies, which were conducted in terms of natural selection theory. New Darwinian concepts reminding orthogenesis were elaborated: "developmental constraint" (Maynard Smith *et al.*, 1985), "generative entrenchment" (Wimsatt, 1986), "epigenetic traps" (Wagner, 1989), "evolutionary ratchet" (Arnold, 1989), "design limitation" (Wake, 1991), "development and design limits" (Arnold, 1992), "spontaneous order" "crystallisation of life" (Kauffman, 1993), "evolutionary channelling", "non-random production of variants" (Schwenk, 1995), "early "morphogenetic" stage of evolution" (Arthur, 1997), etc. Their authors do not consider themselves advocates of orthogenesis, however they claimed the possibility of the self-organised autonomous evolution without selection or significance of some internal forces in evolution. Such hints of orthogenesis dealt mostly either with the hypothetical organisms (modelling without any illustrations by real examples), or with the very remote past (pre-Cambrian), or minute details of the structure of some organisms (e.g. fingers of some salamander species or differences in spot-patterns between segments of Drosophyla body). The directed evolution idea is not allowed to penetrate the foreground of evolutionary theory, but it progressed over the last 150 years. The information collected by its advocates provided backgrounds for decisive conclusions on the evolution directionality.

Table 1. The directed evolution concepts: historical protagonists and milestones[1].

Author	Country	Field of empirical studies	The names and/or the main terms of the evolutionary concepts	The year(s) of the main publications
Carl von Nägeli (1817-1891)	Switzerland Germany	Botany, microanatomy, plant physiology	Theory of perfection Mechanic-physiological theory of evolution	1856-1884
Albert von Köllicker (1817-1905)	Switzerland, Germany	Histology, Zoology	Heterogenesis, theory of perfection	1864-1872
Edward Cope (1840-1897)	USA	Paleontology	Batmism, batmogenesis, Neo-Lamarckism	1868-1897
Karl von Baer (1792-1876)	Russia	Embryology	"Purposeful creation"	1876

[1] Köllicker, 1864, 1872; Nägeli, 1865, 1884; Cope, 1868, 1896; Baer, 1876; Haacke, 1893; Eimer, 1897; Whitman, 1919; Przibram, 1929; Rosa, 1931; Osborn, 1934; Berg, 1922; Sobolev, 1924; Labbé, 1924, 1929; Lwoff, 1944; Teilhard de Chardin, 1950, 1987; Schindewolf, 1947, 1993; Aguirre, 1958; Crusafont Pairó, 1960; Vandel, 1964; Lima de Faria, 1988, 1995; Gutmann, 1994, 1996, 1997; Taylor, 2002; Akhnazarov, 2002; Kawamura, 2003

Wilhelm Haacke (1855-1912)	Germany	Zoology	Orthogenesis	1893
Theodor Eimer (1843-1898)	Germany	Zoology	Orthogenesis	1897
Charles Otis Whitman (1842-1910)	USA	Zoology	Orthogenesis	1919
Daniele de Rosa (1857-1944)	Italy	Zoology	Ologenesis	1898-1931
Henry Osborn (1857-1935)	USA	Paleontology	Aristogenesis	1912-1934
Lev Berg (1876-1950)	Russia	Ichtyology, Geography	Nomogenesis	1922-1926
Dmitry Sobolev (1872-1949)	Russia	Paleontology	Historical biogenetics	1924
William Lang (1878-1966)	Great Britain	Paleontology	Orthogenesis	1910-20s
Hans Przibram (1874-1944)	Austria	Embryology	Apogenesis	1900-1920s
Alphonse Labbé (1869-?)	France	Zoology	Allelogenesis	1920s
André Lwoff (1902-1994)	France	Physiology	Physiological reduction, Orthogenesis	1930-40s
Pierre Teilhard de Chardin (1881-1955)	France	Paleontology	Orthogenesis, Omega principle	1950s
Otto Schindewolf (1896-1971)	Germany	Paleontology	Typostrophism	1930-1950s
Albert Vandel (1894-1980)	France	Zoology	Organicism	1950-60s
Emilio Aguirre	Spain	Paleontology	Orthogenesis	1958
Miguel Crusafont Pairó	Spain	Paleontology	Orthogenesis	1960
Leon Croizat (1894-1982)	Italy, USA, Venezuela	Botany, biogeography	Panbiogeography, Biological synthesis	1960s
Wolfgang Gutmann (1935-1997)	Germany	Animal morphology	Frankfurt school of engineering morphology	1970-1990s
Eduard Akhnazarov (1933-2001)	Russia	Economy	Contours of evolution	1982, 2002
Antonio Lima de Faria	Sweden	Cytogenetics	Autoevolution, Biological periodicity	1980-1990s
Kunio Kawamura	Japan	Chemistry	Subjectivity, self-organization, Shutaisei	1990-2003

3. Directed Evolution Mechanism

The emergence of the directed evolution concepts came from the analysis of the following phenomena: convergence inexplicable in terms of adaptation, constraints on variation, evolution of non-adaptive characters, trends of evolution leading to extinction, the origin of novelties. The mechanism of them could be characterised as follows: individuals inevitably produce copies of themselves, but they are unable to produce the exact self-copies for a long time, that is why organisms evolve inevitable over the change of generations even if they have already reached an adaptive condition; such an inevitable evolution takes place in definite directions because of physical and chemical constraints. In other words: the constraints and variation pressure prevail over selection. This means that the majority of species representatives participate in evolution rather then selected elite, which force out their relatives.

The incapability in exact self-copies production could just be caused by the complexity of processes involved in the organism reproduction. Some faults of the reproduction are inevitable because of such a complexity, and the more complex organic system is, the more numerous are the faults. This means that the period of the reproduction of the approximate self-copies is lesser at the complex organisms. So, the simplest organisms could remain unchanged for a long time, while the highest organisms evolve rapidly irrespective of any other condition.

4. Directed Human Evolution

So far, the attempts to describe the directed human evolution have been rare and inconsistent. At the most cases they reduce to the claim the great importance and uniqueness of the evolutionary trend of the human ancestors and to glorifying the power of human intellect. Usually the researchers of man and mankind do not allow any doubts in the capacity of man to control his own evolution rationally. Even the orthogenesis supporters gave numerous exceptions concerning mankind, not taking into account the laws of evolution common for all species.

Assuming the existence of the mechanism characterised above, the human evolution might take place spontaneously irrespectively of our desire. In a moment the species *Homo sapiens* might be transformed into other(s), or become extinct without any descendent. The second scenario is supported by three evidences: 1. Human species represents the origin of giant specialised if compared to other hominids. The appearance of such forms indicates the forthcoming extinction of a phylogenetic trend. 2. The current hominid evolution stage demonstrates the decrease in the species number up to the minimum. This also takes place at the final stage of a trend. 3. Changes in human environment take place in that directions, which are hardly favourable for the mankind.

5. What Direction is Biosphere Evolving Towards?

Biosphere is replacing for the sphere of human activity, but this activity is hardly reasonable. During the last 40 years the human population doubled, the 20% of topsoil has been lost, wars and other non-rational human activities have not been disappearing. At present a quarter of mankind lives in poverty (Chiarelli, 2003). The negative human impact on nature is not declining. The biosphere care is not the prime interest of those groups of mankind, which operate its activity. Despite great intellectual achievements human species is incapable to use biological resources rationally.

The history of fisheries is a good example of such incapability. The catches of fishes and other water inhabitants have been continuously declining over the last centuries. The scale of such a declining was only recently realised by some specialists. Usually the representatives of one human generation are informed well only on the situation existing several decades before them. They consider that situation to be perfect, while they do not realise that it was much better hundreds years ago. Each human generation states more and more modest tasks undertaking measures for rational control of fisheries. At the moment enthusiasts on ocean wildlife restoration accept a slogan: "shifting baselines", i.e. they appeal to realise the catastrophic decline of biological resources. It would seem, that if the "baselines" are known, it should be very easy to elaborate a rational strategy for the restoration: to restrict the catches, to wait some years, and than to use fish populations rationally in accordance with the well known regularities of the population dynamics. However the mankind is incapable to do it. Even the developed countries cannot stop fisheries and control the groups of mankind, which want to catch fish. The fisheries stop only when the fish population runs out, i.e. when the fishery becomes non-profitable. The only way to solve the problem is to organise the fish breading (The Northwest Salmon Crisis, 1996; Pacific salmon and their ecosystems, 1997; Taylor, 1997). Why do we need to breed fishes, if they could reproduce and grow by themselves? Now only uneducated persons are allowed to express such a viewpoint, while an "intelligent" man is guided by another one: "Why must we prohibit something to ourselves and just to catch fishes, if we can breed them?" The last viewpoint could seem rational, but the problem is, that the fish breeding is very expensive and labour-intensive, and it causes ecological problems. It requires technologies, bureaucracy and scientific researches. Each of such components tends to be more and more complex, requiring new technologies. Moreover, those components appear inevitably, which just pretend to contribute to the fish population restoration, while in reality they just support their own existence. If to spend on the protection of populations and purification of water the huge funds, which are spent directly or indirectly on fish breeding, the ecological and economic effects would be much higher. However it is impossible now. To realise another strategy the self-developing system was already established. It approaches the end of the natural existence of many fish species. Man undermined the populations of numerous land animals thousands years ago, and then was forced to domesticate some of them. Nowadays it should be the fish's turn.

The case of the fish populations undermining shows: human species is incapable to control the activity of the big groups of its representatives; the majority of human species representatives is extremely selfish; human species continuously exterminates all possible kinds of the natural biological resources; solving ecological problems mankind always chooses that strategy, which contributes to the development of technologies.

6. Conclusion

The incapability of organisms to reproduce exact self-copies for a long time causes evolution, which takes place in definite directions because of the physical and chemical constraints. That is why evolution takes place inevitably even in those cases, when organisms reached adaptive condition, or even in non-rational direction. This means that the species *Homo sapiens* will be inevitably transformed into other one(s) or disappear without any descendant. The ability of mankind to control this process rationally is dubious. At least now human species has been incapable even to control the activity of big groups of its representatives. The mankind is increasing unreasonably its own population and destroying the objects of biosphere, which could be used rationally (e.g. fish populations). The human activity indicates on the spontaneous transformation of the whole biosphere in definite direction. The display of such processes is the development of technology. To reproduce and develop technologies human species increases its own population and destroys his environment. The products of technology development tend to substitute the whole biosphere.

Acknowledgements

I would like to express my gratitude to Konrad Lorenz Institute for Evolution and Cognition Research (Altenberg, Austria), which provided me ideal conditions for the work at the final stage of this research, and to the organisers of the N.W. Timofeeff-Ressovsky conference "Modern problems of genetics, radiobiology, radioecology and evolution" (Yerevan, 2005) for the assistance.

References

- 1996, *The Northwest Salmon Crisis. A Documentary History*, Corvallis: Oregon State University Press, Corvallis: 1–374.

Aguirre E., 1958, La orthogénesis y el problema de la evolución biológica, *Arbor*. Abril. MCMLVIII. **39**: 467–502.

Akhnazarov E., 2002, *Kontury evolutsii*. Nedra, Saint-Petersburg, (in Russian. "Contours of evolution").

Arnold S., 1989, How do complex organisms evolve?, in D. Wake,

G. Roth (Eds)*Complex Organismal Functions: Integration and Evolution in Vertebrates*,New York: 403–433.

Arnold S.J., 1992, Constraints on phenotypic evolution, *American Naturalist*. **140** Suppl.,: 85–107.

Arthur W., 1997, *The origin of Animal Body Plans. A study in Evolutionary Developmental biology*. Cambridge University Press.

Baer K.E. v., 1876, Über Darwin'sche Lehre, in Karl Ernst von Baer, *Studien aus dem Gebiete der Naturwissenschaften*. (Verlag der Caiferlichen Hofbuchhandlung H. Schmissdorf, St.-Petersburg: 235–479.

Berg L., 1922, Nomogenes ili evolutsia na osnove zakonomernostey, Gos. Izd., Peterburg, 1922 (in Russian. English translation: Berg, L. 1926, Nomogenesis or evolution determined by law. Constable & Co, Ltd., London.

Chiarelli B., 2003, The 2K3 IVAES Congress and Future of Humankind, *Human evolution* **18** (1-2): 1–8.

Cope E., 1868, On the Origin of Genera, *Proceedings of Academy of Natural Sciences of Philadelphia* **4**: 242–305 (1868).

Cope E., 1896, *Primary factors of organic evolution*. (Open Court Publishing, Chicago,).

Crusafont Pairó M., 1960, La evolución de los Vertebrados. Ortogénesis y Clinogénesis, *Estudios geologicos*. **26**: 1–23.

Croizat L., 1958, *Panbiogeography*, Ed. by the author, Caracas.

Croizat L., 1962, *Space, Time, Form: The Biological Synthesis*, Ed. by the author, Caracas.

Eimer Th., 1897, Orthogenesis der Schmetterlinge. Ein Beweis bestimmt gerichteter Entwickelung und Ohnmacht der naturlichen Zuchtwahl bei der Artbildung. Zugleich eine Erwiderung an August Weisman, W. Engelmann, Leipzig, X,XVI,513 S.

Gutmann W.F., 1994, Konstruktionszwänge in der Evolution: schwimmende Vierfüsser, *Natur und Museum* **124**: 165–189.

Gutmann W.F., 1996, Gibt es Alternativwege für die Entwicklung der Organisation von Lebewesen I, *Natur und Museum* **126**: 250–262.

Gutmann W.F., 1997, Globalisierung-Rueckwirkung auf den Bereich der klassischen Biologie and Paläontologie, *Natur und Museum* **127**: 209–219.

Haacke W., 1893, *Gestaltung und Vererbung. Eine Entwickelungsmechanik der Organismen*. T.O.Weigel Nachfolger, Leipzig: 337 S.

Kauffman S., 1993, *The origins of order: self– organization and selection in evolution*. Oxford University Press, New York.

Kawamura K., 2003, Relativity of the roles of genes, subjectivity and self-organisation for the origin and evolution of life, in Levit G.S., Popov I.Yu., Hossfeld U., Olsson L. & Breidbach O (Eds.) *In the shadow of Darwinism. Alternative evolutionary theories in th 20th century*, Fineday Press, Saint-Petesburg: 220–237. (in Russian).

Köllicker A. v., 1864, Über die Darwin'sche Schopfungstheorie, *Z. wiss. Zool.* **14**: 114–186.

Köllicker A.v., 1872, *Morphologie und Entwickelungsgeschichte des Pennatulidenstammes nebst allgemeinen Betrachtungen zur Descendenzlehre,* Christian Winter, Frankfurt a. M.: 87 S.

Labbé A., 1924, Une conception nouvelle de l'adaptation: l'allélogénèse, *Revue scientifique* **10**: 295–301.

Labbé A., 1929, *Le conflict transformist.* Libraire Félix Alcan, Paris: 1–201.

Lima de Faria A., 1988,*Evolution without selection.* Amsterdam, New York, Oxford.

Lima de Faria A., 1995, *Biological periodicity: Its molecular mechanism and evolutionary implication,* Conn. JAI Press. USA, Greenwich, **22**: 1–366.

Lwoff A., 1943, L'évolution phisiologique. Etude des pertes de fonction chez les microorganisms. *Actual. Sc. Industr.* **970**. 1–308.

Maynard Smith J., Burian R., Kauffman S., Alberch P., Campbell J., Goodwin B., Lande R., Raup D. and Wolpert L, 1985, Developmental constraints and evolution, *The Quarterly Review of Biology* **60**: 265–287.

Nägeli C. v., 1865, *Entstehung und Begriff der naturhistorischen Art.* Muenchen.

Nägeli. C. v., 1884, *Mechanischphysiologische Theorie der Abstammungslehre.* Druck und Verlag von R. Oldenbourg, Munich and Leipzig: 822 S.

Osborn H., 1934, Aristogenesis, the creative principle in the origin of species, *Amer. Nat.* **68**: 193–235.

Popov I., 2002, "Periodical Systems" in Biology (a Historical Issue), *Die Entstehung biologischer Disciplinen **II***: 55–69.

Popov I., 2005, "*Orthogenesis versus Darwinism: a historical issue*" Saint-Petersburg University Press, Saint-Petersburg, (in Russian).

Shoulder D., Bisson P. & Naiman R. (Eds.), 1997, .*Pacific salmon and their ecosystems.* Chapman and Hall, New York: 1–685.

Przibram H., 1929, Théorie apogénétique de l'évolution des organisms, *Revue générale des Sciences pures et appliqués* **40**: 293–299.

de Rosa D., 1931, *L'Ologénèse. Nouvelle théorie de l'évolution et de la distribution géografique des êtres vivants,* Libraire Félix Alcan, Paris: 1–368.

Schindewolf O.H., 1947, *Fragen der Abstammungslehre,* Frankfurt am Main.

Schindewolf O.H., 1993, Basic Questions in Paleontology*: Geologic Time, Organic Evolution and biological Systematics,* The Univ. of Chicago Press: 1–467.

Schwenk K., 1995, An utilitarian approach to evolutionary constraint, *Zoology* **98**: 251–262.

Sobolev D., 1924, *Nachala istoricheskoy biogenetiki*, (Gos. Izd., Simpheropol, (in Russian. "Bases of historical biogenetics").

Taylor J., 1999, *Making Salmon. An Environmental History of the Northwest Fisheries Crisis*. University of Washington Press, Seattle and London: 1–421.

Taylor P. &. Lang P.W.D., 2002, Orthogenesis and the evolution of Cretaceous cribrimorph bryozoans,, *Annals of Bryozoology.* Dublin: 275–299.

Teilhard de Chardin P., 1950, Sur un cas remarquable d'orthogénèse du groupe: l'évolution des siphnéides de Chine, *Paléontologie et transformisme,* Albin Michel, Paris: 169–173.

Teilhard de Chardin P., 1987, *Le Phénomène humain.* (Russian edition, Nauka, Moscow).

Vandel A., 1964, *Biospéologie. La biologie des Animaux Cavernicoles*, Gauthier-Villars Éditeur, Paris: 1–619.

Wagner G., 1989, The origin of morphological characters and the biological basis of homology. *Evolution* **43**, 1157–1171.

Wake D., 1991, Homoplasy: the result of natural selection or evidence of design limitation? *American Naturalist* **138**, 543–567.

Whitman Ch. O., 1919, Orthogenetic evolution in pigeons, *Carnegie Inst. Washington Publ.* **257**, 1–194.

Wimsatt W., 1986, Developmental constraints, generative entrenchment, and the innate-acquired distinction, In: Bechtel, W/. (Ed.) *Integrating Scientific Disciplines*: 185–208.

ASSESSMENT OF RADIATION GENETIC RISK IN MAN

VLADIMIR A. SHEVCHENKO[†]
Vavilov Institute of General Genetics, RAS, Moscow

Abstract. This review covers radiation genetic risk in man, expressed in terms of various genetic diseases, on the basis of the United Nations Scientific Committee on the Effects of Atomic Radiation (UNSCEAR) Reports.

Keywords: genetic risk, hereditary diseases, mutations, chromosomal aberrations, Mendelian diseases, multifactorial diseases, ionizing radiation

1. Introduction

Assessment of radiation genetic effects in human populations is still an open problem. This problem first urgently arose after the atomic bombing of Hiroshima and Nagasaki in 1945. This tragedy motivated scientists from many laboratories all over the world to start studies on the mechanisms of radiation genetic effects in humans or animal models.

Radiation genetic risk is usually determined as the probability of various hereditary diseases occurred in the progeny as a result of radiation-induced mutations in germline parents' cells.

The United Nations Scientific Committee on the Effects of Atomic Radiation (UNSCEAR) was established by the General Assembly of the United Nations in 1955 in order to periodically (every 5 years) summarise the results of studies in this field performed all over the world. The main output of the Committee includes the reports of the General Assembly of the United Nations and comprehensive scientific supplements to these reports. They include analytical surveys of the worldwide achievements in the field of the biological effects of natural and man-made radiation sources.

This paper presents a summary of the methodological basis for the assessment of radiation genetic effects, and also describes the development of the concepts of genetic hazard to men exposed to ionizing radiation.

[†] Dr. Shevchenko past away shortly before the meeting.

A.A. Cigna and M. Durante (eds.),
Radiation Risk Estimates in Normal and Emergency Situations, 219–227.

2. Methodology of Genetic Risk Assessment

The basic principles of radiation genetics risk assessment, based on numerous experimental studies on Drosophilae, mice, primates, as well as on human data available at that time, were summarised in the first UNSCEAR 1958 Report, in the US BEAR 1956 Report, and in the British Medical Research Council (MRC) report in 1956. Some of these principles are summarised below:

- both spontaneous and induced mutations are generally harmful;
- even low doses of ionizing radiation can be associated to a certain genetic risk;
- the number of radiation-induced mutations is proportional to the dose, therefore linear extrapolation from high to low doses is applicable.

Since the methods for human genetic risk assessment were mainly based on the experiments on mice and partially on primates, the Committee made the following provisional assumptions:

- the number of genetic lesions, produced by a certain type of radiation under certain conditions, in human germ cells is assumed to be similar to that in the germ cells of the animal model;
- biological factors (such as sex, development stage, and age of the embryo cells) and physical parameters (radiation quality and dose rate) affect the extent of damage both in man and in experimental animals, from which the extrapolation is made, in the same way;
- a linear relationship between the frequency of genetic effects and the dose is assumed for low doses/dose rates of sparsely ionizing radiations.

There are two basic approaches for quantitative assessment of human radiation genetic risk: the doubling dose (or relative mutation risk) method, and the direct (or absolute risk) method.

Doubling dose is the amount of radiation producing the same number of mutations as spontaneously occurring over one generation. The doubling dose method is normally used to estimate the risk in populations exposed to chronic irradiation over several generations. Hence, genetic risk assessment based on this method takes into account the laws of population genetics.

When applying the doubling dose method, it should be kept in mind that natural mutagenesis in human populations has been evolutionary established on a balanced level, which depends on both the rate of spontaneous mutagenesis, and the rate of induced mutations. The rate of natural selection in modern human populations is greatly reduced, resulting in a high natural human genetic variability level.

In order to employ the doubling dose method it is first necessary to determine the doubling dose (DD) on the basis of experimental data on the frequency of spontaneous and induced mutations. Then, the expected increase in the population equilibrium level is estimated, as the product of the spontaneous mutation frequency (P), the relative mutation risk (1/DD), and the dose absorbed by the population. Finally, the increase in the first generation mutagenesis level is calculated using the above-mentioned equilibrium value.

As noted previously (Shevchenko, 2000), all the human genetic anomalies are divided into Mendelian, chromosomal, and multifactorial diseases. Theoretically, the doubling dose method can be used for risk assessment of different kinds of mutations independently of whether they are classified as either dominant or recessive mutations, or as chromosomal aberrations. However, it best applies to simple dominantly inherited traits, whose equilibrium frequency is presumably proportional to the mutation frequency.

The direct methods are used to estimate absolute genetic risks expected in the progeny on the basis of experimental data on spontaneous and radiation-induced mutations. In spite of the apparent simplicity, there are more difficulties to use this approach in practice as compared to the doubling dose method, because of gaps in knowledge about the human genome structure and frequencies of radiation-induced mutations. As early as the direct methods were started to develop, it was apparent that it was not always possible to overcome the contradiction between the assessments of mutation frequencies and the observed phenotypic manifestations. That is why, for many years, UNSCEAR experts used both methods knowing that either of them had some advantages and shortcomings.

As early as 1962 the UNSCEAR reported the doubling dose of 1 Gy for low-LET ionizing radiations for low dose rates. This value was confirmed in the subsequent Committee Reports up to now. In the case of acute exposures the doubling dose was estimated to reach 0.3-0.4 Gy.

Before 1972 the "man/mouse" model (using the data on spontaneous human variability and on radiation-induced mutations in mice) was mainly used to assess radiation genetic risks in humans. Then for more than 20 years the UNSCEAR Reports used primarily the "mouse/mouse" model (the analysis of spontaneous and induced mutations in mice). The extrapolation methods developed by the Committee and based on the empirical conversion coefficients and data from experiments on mice were utilised to assess radiation risks in humans.

Over recent years, this approach has been revised based on the new extensive data on spontaneous human mutagenesis. Besides, there are some more obstacles to the extrapolation of the risk assessment from mice to man. Taking this into account, in 2001 the Committee chose the "human/mouse" model as the major approach for radiation risk assessment in man.

3. Analysis of Radiation Genetic Risk Assessment in Humans Using the UNSCEAR 1977-2001 Reports

The methodology of radiation genetic risk assessment was developed by the 1970s, therefore we choose the UNSCEAR Reports dated from 1977, 1982, 1986, 1993 and 2001 (UNSCEAR, 1977, 1982, 1986, 1993, 2001) for a comparison.

First of all we analysed the estimates of natural human genetic variability from the above-mentioned reports (Table 1). It is seen that the frequencies of dominant and sex-linked diseases are rather conservative, ranging from 10,000 to 16,500 cases per 1 million new-borns over the 20-year period of time. The number of expected autosomal recessive diseases rose from 1,100 to 7,500. Chromosome diseases have the most conservative estimates – about 4000 over that period. Multifactorial diseases make a major contribution to the growth of natural human genetic variability (in sum from 90,000 to 710,000 cases). These types of diseases account for the overall increase in the assessment of natural genetic variability from 105,000 cases in 1977 to 738,000 in 2001 per 1 million new-borns.

Table 1. Assessment of natural mutagenesis in human populations per 1 million new-borns from the UNSCEAR 1977, 1982, 1986, 1993 and 2001 Reports.

Disease types	UNSCEAR 1977 *	UNSCEAR 1982 *	UNSCEAR 1986	UNSCEAR 1993 **	UNSCEAR 2001 **
Autosomal dominant and sex-linked	10000	10000	10000	10000	16500
Autosomal recessive	1100	2500	2500	2500	7500
Chromosomal	4000	3400	3800	4000	4000
Chronic multifactorial	90000	90000	60000	650000	650000
Congenital			60000	60000	60000
Total	105100	105900	676300	726500	738000

* = data on British Columbia

** - data on Hungary and British Columbia

It is necessary to emphasise that this growth is related to the expansion of our knowledge in this field rather than to increasing natural mutagenesis in human populations. It is possible that the mutation frequency in human populations is rising because of environmental mutagenic factors, but further studies are required to find out the real reasons.

Table 2. Assessment of genetic risks in human populations exposed to low-LET radiation at dose of 1 Gy at low dose rate per 1 million first-generation new-borns from the UNSCEAR 1977, 1982, 1986, 1993 and 2001 Reports.

Disease types	UNSCEAR 1977[×]	UNSCEAR 1982[×]	UNSCEAR 1986	UNSCEAR 1993[××]	UNSCEAR 2001[××]
Autosomal dominant and sex-linked	2000	1500	1500	1500	750-1500
Autosomal recessive	low	low	low	5	low
Chromosomal	3800	240	240	240	included in other types
Chronic multifactorial	500	450	not estimated	not estimated	~ 250-1200
Congenital			not estimated	not estimated	~ 2000
Total	6300	2190	1800	1745	3000-4700
Percentage from current estimate of natural variability	6.0	2.1	0.3	0.2	0.41-0.64

The expected estimate of radiation genetic risk for that period appears to be rather conservative against the background of a 7-fold growth of natural human genetic variability (Table 2). Both absolute and relative estimates of maximum genetic risk are given in the UNSCEAR 1977 Report – 6,300 severe genetic diseases expected in the first generation per 1 million new-borns at a dose of 1 Gy at a low-dose rate, which amounts to 6.0% of natural human genetic variability. The major portion of this risk estimate was shown to result from a high expected frequency of unbalanced products of reciprocal translocations and other types of chromosomal aberrations (3,800 cases). Further studies did not confirm the high expected risk from these chromosomal aberrations.

The estimate of chronic multifactorial diseases and congenital anomalies has been considerably changed now. The risk of these diseases was estimated to be as high as 450-500 cases per 1 million new-borns at a dose of 1 Gy in the UNSCEAR 1977 and 1982 Reports. This estimate was based on the assumption that the mutation component for these types of diseases is as high as 5%. Since then, and until year 2001, no calculations of the risk estimates were performed for these diseases. The UNSCEAR 2001 Report used the concept of potential recovery capacity factor (PRCF), mainly designed by Prof. K. Sankaranarayanan (2000), in order to estimate these risks. This concept was developed using the results of analysis of microdeletion syndrome in humans and mice. Multi-locus mutations, which frequently result from microdeletions, do not always produce a phenotypic manifestation. Apparently, mutations in a small portion only of the genome can show up phenotypically, and therefore a special correction coefficient is required to convert the frequency of mutations on the DNA level into the probability of a mutant phenotype. Certain correction coefficients were determined using this concept. The value of this coefficient for chronic multifactorial diseases was estimated to be in the range of 0.02-0.09. The PRCF concept allowed the risk values for all disease types to be derived (Table 3). It should be mentioned that the UNSCEAR 2001 Report uses the minimum values of the PCRF coefficients, which can increase, as new experimental data become available. Overall, the UNSCEAR 2001 Report gave the highest values of risk estimates in recent years (Figure 1).

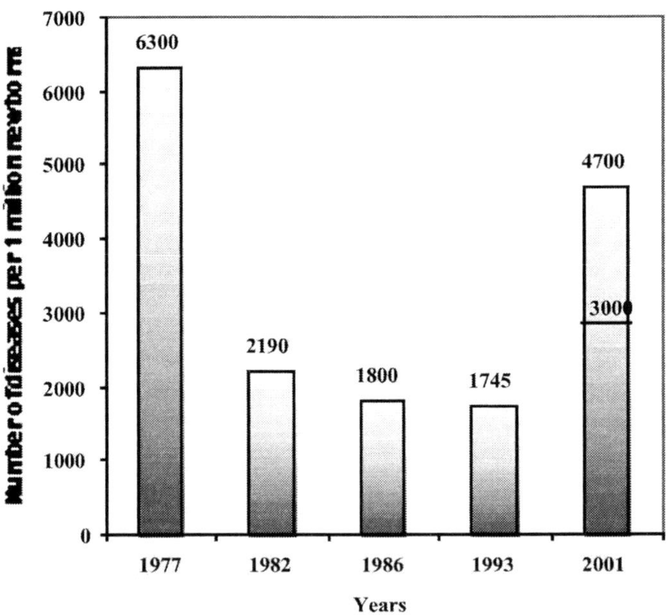

Figure 1. Comparison of general estimates of radiation genetic risk (by all genetic disease types) in human populations exposed to low-LET radiation at dose of 1 Gy at low dose rate per 1 million first-generation new-borns.

Table 3. Assessment of genetic risks in human populations chronic exposed to low-LET radiation at low doses (doubling dose – 1 Gy).

Disease types	Natural genetic variability, per 1 million new-borns	Risks per 1 Gy per 1 million new-borns	
		First-generation	Second-generation
Mendelian			
Autosomal dominant and sex-linked	16500	~ 750 – 1500*	~ 1300 – 2500*
Autosomal recessive	7500	0	0
Chromosomal	4000	Included in other types	Included in other types
Multifactorial			
Chronic multifactorial	650000	~ 250 – 1200	~ 250 – 1200
Congenital	60000	~ 2000	~ 2000
Total	738000	~ 3000 – 4700	~ 3550 – 5700
Genetic risks per 1 Gy (Percentage from natural genetic variability)		~ 0,41 – 0,64	~ 0,48 – 0,77

* Including chromosomal diseases

These estimates are accepted as most adequate, since this Report takes proper account of a broad spectrum of genetic diseases. Besides, the risk estimate for each genetic disease type is more comprehensive and accurate than in the previous Reports.

Radiation exposure of one parental generation results in genetic effects to be observed in several next generations. So far we considered the effects expected in the first generation (in the offspring from exposed parents). Radiation genetic risk for the second generation is estimated to be 1150-3200 additional cases of hereditary anomalies, which amounts to 0.16-0.43% of the natural background of hereditary diseases. The effect expected in the second generation reaches approximately 56% of the effect in the first generation. The effect induced in the parental generation is gradually decreasing as further generations pass by. This decrease, as shown in our previous paper (Shevchenko and Pomerantseva, 1985), can be described by an exponential or power function. Assuming an exponential dependence, it is easy to calculate that the integral genetic effect in 10 generations after a single radiation exposure of the first generation will be as high as 220% of the effect expected in the first generation. This effect over 10 generations is estimated to be about 6600-10340 cases of hereditary anomalies (0.9-1.4% of the natural genetic variability).

Which mutation types will persist for a long time in a population after a single radiation exposure? Multifactorial diseases (88.08%) and congenital anomalies (8.12%) prevail in natural mutagenesis, while the other types of genetic anomalies account for less than 4%. This proportion describes the balanced overall yield of spontaneous mutations. Assuming a parental generation is exposed to radiation, the effect induced in the first generation is classified into three types of genetic diseases: 25.0-31.9% - autosomal dominant and sex-linked anomalies, 8.3-25.5% - multifactorial diseases and 42,6-66.7% - congenital anomalies. The second generation will have a higher fraction of dominant diseases. We believe that the proportion of multifactorial diseases will increase in the further generations. The diseases related with chromosomal damage and expected in the first and second generations are classified to the groups of autosomal dominant diseases and congenital anomalies.

So far radiation-induced mutations and their possible phenotypes have been considered assuming that they occurred in individuals with healthy genotypes and without other mutations. Actually, each of 73.8% individuals normally has a severe mutation (sometimes even two or three), which can be crucial to human vitality. Therefore, in real populations, in 73.8% cases radiation-induced mutations occur in individuals who already harbour genetic deviations. This situation (gene interaction), as known from classical genetics, can often result in an increase in negative effects at the phenotype level. This fact can significantly affect the PRCF factor. It has been so far assumed that a mutation (most often a microdeletion) occurs in a normal genotype, and consequently the gene interaction factor is relatively low. If a microdeletion is induced in a mutant genotype, then the probability of a severe genetic anomaly increases, and, as a result, the PRCF correction factor should be increased

4. Conclusion

In conclusion it should be emphasised that further evolution of the methods of radiation risk assessment in man will apparently be related to the analysis of such genetic disorders as dominant mutations and inheritable cancers, which risks have not been assessed yet. These categories of genetic disorders will certainly increase the overall expected genetic risk from radiation. In the future, the progress in molecular genetics will expand the spectrum of Mendelian diseases, which in turn can also increase the expected risk. As new data on the phenotypic manifestation of the genetic diseases related to radiation-induced microdeletions become available, the PRCF value will be corrected. In the near future, extensive researches are predicted into this field of radiation genetics.

References

BEIR, 1956, *The Biological Effects of Atomic Radiation.* National Academy of Sciences, National Research Council. D.C. Washington.

MRC, 1956, *The Hazards to Man of Nuclear and Allied Radiations.* , Medical Research Council. Cmd 9780, H.M.S.O., London, United Nations.

Sankaranarayanan, K. & Chakraborty, R., 2000, Ionizing radiation and genetic risks. XII. The delineation and use of the concept of "potential recoverability correction factor" (PRCF) to bridge the gap between radiation-induced mutations recovered in mice and the risk of radiation-inducible genetic disease in humans. *Mutat. Res.*, **453**:129–181.

Shevchenko V.A. & Pomerantseva M.D., 1985, Genetic consequences of ionizing radiations. Moscow.: *Nauka*, 1–277.

Shevchenko V.A., 2001, Modern problems of genetic risk assessment of human exposure. Radiation Biology. *Radioecology*. **41**(5): 615–626.

United Nations, 1958, *Report of the United Nations Scientific Committee on the Effects of Atomic Radiation.* Official Records of the General Assembly, Thirteenth session, Suppl. N17 (A.3838), N.Y.

United Nations, 1962, *Report of the United Nations Scientific Committee on the Effects of Atomic Radiation.* Official Records of the General Assembly, Thirteenth session, Suppl. N16 (A.5126), N.Y.

United Nations, 1977, *Sources and Effects of Ionizing Radiation,* United Nations Scientific Committee on the Effects of Atomic Radiation, Report to the General Assembly, E.77.IX.1. United Nations. N.Y.

United Nations, 1982, *Ionizing radiation. Sources and biological effects.* UNSCEAR 1982, Report to the General Assembly United Nations E.82. 1982, United Nations, New York.

United Nations, 1986, *Genetic and Somatic Effects of Ionizing Radiation. United Nations Scientific Committee on the Effects of Atomic Radiation*, Report to the General Assembly, United Nations sales publication E.86.IX.9. United Nations, New York.

United Nations, 1993, *Sources and Effects of Ionizing Radiation*, United Nations Scientific Committee on the Effects of Atomic Radiation, Report to the General Assembly, with scientific annexes. United Nations sales publication E.94.IX.2. United Nations. NY.

TARGETS, HITS AND TRACKS

JÜRGEN KIEFER
University of Giessen, Germany

Am Dornacker 4, S-35435 Wettenberg, Germany
E-mail: juergen.kiefer@strz.uni-giessen.de

Abstract. Target theory was the first successful attempt to describe biological phenomena in the language of theoretical physics. Its basic assumptions are still valid, and their validity can be proven with the help of modern techniques of molecular biology. The theory has also practical implications, e.g. the determination of molecular weights of enzymes which cannot be easily separated to allow standard molecular analysis. For the description of heavy charged particle action the original concept had to be broadened which led to the development of "microdosimetry". Track structure analysis takes into account the spatial distribution of energy deposition at a nanometer scale and thus allowed an understanding of the action of heavy charged particles which is important both for the assessment of radiation risk to humans as well as to subminiature electronic circuits. New developments in radiation biology as, e.g., the "bystander effect" do not call the basic principles into question but ask for a rethinking about the nature of targets and the possible effects following radiation interactions.

Keywords: target theory, track structure, microdosimetry

1. Introduction

Very soon after Röntgen's discovery of X-rays in 1895 their medical potential was realized, both for diagnostic as well as for therapeutic purposes. It took a considerably longer time to develop a theoretical framework for the quantitative understanding of biological radiation effects. The target theory constitutes the first attempt to explain biological phenomena in a similar way as the physicists deal with their experiments. It marks thus the birth of "Theoretical Biology" in analogy to "Theoretical Physics" which celebrated triumphal successes at the turn from the 19th to

A.A. Cigna and M. Durante (eds.),
Radiation Risk Estimates in Normal and Emergency Situations, 229–242.

the 20th century. In effect, the then newly developed "quantum theory" provided the key for an explanation of the interaction of ionizing radiation with biological entities. Right from the beginning, the aim was not a merely formal description of dose-effect curves in the form of "curve fitting" but rather to use radiation as a tool to explore the properties of biological systems and the way they work. Radiation biology has thus laid the foundation for the developments of modern biology, particularly to molecular genetics.

The historical landmarks have been described elsewhere (Kiefer, 2001). Target theory started with the first formulations by Dessauer (1922) and Crowther (1924), was taken up and refined by Timofeeff-Ressovsky, Delbrück and Zimmer, particularly in their famous "green pamphlet" (1935) whose influence reached as far as Francis Crick, one of the founder fathers of modern molecular biology. But the ideas of these early attempts should not only be seen with an historical perspective as they are still important parts of even contemporary quantitative radiation biology (see also Zimmer 1961 for some special applications). It seems thus worthwhile to follow some of the further developments in order to see what kind of new insights were gained.

1. Target Theory and Consequences

Target theory in its original form is based on two simple principles:

1. Radiation interaction with matter is of quantal nature, i.e. discrete and hence not continuous. This implies that local energy absorption cannot be infinitesimally small. The interaction events are called "hits"
2. A certain number of hits received in the biological target will lead to its inactivation.

The mathematical formalism of target theory is quite simple: If N is the number of undamaged targets after an exposure of N_0 targets to a radiation dose D, ε the deposited energy per hit, m the target mass and k the number of hits necessary for target inactivation one has

$$N = N_0 \sum_{0}^{k-1} e^{-mD/\varepsilon} (mD/\varepsilon)^i / i! \qquad (1)$$

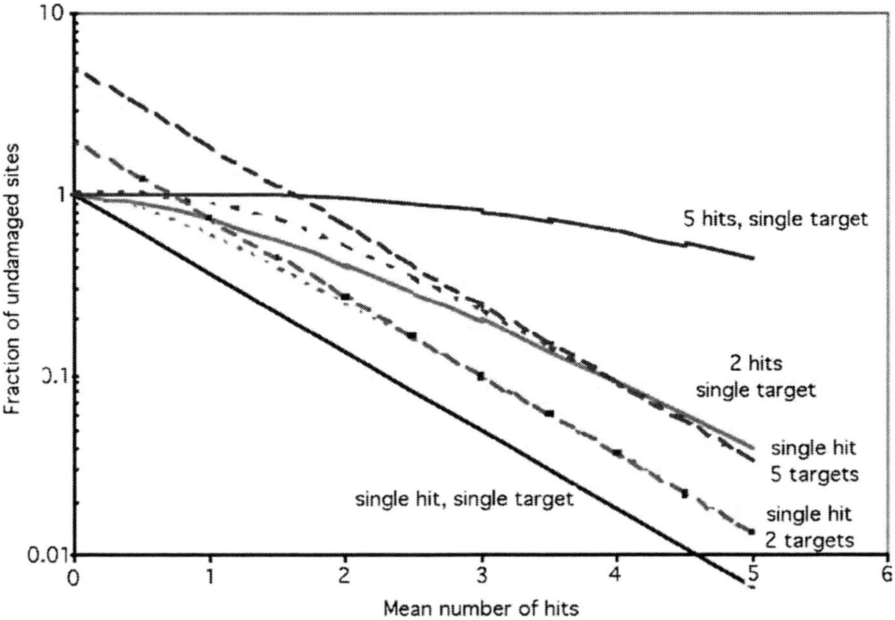

Figure 1. Examples of theoretical survival curves after radiation action. Shown are multi-hit-single target and single-hit-multi-target types. Note that curves for a single hit but multiple targets are closer together than curves for a single target but multiple hits.

The number of hits follows a Poison distribution indicating that the mean number is small. Fig. 1 gives examples of multiple hit curves.

A modification is the assumption that a site contains several sensitive targets which all have to be hit to inactivate the whole site. In the case of a single hit per target and n targets one has the formula

$$N=N_0[1-(1-e^{-mD/\varepsilon})^n] \tag{2}$$

There are also examples for this expression in Fig. 1. One sees an important difference: the number of necessary hits has a much stronger influence on the survival curve than the number of targets.

Expression (2) has been very popular over a long time as many survival curves (mainly with microorganisms) could be rather easily fitted. Sometimes the targets were identified with intracellular sites, e.g. chromosomes. All this broke down when it was realized that survival curve shapes can be changed by various experimental conditions. Consequently m was termed just "extrapolation number" as it is obtained if the terminal exponential part of the survival curve is back-extrapolated to the ordinate (Alper, Elkind and Gillies 1960). When the interest of radiobiology shifted to mammalian cells it was realized that the multi-target type of curves were not appropriate to describe their survival. The currently popular version is the "linear-quadratic" expression which was originally based on microdosimetric considerations (see below). It is not the place here to follow the history of dose-effect curve expressions but just to demonstrate the impact which the initial approach had on the further developments.

The physical foundation of target theory is still true, sites are hit according to a Poisson distribution but the biological assumptions are much too rigid. It is postulated that the physical event of energy absorption (the "hit") determines unequivocally the further fate. This view has been completely shattered by the discovery of repair processes, which govern present days interest in cellular radiation biology. The "shoulder" in the survival curves can no longer be interpreted in term of multiple hits or targets but they must be ascribed to the dependence of repair capacity on the number of lesions formed as a result of radiation exposure.

Also, biological units are generally not inactivated by a fixed number of hits. This is even true if the survival curve is of the "single-hit" type, i.e. strictly exponential. The literal meaning would be that only a unit without any hit is able to survive. Exponential curves can be interpreted also in a different way (Kiefer, 1971): If one assumes that a hit leads with a probability q to inactivation the survival can be expressed as

$$N = N_0 \sum_{0}^{\infty} e^{-mD/\varepsilon}(1-q)^i (mD/\varepsilon)^i / i!$$

$$\ldots = N_0 e^{-qmD/\varepsilon} \qquad (3)$$

A "single-hit" curve can hence be interpreted as the unit being able to survive even with many hits although every one received lowers the survival probability.

"Hits" in the original concept described interaction processes in a rather abstract form. They could not be measured and certainly not be seen. This situation has now changed in some respect. Radiation absorption leads to the formation of double strand breaks (DSB) in the DNA of the affected cells which are the most important biologically significant lesions. With the help of immunofluorescence techniques it is now possible to make them visible and determine their distributions down to X-ray doses of a few mGy (Rothkamm and Loebrich 2003). Figure 2 shows an example from their work. The number of DSB per cell nucleus follows very nicely the predicted Poisson distribution. It must be pointed out at this stage, however, that this is typical only for sparsely ionizing radiations like X- or gamma-rays, the situation is different with densely ionizing particles as discussed later.

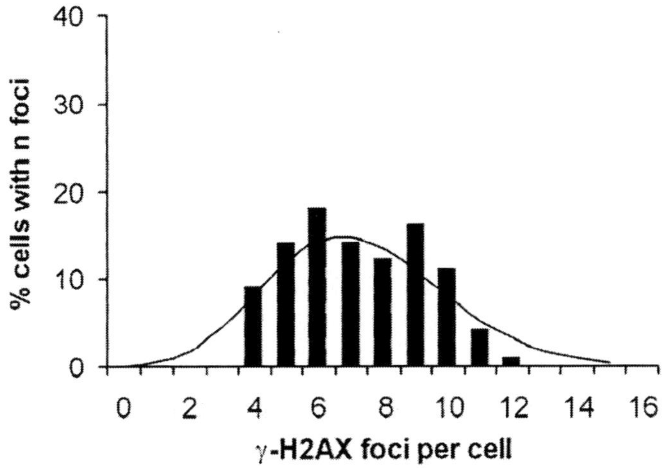

Figure 2. Number of fluorescent foci as markers of DNA double strand breaks in nuclei of human cells after exposure to 200 mGy X-rays (Rothkamm and Löbrich 2003). The line shows Poisson probabilities for a mean yield of 7 foci/cell. Courtesy of M. Löbrich.

2. Radiation as Analytical Tool: Determination of Molecular Masses

While the success of target theory to explain cellular radiation action was only rather limited it developed to a useful tool for the analysis of molecular masses. There are, of course, quite a number of standard techniques for their determination (ultracentrifugation, electrophoresis etc.) but they require the isolation or separation of the molecules and it cannot be excluded that the manipulation alters the conformation and even the size compared to the functional environment. This is, for instance, the case with membrane bound enzymes. If they can be inactivated by radiation the dose dependence is a function of their size *in situ*.

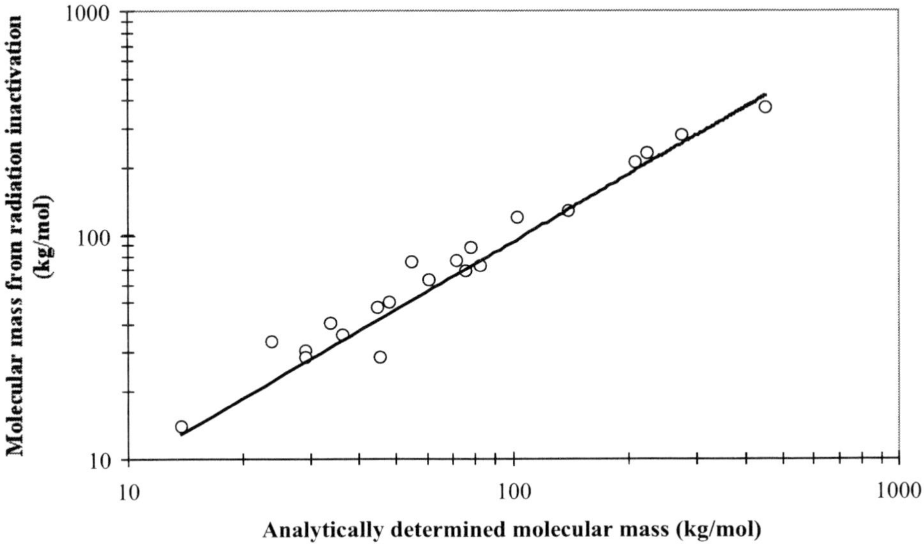

Figure 3. Relationship between analytically determined molecular weights and those measured by radiation inactivation (data from Kempner and Schlegel 1979).

In the case of single hit processes which are found experimentally in most cases equation (1) reduces to

$$N=N_0e^{-mD/\varepsilon}\qquad(4)$$

which can be directly applied. Kempner and Schlegel (1979) compared molecular weights obtained by sedimentation and radiation analysis and found a linear dependence over two orders of magnitude as summarized in Fig. 3. The topic was followed up in a number of more recent papers (Kempner 1988, 1993, 1999). It was also shown that radiation target analysis is more reliable than centrifugation methods as it is insensitive to changes in pH that may alter sedimentation behavior (Osborne *et al.*, 2000).

3. Radiation Quality

The original target analysis was concerned with sparsely ionizing radiations. In this case ionization events are stochastically distributed if the exposed sites are not too small and the doses not too low. This is the situation described in the previous section so that the method is applicable with X- or gamma-rays. If one deals with densely ionizing particles like α–particles or neutrons one has to take into account that they are able to deposit comparatively high amounts of energy even by single traversals. This is described quantitatively by the "linear energy transfer" (LET) which gives the "energy deposited *locally* per unit pathlength"- The common unit is "keV/μm". The term "locally" refers to track structure which will be discussed below.

With densely ionizing particles ionizations are concentrated largely in their path, contrary to e.g. X-rays they leave tracks in matter. To be strictly correct this is also the case with the electrons which are liberated by photon radiations but here ionizations are much further apart. A direct consequence of high LET-values is that for a given dose the number of particles per unit area (the "fluence") decreases as dose D and fluence Φ are linearly related through the LET L:

$$D = L \cdot \Phi \quad \text{.......................................(5)}$$

In the case of smaller (e.g. subcellular) sites there is no longer an evenly distributed energy deposition because it is governed by the probability p to be hit by the particle. It depends both on the site area and the fluence but it is still Poissonian:

$$P(n) = e^{-a\Phi} \, (a\Phi)^n/n! \text{......................................(6)}$$

where p(n) is the probability of n hits to a site of area a.

The energy deposited per mass (the dose) varies thus from site to site and thus loses its unequivocal meaning in smaller sites. In order to take this into account a new quantity was introduced, the "specific energy" z (also defined as energy deposited per mass) as the microscopic counterpart to the macroscopic dose. Specific energy is a stochastic variable characterized by a distribution p(z). Dose can now be defined in a more stringent way as the expectation of specific energy:

$$D = \int zp(z)dz \text{.........................(7)}$$

Energy deposition in small sites, their variation and quantitative description as well as the interpretation of biological effects is the realm of a new sub-branch of radiation science, namely "microdosimetry" (Rossi and Zaider, 1996).

The inactivation of a biological target depends on z. In the case of single inactivation events one has with z_1 the specific energy per single traversal.

This is an exponential relationship on fluence and, because of eq. (5), also on dose. The factor attached to the fluence is usually called the "action cross section", or in this case, the "inactivation cross section" as it has the dimension of an area. If one denotes it with σ one sees that

$$\sigma = a(1 - e^{-z_1/\varepsilon}) \qquad (8)$$

in other words: the inactivation cross section increases exponentially with the energy deposited by a single particle traversal and reaches a saturation value equal to the geometrical cross section of the site. The specific energy deposited by a single traversal z_1 is proportional to LET

$$z_1 = L/\rho\, v \qquad (9)$$

with v as the site volume and ρ the density.

One can thus conclude that with very high ionization density biological targets can be inactivated by a single particle traversal.

$$N = N_0 \sum_n p(n) e^{-nz_1/\varepsilon}$$
$$\ldots = N_0 \sum_n (e^{-z_1/\varepsilon})^n e^{-a\Phi} (a\Phi)^n / n!$$
$$\ldots = N_0 \exp[-a\Phi(1 - e^{-z_1/\varepsilon})] \qquad (10)$$

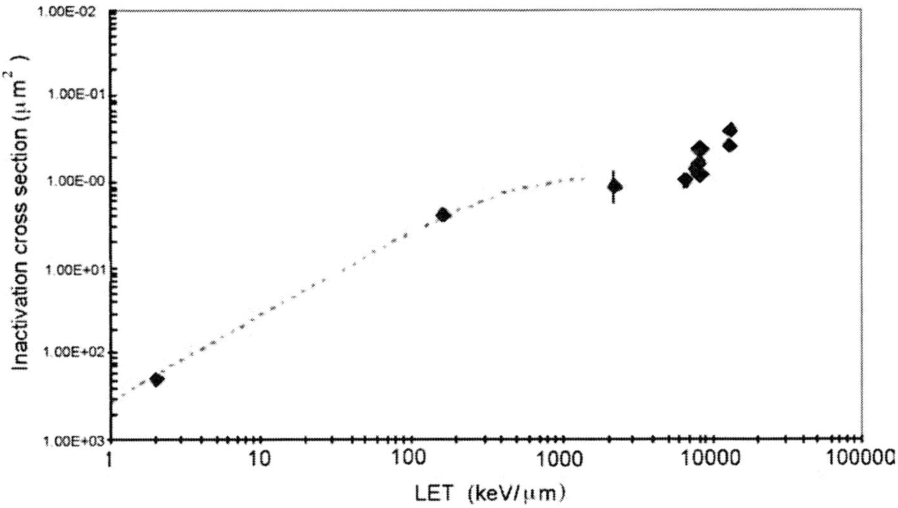

Figure 4. Inactivation cross section for haploid yeast cells after exposure to heavy ions of different LET and energies below 3 MeV/u. Data from Kiefer *et al.* (1983).

The validity of the theoretical approach described was tested experimentally by measuring the inactivation of haploid yeast exposed to low energy heavy ions. Results are shown in Fig. 4.

One sees that the experimental points can be fitted by the theoretical expression according to eq. (8). It has to be pointed out, however, that this holds true only in the case of low energy ions. If particles of higher energy are also included one obtains the relationship shown in Fig. 5. It is obvious that the unequivocal dependence on LET breaks down. The reason is that ion tracks cannot be considered as ultrathin "pencil beams" but possess lateral extensions formed by secondary electrons which contribute to inactivation. The "LET" approach is thus only applicable in the case of negligible energy depositions at some distance from the track core. This is the meaning of "locally deposited" in the definition given above.

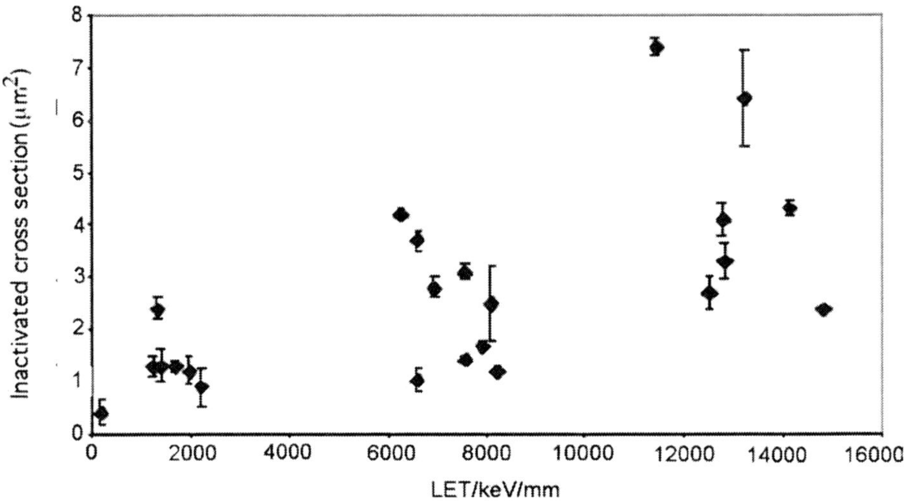

Figure 5. As in figure 4 but including all ion energies up to 12 MeV/u. Data from Kiefer *et al.* (1983).

4. Track Structure

At the point of collision electrons are ejected into different directions and with different velocities and angles. They form an "ionization cone" whose radius depends on ion velocity.

Track structure has been studied extensively, both theoretically as well as by experimental measurements. Fig. 6 depicts the track of a heavy charged particle in a very simplified schematic way: The impinging ion liberates electrons with energies which depend on ion velocity, their number increases essentially as a function of the ion's effective charge. As a first approximation the situation can be described by classical collision dynamics (Kiefer and Straaten 1986). This yields the following expressions:

1. *Energy density* $\rho_\varepsilon(x)$. This is the energy per mass deposited as a function of distance x from the track center

$$\rho_e(x)=1.25(10^{-4})\frac{z^{*2}}{\beta^2}1/x^2 \qquad (11)$$

z* is ion effective charge and ß ion velocity relative to that of light *in vacuo*

2. Penumbra radius x_p: This is the maximum radial range of secondary electrons

$$x_p=6.16(10^{-2})(E/m)^{1.7} \quad \text{....................(12)}$$

E is ion energy, m its mass. E/m is proportional to ion velocity.

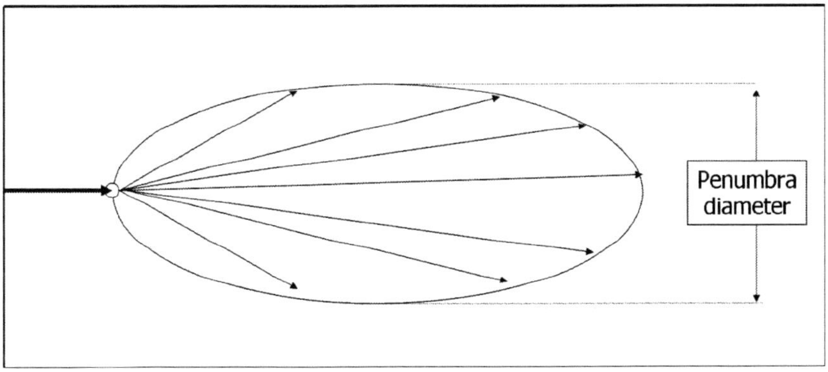

Figure 6. Simplified scheme of an ion track.

If track structure is taken into account the inactivation of a site by particles is more complex. As energy depositions by secondary electrons are possible at considerable distances from the track core the inactivation cross section may be larger than its geometrical cross section. Whether this occurs depends on the electron density in the penumbra (i.e. on ion charge) and also the penumbra extension (i.e. on ion velocity). The increase in inactivation cross section is, therefore, expected to be most expressed with ions of high charge and high velocity. This leads to a consequence which is contrary to conventional expectations as seen in Fig. 5: At high LET-values which are obtainable only by very heavy ions one finds an increase of the inactivation cross section with *decreasing* LET. This so-called *hook* structure is simply the reflection of the fact that in this region the increase in ion velocity (larger penumbra radius, smaller LET) is more important than the LET.

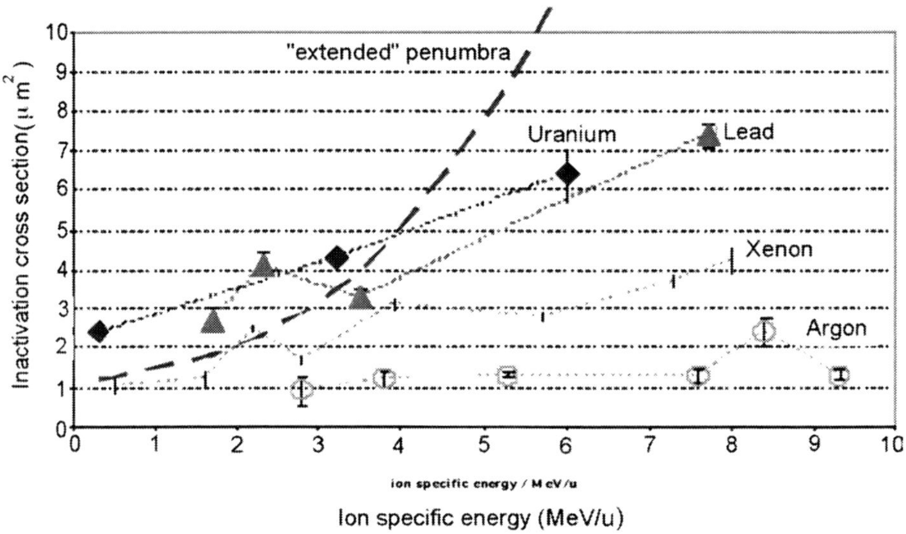

Figure 7. Inactivation cross section for haploid yeast cells after exposure to various heavy ions and its dependence on their energy.

Figure 7 shows inactivation cross section versus ion energy for various ions. The increase is obviously more marked with heavier ions, underlining what was said above. To give a more quantitative comparison the area of "extended" penumbras are also shown. They were calculated as circles where the radius is the sum of the estimated site radius and the penumbra radius according to eq. 12. With lighter ions the experimental action cross section is always smaller than that of the extended penumbra, and the difference is less for heavier ions as expected. A very interesting case, however, is seen with heavy ions at lower energies (lead and uranium): The inactivation cross sections are clearly and significantly higher than that of the extended penumbra. This result could not be explained when it was first noticed, it is tempting to attribute it to the "bystander effect". Although this has until now only been demonstrated for protons and alpha particles there is no reason to assume that it would not also occur with heavier ions.

5. Conclusions

Quantitative radiation biology has gone a long way. It started successfully nearly a century ago with the invention of the target theory whose principles still form the cornerstones of any attempt to explain and understand irradiation action to microscopic sites. The applicability is not limited to biological effects although they are still in the center of interest but target theory and its further development can also be used to understand e.g. damage by space radiation to subminiature electronic circuits in satellites. New developments in radiation biology, e.g. the bystander effect (see e.g. Prise *et al.*, 2005 for a recent review), do not invalidate the basic principles but will help to remind us not to think in too simplified ways. For a long time targets were only identified with DNA where "lethal lesions" lead immediately to cell death. We know today that biological processes are much more complex and indirect effects and the influence of neighboring cells as well as signal transduction have to be taken into account. Nevertheless, all kinds of radiation action start with the absorption of quantum or particle energy in a sensitive site, and this very basic mechanism still follows the "old" principles of target theory.

On the other hand the approaches described in this paper to understand in a quantitative way the action of heavy charged particles on biological entities will be equally useful in the context of heavy ion radiotherapy and in radiation risk assessment for astronauts in space.

Acknowledgements

I thank Markus Löbrich for continuous fruitful discussion and for the permission to use his data for this paper.

References

Alper, T., Gillies, N., Elkind, M.M., 1960, The sigmoid survival curve in radiobiology, *Nature*. **186**:1062–1063.

Crowther, J.A., 1924, Some considerations relative to the action of X-rays on tissue cells. *Proc. Royal Soc.* (London) **B96**, 207–211.

Dessauer, F., 1922, Über einige Wirkungen von Strahlen. *Z. Physik* **12**: 38–47.

Kempner E.S., Schlegel W., 1979, Size determination of enzymes by radiation inactivation. *Anal Biochem*. **92**,:2–10.

Kempner E.S., 1999, Advances in radiation target analysis. *Anal Biochem*. **276**:113–23.

Kempner E.S., 1988, Molecular size determination of enzymes by radiation inactivation. *Adv Enzymol Relat Areas Mol Biol.*; **61**:107–147.

Kempner E.S., 1993, Novel predictions from radiation target analysis. *Trends Biochem Sci.* **18**:236–239.

Kiefer J., Straaten H., 1986, A model of ion track structure based on classical collision dynamics. *Phys Med Biol.* **31**:1201–1209.

Kiefer, J., 1971, Target theory and survival curves. *J Theor Biol.* **30**:307–317.

Kiefer, J., Rase S., Schopfer F., Schneider E., Weber K., Kraft G., 1983, Heavy ion action on yeast cells: inhibition of ribosomal-RNA synthesis, loss of colony forming ability and induction of mutants. *Adv Space Res* **3**(8):115–125.

Kiefer, J., 2001, Radiation Biology – Glory of the past. Chances and challenges of the future. In: Korogodin V.I. (Ed.), *Modern problems of Radiobiology, Radioecology and Evolution.* Proc. of the Intern. Conf. dedicated to the Centenary of the Birth of N.W. Timofeeff-Ressovsky, Dubna, 6-9 September 2000, Dubna: JINR: 130–139.

Osborne J.C. Jr., Miller J.H., Kempner E.S., 2000, Molecular mass and volume in radiation target theory. *Biophys J.* **78**:1698–1702.

Prise K.M., Schettino G., Folkard M., Held K.D., 2005, New insights on cell death from radiation exposure. *Lancet Oncol.* **6**:520-528.

Rossi, H.H., Zaider, M., 1996, *Microdosimetry and its application* New York: Springer.

Rothkamm, K., Lobrich, M., 2003, Evidence for a lack of DNA double-strand break repair in human cells exposed to very low x-ray doses. *Proc Natl Acad Sci U S A.* **100**: 5057–5062.

Timofeeff-Ressovsky, N.W., Zimmer, K.G., 1947, *Das Trefferprinzip in der Biologie.* S. Hirzel Verlag Leipzig.

Timofeeff-Ressovsky, N.W., Zimmer, K.G., Delbrück, M., 1935, Über die Natur der Genmutation und der Genstruktur, *Nachr. Ges. Wiss. Göttingen* **VI** (1): 189–245.

Zimmer, K.G., 1961, *Studies on quantitative radiation biology,* Edinburgh and London: Oliver and Boyd.

RADIATION CYTOGENETICS: THE COLOR REVOLUTION

MARCO DURANTE
Department of Physics,
University Federico II, Naples, Italy
E-mail; durante@na.infn.it

Abstract. Analysis of radiation-induced chromosomal aberrations has long been a powerful tool to understand the mechanisms of radiation action in living cells. Early cytogenetics was based on observations of solid-stained chromosomes, although banding techniques were soon developed for karyotyping human cells. Banding is a complex, error-prone, and time-consuming methodology, especially when applied to radiation-induced aberrations that, unlike genetic syndromes, are randomly induced in the genome and in the cellular population. The "color revolution" occurred in the 80's with the introduction of fluorescence in situ hybridization (FISH) The first great improvement provided by FISH-painting was the opportunity to analyze symmetrical, transmissible aberrations (such as translocations and inversions) simply and rapidly, whereas dicentrics had long been the main endpoint analyzed by solid staining. Multi-color painting demonstrated that radiation-induced rearrangements are much more complex than previously thought, and even low doses of densely ionizing radiation produce mostly complex-type exchanges. The impact of multi-color painting on the understanding of radiation-induced genetic effects will be discussed here.

Keywords: Chromosomal aberrations; FISH; densely ionizing radiation

1. Introduction

When in 1935 Timofeeff-Ressovsky introduced the concept of genetic target of radiation (Timofeeff-Ressovsky *et al.,* 1935), it soon became clear that analysis of chromosomes could be used as a powerful tool to test target theory. The work on plant and insect chromosomes performed in those years (reviewed in Lea, 1946) is of paramount importance for understanding the mechanisms of radiation action. The microscopic

A.A. Cigna and M. Durante (eds.),
Radiation Risk Estimates in Normal and Emergency Situations, 243–252.

observations on structural rearrangements occurring in chromosomes exposed to ionizing radiation were explained assuming that interphase chromosomes were physically broken by electron or ion tracks. In most cases, the broken ends would rejoin correctly to restitute the original chromosome (*eurepair*). However, if two chromosome breaks are close enough together in time and space, exchanges can occur to form the structural rearrangements visible at mitosis (*misrepair*). Finally, a subset of initial breaks may not rejoin at all, producing terminal deletions (*unrepair*). The induction of chromosome breaks will be proportional to the number of tracks, i.e. to the dose, whilst if two interacting breaks are induced by separate tracks, exchanges are proportional to the square of the dose.

This simple picture is known as *classical* or *breakage-and-reunion* theory. Several observations of great interest in radiobiology were readily interpreted within this framework. For instance, if a given dose D is delivered in a longer time by protracted irradiation (low dose rate) or by giving two or more smaller doses separated in time (fractionation), the same number of breaks produced by an acute dose D will be induced in the cells, but breaks occurring early in the protracted or fractionated exposure may restitute and thus be unavailable for interaction to form an exchange with breaks occurring late in the chronic treatment. Therefore, less exchanges will be produced at low dose rate or after fractionation (Sax, 1939). Early observation on increased effectiveness of high-LET radiation compared to photons (Zimmer and Timofeeff-Ressovsky, 1939) could also be explained. In fact, single tracks of densely ionizing radiation can produce two or more chromosome breaks in close proximity, which can interact to form exchanges. Most of the exchanges produced by heavy charged particles will be formed along the ion track (intra-track), while the interactions of breaks produced by independent electrons is necessary for the formation of exchanges induced by photons (inter-track). Therefore, linear dose-response curves for the induction of exchanges will be observed after exposure to densely ionizing radiation, and the effectiveness will be highly increased as compared to sparsely ionizing radiation especially at low doses, where photons are relatively ineffective in forming exchanges (Lea, 1946).

The classical theory has been challenged by several other models. The most significant alternative models are perhaps the *exchange hypothesis* (Revell, 1974), where it is assumed that all aberrations are produced by pairwise interactions or initial lesions, that are not true breaks, and the *molecular theory* (Chadwick and Leenhouts, 1981), where it is assumed that exchanges are produced by the interaction of breaks with unbroken chromosomes. With the introduction of fluorescence in situ hybridization, about twenty years ago, radiation cytogenetics found the tools to clarify this issue.

2. FISH

Studies in radiation cytogenetics have been performed for over half century using Giemsa-staining. Although G-banding (Drets and Shaw, 1971) was in principle able to reveal all kinds of symmetrical and asymmetrical rearrangements in the human genome, the method can hardly be applied in large experiments, is extremely time-consuming, and error-prone (Savage, 1977). In fact, dicentrics, that are easily visualized by solid Giemsa staining, were virtually the only aberration analyzed in human cells until the middle eighties, when the Lawrence Livermore National Laboratory introduced the technique of fluorescence in situ hybridization (FISH) with whole chromosome probes (Pinkel *et al.*, 1986). The basic idea underlying the method is to label DNA probes originating from a single chromosome with a fluorescent probe, and then hybridize the labeled probe to the chromosomes attached to a microscope slide.

Soon after the introduction of FISH, cytogenetists were attracted by the possibility to score reciprocal translocations. Symmetrical interchanges are more stable than dicentrics, and can therefore provide useful indications for retrospective biodosimetry. In fact, FISH translocations could be measured in A-bomb survivors and the results correlated with the calculated dose absorbed by the subjects over 40 years before (Lucas *et al.*, 1992; Nakano *et al.*, 2001).

FISH could be also used to elucidate mechanisms of formation of chromosomal aberrations. For instance, the Revell's exchange theory predicts pairwise lesion interaction only, and therefore a multi-color DNA fragment with material from three different chromosomes cannot occur. However, these 3-color chromosomes were soon observed by FISH (Lucas and Sachs, 1993), supporting the classical theory.

FISH could also be used to paint prematurely condensed chromosomes (PCC), that had been used to characterize initial chromosome damage (reviewed in Cornforth and Bedford, 1993). FISH-painting of PCC fixed at different times after exposure could provide insights on the kinetics of formation of chromosomal aberrations. Both in human fibroblasts (Brown *et al.*, 1993) and lymphocytes (Durante *et al.*, 1996) exchange formation starts within a few minutes from irradiation, but interchanges first appear in incomplete forms, and are then completed within a few hours when the second broken ends also misrejoin. The view of "floating broken ends" seeking partners for enzyme-mediated rejoining can be easily modeled (Wu *et al.,* 1996), and is perfectly coherent with the classical theory.

3. Multi-fluor FISH

One further striking observation soon noted by FISH was that many exchanges were "complex" (Lucas and Sachs, 1993; Savage and Simpson, 1994). Complexes are defined as those exchanges involving more than two breaks in at least two chromosomes (Savage, 2002), and can be basically divided into insertions and non-reciprocal exchanges. Exchanges involving several breaks and chromosome arms had been described in Giemsa-stained samples (Savage, 1976), but they were thought to be relatively rare as compared to two-way symmetrical or asymmetrical exchanges. The abundance of complex-type exchanges became clear when, toward the end of the last century, the number of colors in FISH-painting could be increased. In fact, the interferometer-based spectral karyotyping (SKY) system (Schrok *et al.*, 1996) and the filter-based multi-fluor FISH (mFISH) approach (Speicher *et al.*, 1996) made possible analyzing full karyotypes, where each chromosome pair has a different color. Radiation-induced complex-type exchanges were soon observed using mFISH (Greuclich *et al.*, 2000), and then described in detail by Cornforth (2001). Again, the classical breakage-and-reunion theory explains the existence of these events, whereas other models unavoidably fail (Sachs *et al.*, 2000). However, the current models, based on computer simulation, tend to underestimate the number of chromosomes that can be involved in exchanges (Vazquez *et al.,* 2002). Another important improvement to the classical model is the origin of the curvature in the dose-response curve for the induction of interchanges. It is now clear that most of the curvature in the dose-response curve for the induction of dicentrics is caused by the occurrence of complex-type exchanges at high doses, thus suggesting that most simple interchanges are induced in intra-track mode (Loucas and Cornforth, 2001). However, other authors contend that the presence of a β-term in the dose-response curve for dicentrics even at low doses indicates that simple exchanges are not linear with dose (Edwards *et al.*, 1999).

Whilst complexes are generally induced by sparsely ionizing radiation at doses > 2 Gy only, even single tracks of densely ionizing radiation can produce complex rearrangements. In fact, mFISH studies with α-particles (Anderson *et al.*, 2002), and heavy ions (Durante *et al.*, 2002) point to a very high RBE of high-LET radiation in the induction of complexes at low doses. Transmissible complexes have also been observed in vivo in lymphocytes from subjects exposed to high doses of Pu α-particles many years before the test (Anderson *et al.*, 2005; Hande *et al.*, 2005), and in radiotherapy patients treated with accelerated carbon ions (Yamada *et al.*, 2000).

4. Multi-color Banding

FISH studies with whole-chromosome probes are ideal tools for scoring interchanges, but intra-chromosomal rearrangements will escape the analysis. Arm-specific DNA probes have been used to detect pericentric inversions (Natarajan *et al.*, 1996). Intra-arm rearrangements can be detected using high-resolution multi-color banding (mBAND), where single chromosome pairs are painted in 23 different combinatorial colors (Chudoba *et al.*, 1999).

Theoretical models based on the classical theory predict that the ratio inter-/intra-chromosomal exchanges (*F*-ratio) decreases by increasing radiation LET (Brenner and Sachs, 1994). The variation of the *F*-ratio with radiation quality, as well as of other aberration ratios, is strongly dependent on the nuclear architecture in interphase and on the movement of the chromosome domains. Evidence of a decreased *F*-ratio has been provided by mBAND analysis of former radiation workers overexposed to either α-particles or γ-rays (Hande *et al.,* 2003). However, in vitro results generally failed to detect significant changes, and point to an excess of interchanges compared to intrachanges (Johannes *et al.*, 2004).

Multicolor banding can also be observed in the whole genome when gibbon DNA probes labeled with three different fluorochromes are hybridized to human chromosomes. Due to the extensive DNA homology between apes and man, and to the many rearrangements that occurred during evolution, the hybridization results in approximately 90 visible bands in the human karyotype in five different colors. The method is known as cross-species color segmenting, or rainbow FISH (RxFISH) (Muller *et al.*, 1997). Although the resolution is much lower than in G-banding or mBAND, it allows a simultaneous analysis of intra- and inter-changes in the whole genome. The method has been recently applied to human lymphocytes exposed to Fe-ions or γ-rays and harvested 144 h after stimulation to in vitro growth (Durante *et al.*, 2006).

5. Nuclear Architecture

FISH-painting has been used to elucidate the position of the chromosome domains in interphase. FISH has clearly shown that chromosomes are compartmentalized into discrete territories, and the topology of the nucleus influence gene expression and regulation (reviewed in Cremer and Cremer, 2001). The 3D position of the genes strongly affects cell function in a developmental and tissue-specific way during the cell-cycle. Gene regulation by conformational changes in the chromatin is now referred to as "chromosomics" (Claussen, 2005). The topology of the interphase nucleus has obviously large impact on the formation of chromosomal aberrations.

First, the classical theory will predict that rearrangements occur more easily for genes in close proximity. Using interphase FISH, Nikiforova *et al.* (2000) elegantly showed that the RET and H4 genes are often juxtaposed in thyroid cells, but not in mammary human cells. Papillary thyroid cancer is characterized by inversion in chromosome 10 leading to a recombination between RET and H4. This experiment clearly shows that "proximity matters" (Savage, 2000) in the formation of specific chromosomal aberrations that can lead to cancer. Close proximity in interphase of oncogenes whose rearrangement is necessary for B-cell lymphoma (Roix *et al.*, 2003) or chronic myeloid leukemia (Kozubek *et al.*, 1997) has been reported.

Second, DNA double-strand breaks can move in the interphase nucleus and interact with other open ends, eventually leading to misrepair. Movements of DNA breaks can be now visualized by immunostaining of the histone H2AX, which is phosphorylated in the site of the DNA breaks. Using this method, Aten *et al.* (2004) have shown that DNA breaks move within minutes form their formation, and can cluster in specific areas where they are apparently repaired, in a process involving Mre11. Again, this mechanism is coherent with the breakage-first hypothesis.

Finally, early FISH results pointed to large variations in individual sensitivity of different chromosomes to radiation breakage, i.e. to a nonrandom involvement of different chromosomes in exchanges (e.g. Knehr *et al.*, 1994). However, these observations were probably mainly caused by technical difficulties. In fact, multi-fluor FISH analysis demonstrate that only small deviations from randomness are observed in exchanges induced by either sparsely (Cornforth *et al.*, 2002) or densely (Durante *et al.*, 2002) ionizing radiation. The deviations from random behavior (Arsuaga *et al.*, 2004) reflect the nuclear architecture: gene-rich chromosomes are mainly located in the centre of the nucleus (Cremer and Cremer, 2001), and will then be on average closer to each other than randomness would predict. Some chromosomes will then preferentially exchange their genetic material with the neighbors, notwithstanding a considerable randomness of chromosome-chromosome juxtapositions.

6. Conclusions

In the past 20 years, the introduction of FISH-painting has certainly changed cytogenetics. Multi-color painting of the whole genome, or of some specific genes or subchromosomal domains, is commonly used today to elucidate basic questions in molecular biology and radiation biophysics. It is somehow surprising that these new techniques have substantially confirmed the classical theory of radiation action that was developed over 70 years ago using technologies that no one would even consider today. The "color" revolution did not lead to a conceptual revolution with respect to the ideas elegantly expressed so many years ago by Timofeeff-Ressovsky, Zimmer, Sax, and Lea. The basic frame of the breakage-first theory, although challenged so many times, remains substantially valid. Multi-color painting of the human chromosomes is now substantially contributing to human genomics, disease and cancer genetics, chromosome evolution and the relationship of nuclear structure to function (Trask, 2002). As to radiation effects in chromosomes, the search for biomarkers of radiation quality and risk is still ongoing and particularly promising. Transmission of radiation-induced aberrations, and the appearance of delayed chromosomal instability (Sieber *et al.*, 2003) are indeed responsible for late effects. The shift from early events, so well described by the classical theory, to the late consequences of the chromosomal rearrangements may lead in future years to a better understanding of the carcinogenic pathway and suggest novel therapeutic approaches.

References

Anderson, R.M., Stevens, D.L., and Goodhead, D.T., 2002, M-FISH analysis shows that complex chromosome aberrations induced by alpha -particle tracks are cumulative products of localized rearrangements, *Proc. Natl. Acad. Sci USA*. **99**: 12167–12172.

Anderson, R.M., Tsepenko, V.V., Gasteva, G.N., Molokanov, A.A., Sevan'kaev, A.V., and Goodhead, D.T., 2005, mFISH analysis reveals complexity of chromosome aberrations in individuals occupationally exposed to internal plutonium: a pilot study to assess the relevance of complex aberrations as biomarkers of exposure to high-LET alpha particles. *Radiat. Res.* **163**: 26–35.

Arsuaga, J., Greulich-Bode, K.M., Vazquez, M., Bruckner, M., Hahnfeldt, P., Brenner, D.J., Sachs, R.K. and Hlatky, L., 2004, Chromosome spatial clustering inferred from radiogenic aberrations. *Int. J. Radiat. Biol.* **80**: 507–515.

Aten, J.A., Stap, J., Krawczyk, P.M., van Oven, C.H., Hoebe, R.A., Essers, J. and Kanaar, R., 2004, Dynamics of DNA double-strand breaks revealed by clustering of damaged chromosome domains, *Science* **303**: 92–95.

Brenner, D.J. and Sachs, R.K., 1994, Chromosomal "fingerprints" of prior exposure to densely ionizing radiation. *Radiat. Res.* **140**: 134–142.

Brown, J.M., Evans, J.W. and Kovacs, M.S., 1993, Mechanism of chromosome exchange formation in human fibroblasts: insights from "chromosome painting". *Environ. Mol. Mutagen.* **22**: 218–224.

Chadwick, K.H. and Leenhouts, H.P., 1981, *The Molecular Theory of Radiation Biology*, Springer Verlag, Berlin.

Claussen, U., 2005, Chromosomics, *Cytogenet. Genome Res.* **111**: 101–106.

Cornforth, M.N., 2001, Analyzing radiation-induced complex chromosome rearrangements by combinatorial paining, *Radiat. Res.* **155**: 643–659.

Cornforth, M.N. and Bedford, J.S., 1993, Ionizing radiation damage and its early development in chromosomes. *Adv. Radiat. Biol.* **17**: 423–496.

Cornforth, M.N., Greulich-Bode, K.M., Loucas, B.D., Arsuaga, J., Vazquez, M., Sachs, R.K., Bruckner, M., Molls, M., Hahnfeldt, P., Hlatky, L. and Brenner, D.J., 2002, Chromosomes are predominantly located randomly with respect to each other in interphase human cells, *J. Cell Biol.* **159**: 237–244.

Cremer, T. and Cremer, C., 2001, Chromosome territories, nuclear architecture and gene regulation in mammalian cells. *Nat. Rev. Genet.* **2**: 292–301.

Drets, M.E. and Shaw, M.W., 1971, Specific banding patterns of human chromosomes. *Proc. Natl. Acad. Sci. USA* **68**, 2073–2077.

Durante, M., George, K., Wu, H.-L., and Yang, T.C., 1996, Rejoining and misrejoining of radiation-induced chromatin breaks. I. Experiments in human lymphocytes, *Radiat. Res.* **145**: 274–280.

Durante, M., George, K., Wu, H.-L. and Cucinotta, F.A., 2002, Karyotypes of human lymphocytes exposed to high-energy iron ions, *Radiat. Res.* **158**: 581–590.

Durante, M., George, K. and Cucinotta, F.A., 2006, Chromosomes lacking telomeres are present in the progeny of human lymphocytes exposed to heavy ions, *Radiat. Res.* **165**: 51–58.

Edwards, A.A., Hone, P.A., Moquet, J.A. and Lloyd, D.C., 1999, Simple chromosome exchanges are not linear with dose, *Int. J. Radiat. Biol.* **75**: 1113–1117.

Greulich, K.M, Kreja, L., Heinze, B., Rhein, A.P., Weier, H.G., Bruckner, M., Fuchs, P. and Molls, M., 2000, Rapid detection of radiation-induced chromosomal aberrations in lymphocytes and hematopoietic progenitor cells by mFISH, *Mutat. Res.* **452**: 73–81.

Hande, M.P., Azizova, T.V., Geard, C.R., Burak, L.E., Mitchell, C.R., Khokhryakov, V.F., Vasilenko, E.K. and Brenner, D.J., 2003, Past exposure to densely ionizing radiation leaves a unique permanent signature in the genome, *Am. J. Hum. Genet.* **72**: 1162–1170.

Hande, M.P., Azizova, T.V., Burak, L.E., Khokhryakov, V.F., Geard, C.R. and Brenner, D.J., 2005, Complex chromosome aberrations persist in individuals many years after occupational exposure to densely ionizing radiation: an mFISH study, *Genes Chromosomes Cancer* **44**: 1–9.

Johannes, C., Horstmann, M., Durante, M., Chudoba, I. and Obe, G., 2004, Chromosome intrachanges and interchanges detected by multicolor banding in lymphocytes: searching for clastogen signatures in the human genome, *Radiat. Res.* **161**: 540–548.

Kozubek, S., Lukasova, E., Ryznar, L., Kozubek, M., Liskova, A., Govorun, R.D., Krasavin, E.A. and Horneck, G., 1997, Distribution of ABL and BCR genes in cell nuclei of normal and irradiated lymphocytes, *Blood* **89**: 4537–4545.

Knehr, S., Zitzelsberger, H., Braselmann, H. and Bauchinger, M., 1994, Analysis for DNA-proportional distribution of radiation-induced chromosome aberrations in various triple combinations of human chromosomes using fluorescence in situ hybridization. *Int. J. Radiat. Biol.* **65**: 683–690.

Lea, D.E., 1946, *Actions of Radiations on Living Cells*, Cambridge University Press, London.

Loucas, B.D. and Cornforth, M.N., 2001, Complex chromosome exchanges induced by gamma rays in human lymphocytes: an mFISH study. *Radiat. Res.*, **155**: 660–671.

Lucas, J.N., Awa, A.A., Straume, T., Poggensee, M., Kodama, Y., Nakano, M., Ohtaki, K., Weier, H.-U., Pinkel, D., Gray J.W. and Littlefield, G., 1992, Rapid translocation frequency analysis in humans decades after exposure to ionizing radiation, *Int. J. Radiat. Biol.* **62**, 53–63.

Lucas, J.N. and Sachs, R.K., 1993, Using three-color chromosome painting to test chromosome aberration models, *Proc. Natl. Acad. Sci. USA* **90**:1484–1487.

Muller, S., O'Brien, P.C., Ferguson-Smith, M.A. and Wienberg, J., 1998, Cross-species colour segmenting: a novel tool in human karyotype analysis. *Cytometry* **33**: 445–452.

Nakano, M., Kodama, Y., Ohtaki, K., Itoh, M., Delongchamp, R., Awa, A.A. and Nakamura, N. (2001) Detection of stable chromosome aberrations by FISH in A-bomb survivors: comparison with previous solid Giemsa staining data on the same 230 individuals. *Int. J. Radiat. Biol.* **77**: 971–977.

Natarajan, A.T., Boei, J.J., Vermeulen, S. and Balajee, A.S., 1996, Frequencies of X-ray induced pericentric inversions and centric rings in human blood lymphocytes detected by FISH using chromosome arm specific DNA libraries, *Mutat. Res.* **372**: 1–7.

Nikiforova, M.N., Stringer, J.R., Blough, R., Medvedovic, M., Fagin, J.A. and Nikiforov, Y.E., 2000, Proximity of chromosomal loci that participate in radiation-induced rearrangements in human cells, *Science* **290**: 138–141.

Pinkel, D., Straume, T. and Gray, J.W., 1986, Cytogenetic analysis using quantitative, high-sensitive fluorescence hybridization. *Proc. Natl. Acad. Sci. USA* **83**: 2934–2938.

Revell, S.H., 1974, The breakage-and-reunion theory and the exchange theory for chromosomal aberrations induced by ionizing radiation: a short history, *Adv. Radiat. Biol.* **16**: 77–124.

Roix, J.J., McQueen, P.G., Munson, P.J., Parada, L.A. and Misteli, T., 2003, Spatial proximity of translocation-prone gene loci in human lymphomas. *Nat. Genet.* **34**: 287–291.

Sachs, R.K., Rogoff, A., Chen, A.M., Simpson, P.J., Savage, J.R.K., Hahnfeldt, P. and Hlatky, LR., 2000, Underprediction of visibly complex chromosome aberrations by a recombinational-repair ('one-hit') model, *Int. J. Radiat. Biol.* **76**:129–148.

Schröck, E., du Manoir, S., Veldman, T., Schoell, B., Wienberg, J., Ferguson-Smith, M.A., Ning, Y., Ledbetter, D.H., Bar-Am, I., Soenksen, D., Garini, Y., nd Ried, T., 1996, Multicolor spectral karyotyping of human chromosomes, *Science* **273**: 494–497.

Savage, J.R.K., 1976, Classification and relationships of induced chromosomal structural changes. *J. Med. Genet.* **13**: 103–122.

Savage, J.R.K., 1977, Assignment of aberrations breakpoints in banded chromosomes, *Nature* **270**, 513–514.

Savage, J.R.K., 2000, Cancer: proximity matters, *Science* **290**: 62–63.

Savage, J.R.K., 2002, Reflections and meditations upon complex chromosomal exchanges. *Mutat. Res.* **512**: 93–109.

Savage, J.R.K. and Simpson, P.J., 1994, FISH "painting" patterns resulting from complex exchanges, *Mutat. Res.* **312**: 51–60.

Sax, K., 1939, The time factor in X-ray production of chromosome aberrations. *Proc. Natl. Acad. Sci. USA* **25**, 225–233.

Sieber, O.M., Heinimann, K. and Tomlinson, I.P., 2003, Genomic instability-the engine of tumorigenesis? *Nat. Rev. Cancer* **3**: 701–708.

Speicher, M.R., Gwyn Ballard, S. and Ward, D.C., 1996, Karyotyping human chromosomes by combinatorial multi-fluor FISH, *Nat. Genet.* **12**:368–375.

Timofeeff-Ressovsky, N.W., Zimmer, K.G. and Delbrück, M., 1935, Über die Natur der Genmutation und der Genstruktur, *Nachr. Gess. Wiss. Göttingen. Biologie.* **1**:189–245.

Trask, B.J., 2002, Human cytogenetics: 46 chromosomes, 46 years and counting, *Nat. Rev. Genet.* **3**: 769–778.

Vazquez, M., Greulich-Bode, K.M., Arsuaga, J., Cornforth, M.N., Bruckner, M., Sachs, R.K., Hlatky, L., Molls, M. and Hahnfeldt, P., 2002, Computer analysis of mFISH chromosome aberration data uncovers an excess of very complicated metaphases, *Int. J. Radiat. Biol.* **78**: 1103–1115.

Wu, H.-L., Durante, M., George, K., Goodwin E.H. and Yang, T.C., 1996, Rejoining and misrejoining of radiation-induced chromatin breaks. II. Biophysical model, *Radiat. Res.* **145**: 281–288.

Yamada, S., Durante, M., Ando, K., Furusawa, Y., Kawata, T., Majima, H. and Tsujii, H., 2000, Complex-type chromosomal exchanges in blood lymphocytes during radiation therapy correlate with acute toxicity. *Cancer Lett.* **150**: 215–221.

Zimmer, K.G., Timofeeff-Ressovsky, N.W., 1939, Note on the biological effects of densely ionizing radiation. *Phys. Rev.* **55**, 411.

THE MINIMUM DETECTABLE DOSE BY BIODOSIMETRY IN A RADIATION OVEREXPOSURE*

D.C. LLOYD*, A.A. EDWARDS and M. SZŁUIŃSKA
Health Protection Agency, Radiation Protection Division
Chilton, Didcot, OX11 0RQ. UK

*To whom correspondence should be addressed.

Abstract. Chromosomal aberration analysis is the most sensitive biological method to indicate exposure to ionising radiation. This paper will distinguish between the *detection* and the *measurement* of low doses by aberration analysis and show how to quantify their limits. Worked examples will be presented using the lymphocyte dicentric assay and data typical of Co-60 gamma rays. The principles illustrated can be applied to other aberration types and other radiation qualities. Two situations will be considered: conventional by eye scoring of 1,000 metaphases from a suspected low exposure patient and scoring more metaphases with computer assisted microscopy. Low dose quantification is ultimately limited by the uncertainty on the assumed background level of dicentrics. With conventional scoring the Poisson uncertainty on the patient's observed dicentric frequency is the major component to the uncertainty on low dose estimates. With increased scoring, assisted by computer, this is reduced but the standard error on the linear calibration coefficient becomes more important. The optimum is reached where both components contribute equally to the overall uncertainty. A dose estimate may be considered as a *measurement* when its lower 95% confidence limit falls above zero. A dose can be regarded as having been *detected* when the dicentric frequency is above an assumed background but the lower 95% confidence limit includes zero. Conventional scoring of 1,000 metaphases will permit a measurement lower limit of about 100mGy of gamma radiation. This can be reduced by scoring many more cells (~10,000) to about 70mGy. Further improvement is unlikely due to the background 'noise' in the assay.

Keywords: biological dosimetry, dicentric, low dose detection

* Note: A full text version of this presentation is in press (Lloyd *et al.*, International Journal of Low Radiation). Therefore for the purpose of this conference proceedings an Extended Abstract is provided.

A.A. Cigna and M. Durante (eds.),
Radiation Risk Estimates in Normal and Emergency Situations, 253–258.

1. Introduction

Chromosomal aberration analysis is the most sensitive biological method available to indicate a person's overexposure to ionising radiation. For many years this has been done by detecting dicentric aberrations in cultured lymphocytes taken from a small blood sample. Other techniques, applied also to lymphocytes, include assays for micronuclei or translocations although in practice these tend to be used less frequently. For convenience the remarks in this paper will be confined to the dicentric assay although the principles also can be applied to the others.

Fortunately most actual or suspected accidental exposures to radiation involve small doses. These therefore comprise the major part of the case load referred to biodosimetry laboratories. Whilst low doses may incur low risks to health, and possibly zero risk in some peoples' opinion, they are nevertheless classed as overexposures in the administrative sense. The recommended occupational and public dose limits are set very low (ICRP-60) and so there is often pressure on biodosimetry laboratories to quantify exposures at levels below the capability of the dicentric assay.

Thus the question is frequently asked: 'What is the low dose detection limit by biodosimetry?'

This is a simple question to ask but the reply needs much consideration. This paper explores some of the issues involved.

2. Low dose Quantification

The first point to consider is whether the questioner really means low dose detection or measurement and the difference between these is dependent on the statistical uncertainties associated with establishing the dicentric frequency and interpreting it in terms of absorbed radiation dose.

The usual simple approach is to calculate the mean observed dicentric frequency with upper and lower confidence limits. These values are then converted to doses by reference to an in vitro dose response calibration curve. This curve consists of a background level (c) with linear (α) and square law (β) coefficients. All of these coefficients carry uncertainties expressed as standard errors. So, what is the lowest dose that can be reasonably measured and what is the lowest that can be detected? How does one set about deriving such values?

In most situations the Poisson uncertainty on the observation made from typically 1,000 of the patient's lymphocytes is much greater than the uncertainty on the calibration coefficients c and α and so the calibration's contribution to uncertainty on the dose estimate is overlooked. Moreover, at low doses the contribution to the yield from the coefficient β is so small that it and its standard error can be safely ignored.

The question of whether a dose has been measured or detected depends on its confidence limits and, as 95% limits are the generally accepted practice, these have been chosen for the illustrations used here. It is suggested that an exposure can be regarded as having been *detected* whenever the dose estimate is positive but its lower confidence limit is zero or less. If the lower limit is above zero it is suggested that the dose estimate may be regarded as a *measurement*. This is illustrated in table 1 which shows, for a typical linear/quadratic dose response equation for cobalt-60 (where Y = the yield of dicentrics per cell and D = dose in Gy):

$$Y = 0.001 + 0.02D + 0.06D^2$$

that for scoring 1,000 cells, the lower limit of measurement occurs when 4 dicentrics are seen and this corresponds to a dose of about 100 mGy. The observations of 2 and 3 dicentrics lead to smaller detected doses but their lower confidence limits are zero or less. A limit less than zero, implying a negative value in Gy, is of course a scientific absurdity and therefore is shown in the table as zero.

Table 1. Dose estimates and their 95% confidence limits in Gy for various numbers of dicentrics scored in 1,000 cells with an assumed background of 1/1,000 cells. The yield equation is $Y=0.001 + 0.02D + 0.06D^2$. Yields and resultant doses regarded as measurements are shown in bold.

Observed no. of dicentrics	Lower Limit	Mean Dose	Upper Limit
0	-	0	0.10
1	0	0	0.16
2	0	0.04	0.20
3	0	0.08	0.23
4	**0.0004**	**0.11**	**0.26**
5	**0.03**	**0.14**	**0.29**

The scoring of 1,000 cells is typical of the analysis effort that can be deployed on a case investigation by a microscopist seated at a conventional microscope. However computer assisted microscopy is now a reality and automatic metaphase finding has been available for many years (Finnon *et al.*, 1986). Currently the metaphase images are passed to an operator for review although the prospects for a fully hands-off "dicentric hunter" seem promising (Roy *et al.*, 2003). If, in the near future, a laboratory could deploy such a machine for biological dosimetry it would enable many more cells to be scored and intuitively this could reduce the minimum measurable or detectable limits.

2.1 BACKGROUND NOISE

If automation can remove the economic constraints on the number of cells scored consideration has to be given to what is an optimal number to analyse. Using the same parameters as in table 1 but increasing the scoring to, as an extreme example, one million cells the lowest *measurable* dose is reduced to 3.3 mGy (95% conf. limits: 0.1 and 6.4 mGy). However this is giving a deceptive impression of accuracy because the background level of dicentrics is not well known.

In the absence of a pre-accident control sample from the patient it is necessary to assume a generic value based on historical data from appropriate control subjects. The background dicentric level is normally assumed to be around 1/1,000 cells (IAEA, 2001) and this is represented by the control coefficient (0.001) in the yield equation shown above. However there is some inter-laboratory variability in published control frequencies (Lloyd *et al*., 1980) and in practice it could lie in the range 1/500 to 1/2,000.

An observation of 2000 dicentrics observed in one million cells would correspond to zero dose at the higher background frequency whilst at the lower end of the frequency range (I/2000 cells) it leads to a measured dose of 63 mGy (conf. limits; 60 and 66mGy). Given that for the patient being examined, the personal 'noise' is unknown, 63 mGy is really not much different from the 100 mGy lowest measurable dose shown in table 1. Thus scoring one million cells is far too many and will lead to a spurious impression of low dose detection accuracy.

2.2 THE OPTIMUM NUMBER OF CELLS TO SCORE

What is the cut-off point between scoring more cells to improve the real measurable lower limit rather than simply getting the spurious sensitivity as discussed above? The answer depends on the accuracy with which a yield may be converted to dose, i.e. the accuracy of the linear calibration coefficient (α). The 95% confidence limits shown in table 1 are based solely on the uncertainty that is associated with the patient's data; they take no account of the uncertainty in the calibration coefficients. However with more cells scored the uncertainty on the observation is substantially decreased and it is no longer justified to ignore the calibration errors. An examination of the published gamma-ray calibration curves from the leading laboratories shows that good curves have uncertainty on the linear coefficient of about 20% (1 standard error). In order to achieve this a substantial effort has to be made to produce a lot of calibration data below 0.5 Gy (IAEA, 2001). Since this is the limiting accuracy on a conversion to dose then little is gained by measuring a patient's dicentric frequency to better than 20 %, that is scoring about 25 dicentrics.

Table 2 shows a calculation for scoring 10,000 cells. The background value and calibration coefficients are as used previously in table 1. The minimum measurable dose, 67 mGy, occurs when the observation is 26 dicentrics and the standard error on this (√26) is about 20%. Thus in this example scoring 10,000 cells by machine is about the cut-off beyond which further scoring produces a progressively diminishing improvement in accuracy.

Table 2. Dose estimates and 95% confidence limits for various numbers of dicentrics scored in 10,000 cells with an assumed background of 1/1,000 cells. The yield equation is $Y=0.001 + (0.020 \pm 0.003)D + (0.060 \pm 0.002)D^2$. Confidence limits include uncertainties in the calibration curve. The yield and resultant dose estimate regarded as measurement are shown in bold.

Observed no. of dicentrics	Lower limit	Mean Dose	Upper limit
22	0	0.053	0.109
24	0	0.060	0.113
26	**0.006**	**0.067**	**0.117**

3. Conclusions

This paper has addressed the question of the ability of the dicentric assay to discriminate low doses of radiation. It is suggested that the lowest measurable dose is that value where its lower confidence limit is above zero. Lower derived dose values associated with dicentric frequencies in excess of background, but where the lower confidence limit is zero or less, can be considered as being detected but not measured.

The worked examples have used background dicentric frequencies and gamma ray calibration yield coefficients typical of several leading laboratories. Any individual laboratory may of course substitute its own data into the calculations to derive its own ability to discriminate low doses. Similarly one may substitute yield coefficients for other qualities of radiation.

It is concluded that by the traditional method of scoring 1,000 metaphases the lowest measurable γ-ray dose is around 100mGy. In this situation the uncertainty on the dose estimate is driven by the Poisson uncertainty on the observed number of dicentrics and the need to assume a generic background frequency. The uncertainty on a good dose response curve is relatively so small that it can be ignored.

As more cells are scored, with the aid of automated microscopes, the sensitivity of the technique is improved but, perhaps surprisingly, not greatly. This is because as the statistics on the individual dicentric frequency improve it is no longer justified to ignore the calibration uncertainty. Freed from the economic constraints of microscopist time, it is suggested that the optimum number of cells to analyse is about 10,000 at which point the uncertainties from the sample and from the linear yield coefficient contribute more-or-less equally to the overall error on the dose estimate. Substantially more cells than this leads to no real improvement because of dicentric background 'noise'. The optimum use of automation might permit the lowest measurable dose of 100mGy of γ-rays achievable by conventional scoring to reduce to around 70mGy.

References

Finnon P., Lloyd D.C., Edwards A.A., 1986, An assessment of the metaphase finding capability of the Cytoscan 110. *Mutation Research* **164**: 101–108.

IAEA., 2001, *Cytogenetic Analysis for Radiation Dose Assessment. A Manual.* Technical Report Series 405., Vienna: International Atomic Energy Agency.

ICRP, 1991, 1990 Recommendations of the International Commission on Radiological Protection. ICRP Publication 60, *Annals of the ICRP* **21** (1–3).

Lloyd D.C., Purrott R.J., Reeder E.J., 1980, The incidence of unstable chromosome aberrations in peripheral blood lymphocytes from unirradiated and occupationally exposed people. *Mutation Research* **72**: 523–532.

Lloyd D.C., Edwards A.A. and Szluinska M. The minimum detectable dose by biodosimetry in a radiation accident. *International Journal of Low Radiation,* (in press).

Roy L., Delbos M., Paillole N., Durand V., Voisin P., 2003, Comparison de systèmes d'analyse d'images cytologiques en dosimétrie biologique. *Radioprotection* **38**:323–340.

A SIMPLE METHOD FOR THE EVALUATION OF SIDE DOSES IN RADIOTHERAPY

ARRIGO A. CIGNA
Past President Union Internationale de Radioécologie
Strada Bottino 2, Frazione Tuffo
I-14023 COCCONATO (Asti) Italy
E-mail: arrigocigna@tiscali.it

Abstract. Radiation protection is taken into account in radiotherapy practice from the point of view of the medical aspect in order to achieve the best ratio between the tumour control probability (TCP) and the normal tissue complication probability (NTCP). But any radiotherapy treatment implies a dose delivered to the patient's body also outside the beams. Therefore, there is a certain interest important to quantify that which can be considered a negative impact of radiotherapy also if it cannot absolutely be avoided and, in any case, the benefit of radiotherapy should prevail.

The dose delivered to the patient by the radiation scattered outside the direct beams has been measured in general in phantoms. This procedure has the advantage of avoiding the direct involvement of the patient but, on the other hand, a phantom implies a certain degree of approximation in the reproduction of the real characteristics of a human body.

In particular rather inaccurate phantoms are normally used. This simple solution supplies reliable results to some specific questions as, e.g., the measurement of the collimator scatter or the dose distribution in the penumbral region. But when the dose to the whole body or to a single organ has to be evaluated, such phantoms are totally unreliable. In fact the dose distribution within the body varies notably with the distance from the target volume and the average value can be identified with difficulty if only one value is available.

The scope of this study is the description of a simple method for the evaluation of the dose, outside the beams, delivered to the body of a patient submitted to external radiotherapy by means of direct measurements. The radiation scattered by the target was monitored by means of pen dosimeters distributed along the body. A virtual model was obtained by dividing the body into about 800 small volumes where the doses could be assumed uniform. The power functions obtained by the best fit of the experimental values were used to calculate the effective dose delivered to the body and single organs due to scattering only.

A.A. Cigna and M. Durante (eds.),
Radiation Risk Estimates in Normal and Emergency Situations, 259–268.

Keywords: radiotherapy, scattered radiation, side doses evaluation

1. Introduction

The dose delivered to the patient in radiotherapy practice by the radiation scattered outside the direct beams has been measured in general in phantoms (Ing *et al.*, 1982; ICRP, 1985; Van Gasteren *et al.,* 1991; Kron *et* al., 1993; Miah *et al.*, 1998; Lope Lope *et al.*, 2001; Roy *et al.*, 2001; Sung, 2001). This procedure has the advantage of avoiding the direct involvement of the patient but, on the other hand, a phantom implies a certain degree of approximation in the reproduction of the real characteristics of a human body.

In particular rather inaccurate phantoms are normally used. This simple solution supplies reliable results to some specific questions as, e.g., the measurement of the collimator scatter (Van Gasteren *et al.*) or the dose distribution in the penumbral region. But when the dose to the whole body or to a single organ has to be evaluated, such phantoms are totally unreliable. In fact the dose distribution within the body varies notably with the distance from the target volume and the average value can be identified with difficulty if only one value is available.

An interesting improvement was proposed by Diallo *et al.* (1996) with a dose calculation programme, which uses a patient phantom based on sex gender, and height. Such a phantom contains 151 anatomical points of interest and therefore supplies results somewhat more accurate than those provided by simple and rough phantoms.

Since a recent generation of conformal radiotherapy, the Intensity-modulated radiotherapy (IMRT), could imply a still higher dose to the whole body from the scattered radiation, the evaluation of the dose delivered by scattered radiation become still more important.

Since the author had the chance to be treated for a prostate cancer by radiation therapy, the radiation scattered by the target was monitored by means of pen dosemeters distributed along the body (Cigna *et al.*, 2004). This very simple method is described.

2. Material and Method

The dose delivered by the radiation scattered by the target was measured by means of pen dosemeters manufactured by the Bendix Corporation, CD V-736 (0-200 mR) and CD V-746 (0-600 R) were used (Fig. 1).

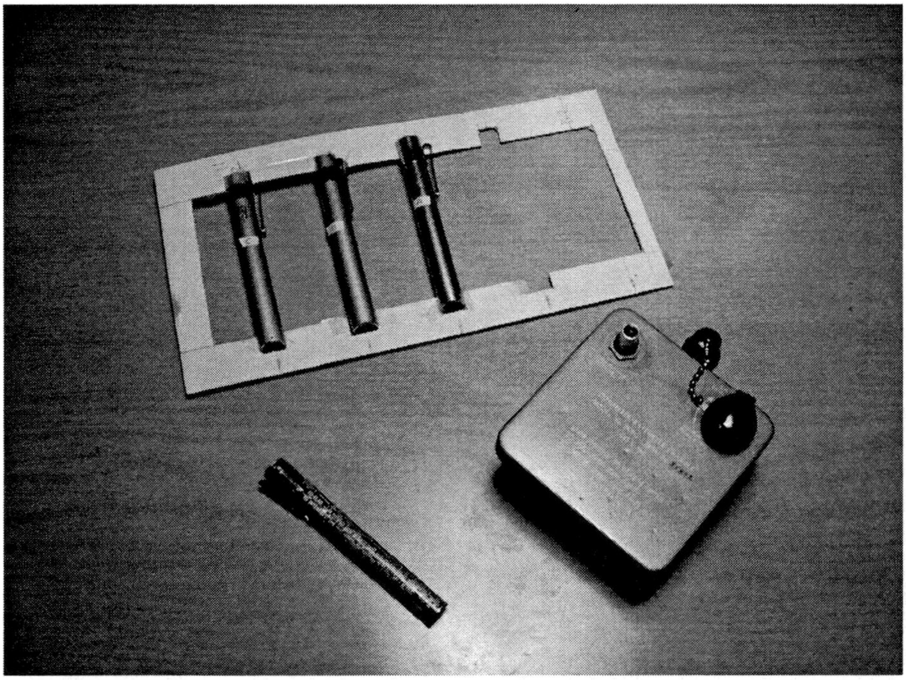

Figure 1. The pen dosemeters used to measure radiation scattered by the target. In the array three 0-200 mR dosemeters; lower left a 0-600 R dosemeter and lower right the charger.

The former was placed at different points along the body of the patient, while the latter was placed 2 cm from the each edge of the beam.

Since the condenser ionisation chambers have been calibrated in roentgens, a conversion factor, D/X, from the exposure in roentgens to the absorbed dose in water in milligrays, was calculated. By assuming a conversion factor of 8.73 mGy/R for the absorbed dose in air (Johns and Cunningham, 1983) and the mass absorption coefficients reported by Hubbel (1982) the conversion factor for the absorbed dose in water (mGy/R) ranges through 9.69 at 0.2 MeV, 9.71 at 1 MeV and 9.65 at 4 MeV. Therefore, it was assumed a rounded conversion factor

$$D/X = 9.7 \text{ mGy/R}$$

The radiation weighting factor for photons was assumed to be unity and, therefore, equivalent doses in mSv are here reported.

The dosemeters were zeroed by means of a CD V-756 dosemeter charger, and proved quite stable and reliable during the whole procedure. It must be emphasised that the pen dosemeters were placed along the body of the patient by keeping the same orientation with respect to the beam (Figs. 2 and 3).

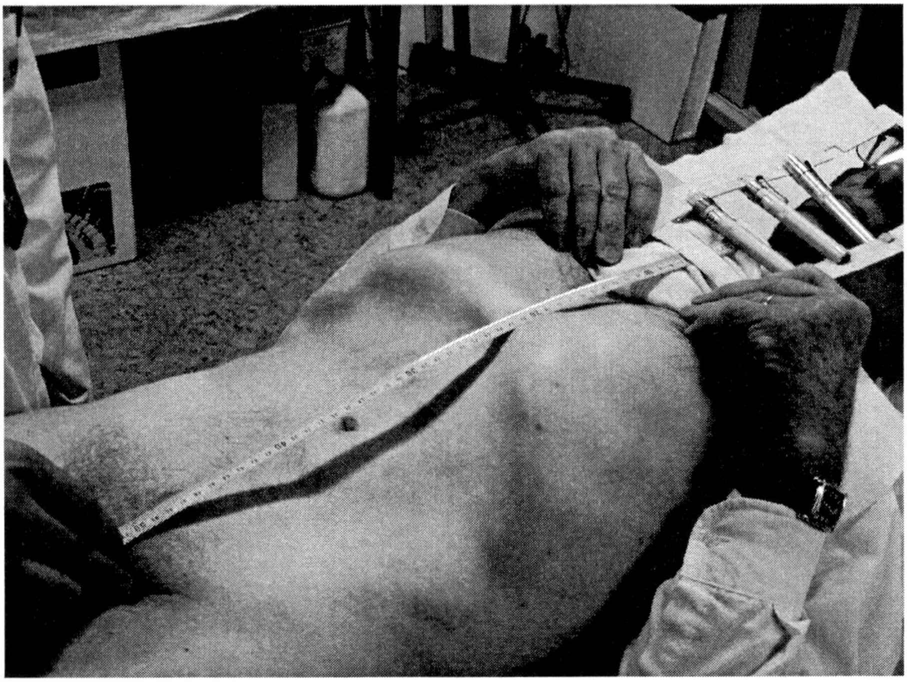

Figure 2. Positioning of the pen dosemeters towards the head.

The performance against a secondary standard was verified at the Calibration Laboratory of the Joint Research Centre of Ispra (Italy) by keeping the same geometry. The agreement was rather good, within a few percent.

Therefore, notwithstanding a generally poor angular and energy response of pen dosemeters, in the present study a standard deviation of 15% for each measurement could be considered, in order to take into account also other sources of uncertainty as the measurement of the distance from the centre of target volume.

The radiation scattered outside the beams is due to an external component produced by any device (collimators, couch, etc.) in the vicinity of the beam and an internal one produced within the patient. In the case described here, the internal component prevails with respect to the external one because measurements above and below the body had no detectable difference, the dose depending on the distance from the centre of target volume only.

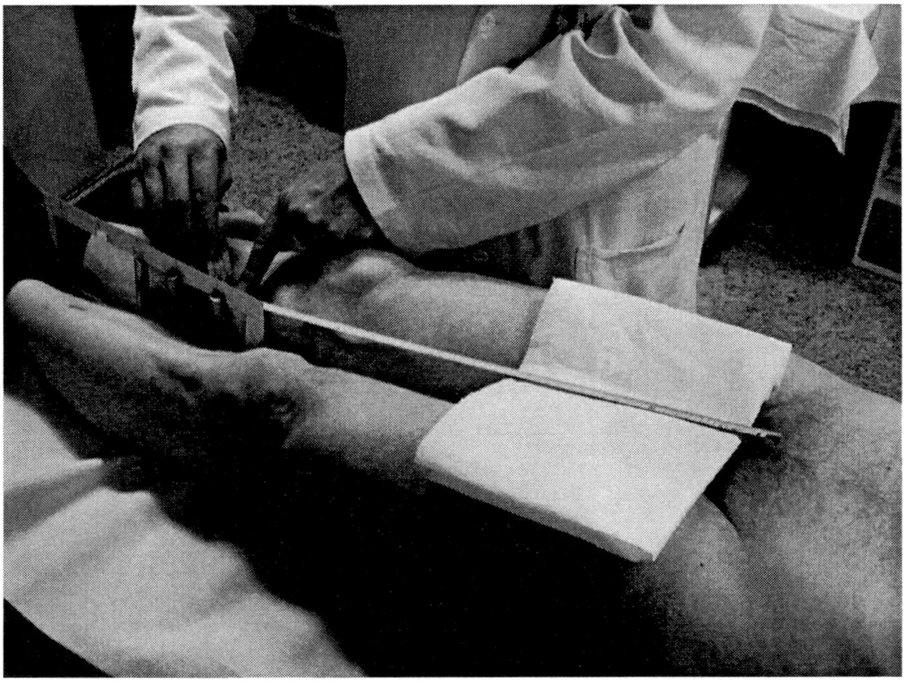

Figure 3. Positioning of the pen dosemeters towards the feet.

The radiation scattered outside the beams is due to an external component produced by any device (collimators, couch, etc.) in the vicinity of the beam and an internal one produced within the patient. In the case described here, the internal component prevails with respect to the external one because measurements above and below the body had no detectable difference, the dose depending on the distance from the centre of target volume only.

Because of the conceptual difference between therapy and radiation protection and of the different biological mechanisms involved, only the regions not affected by the beams were considered.

A virtual model of the patient body was designed and subdivided into slices 1 cm thick. To avoid a too large approximation, slices of "Chest and abdomen" were subdivided into 8 parts, 9 cm x 9 cm each, and slices of "Thigh" into two halves, external and internal respectively. As a result of this procedure, 736 and 744 slices for the first and the second beam arrangement, small enough to be considered homogenous from the point of view of the dose (Fig. 4), were assumed.

On the basis of the treatment plan, CT and normal radiography, the position of each organ could be determined with reference to the centre of the target volume. Then each organ was also divided into slices 1 cm thick by applying the same procedure adopted for the whole body. A total of 251 slices for 12 organs were assumed.

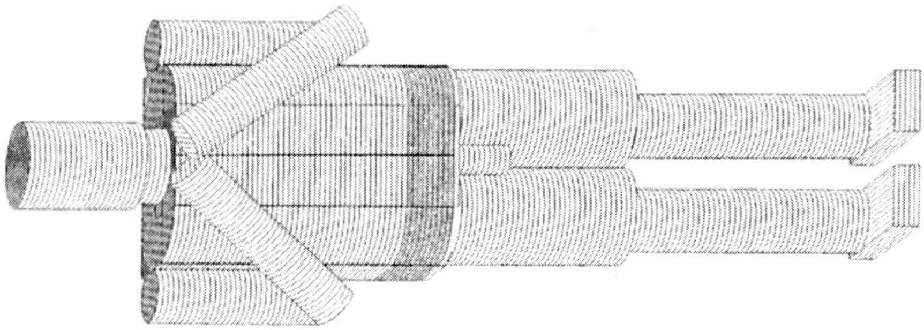

Figure 4. Virtual model of the patient. The darker band on the abdomen corresponds to the area interested by the beams and not considered in the calculation of the mean whole body dose delivered by the scattered radiation.

In principle, the attenuation of the scattered radiation inside the body has two components. One is purely geometrical and is the inverse square law attenuation. The other is due to the interaction processes with matter and is exponential. But, on account of the geometry involved in the present study the size of the source (target volume) is of the same order of magnitude as the distances where the scattered radiation was measured and where the radiation is subject to multiple scattering. Therefore it was assumed to be unrealistic to apply rigorously both the inverse square law attenuation and the physical attenuation for a narrow beam.

For the reason reported above, empirical functions, as a best fit interpolating the experimental measurements, are searched. Such functions must be found for each irradiation arrangement and to cover each part of the body. E.g. in the case of the prostate radiotherapy (Cigna *et al.*, 2004), if D is the dose (mSv) delivered by the scattered radiation at the distance d (cm) from the centre of the target volume and d is defined as:

$$d = \{(x-x_0)^2 + (y-y_0)^2 + (z-z_0)^2\}^{1/2}$$

when x_0, y_0, z_0 are the co-ordinates of the centre of the target volume, and A is a constant, three functions of the type

$$D = Ad^{-f}$$

were found:

1 - First beams arrangement (Box: 100 cm SSD). Dose delivered by the radiation scattered by the target vz. distance d from the target centre (both towards the feet and the head) (Fig. 5).

2 - Second beams arrangement (Boost: 100 cm Isocentric). Dose delivered by the radiation scattered by the target towards the feet vz. distance d from the target centre.

3 - Second beams arrangement (Boost: 100 cm Isocentric). Dose delivered by the radiation scattered by the target towards the head vz. distance d from the target centre.

These empirical functions agree rather well with the relationships expected for the dose due to scattered radiation.

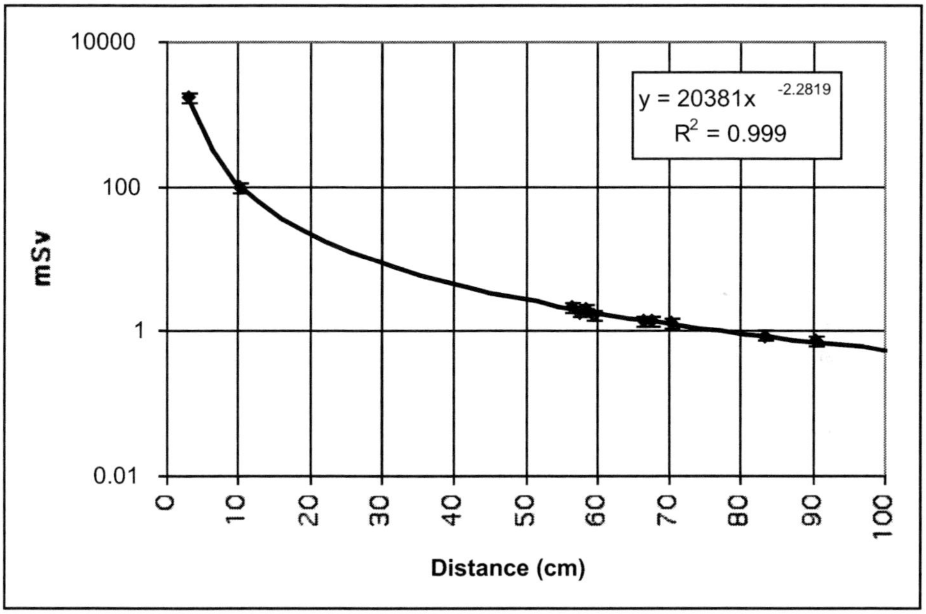

Figure 5. Example of an empirical function. Dose delivered by the radiation scattered by the target vz. distance from the target centre (Cigna *et al.*, 2004).

3. Results

In an Excel spreadsheet each slice of the virtual body described above was reported and its specific data were calculated. In Table 1 the slices are summarised. Obviously this method can be easily adapted to any other irradiation arrangement. Its advantage is the great simplicity since the hardware is reduced to a minimum and the software is available in any computer.

Table 1. Total body: number of slices.

Part of the body	Number of slices
Head	23
Neck	5
Upper arm	33 (x 2 = 66)
Forearm	40 (x 2 = 80)
Chest and abdomen	46 (x 8 = 368)
Gonads	10
Thigh	35 (x 2 = 70)
Leg	47 (x 2 = 94)
Foot	5 (x 2 = 10)

For each slice the following parameters were calculated:

Co-ordinates (x,y,z with reference to the origin assumed to be in the centre of the target volume)
Distance (from the origin)
Volume (cm^3)
Dose (mSv)

A similar procedure was also adopted to evaluate the dose to single organs as summarised in Tables 2 and 3.

Table 2. Organs: number of slices.

Organ	Number of slices
Brain	9
Thyroid	2
Lung	29
Spleen	5
Oesophagus	31
Stomach	6
Liver	12
Colon	13
Kidneys	16
Testes	6

Table 3. Bone marrow (red)(*): number of slices.

Bone marrow (red)	Number of slices
Cranium	12
Mandible	4
Scapulae	32
Clavicles	4
Sternum	21
Ribs	24
Cervical vertebrae	14
Thoracic vertebrae	24
Lumbar vertebrae	10
Humeri (upper half)	17

(*) Excluding the section partially in the beam (which corresponds to about 40% of the total).

A copy of the Virtual Model is available from the author.

Acknowledgements

The author is grateful to Prof. G. Badino for the fruitful discussion, to Prof. E. Casnati for his most useful suggestions, to Mr. D. Walker for the verification of the calibration of the dosemeters. The support by the whole staff of the Div. of Radiotherapy, ASL 19, Asti, Italy for the careful co-operation during the treatment, is here gratefully acknowledged.

References

Cigna A.A., Nassisi D., Masenga D., Raffo R. and Rotta P., 2004, Dose due to scattered radiation in external radiotherapy: a prostate cancer history. *Radiation Prot. Dosim.*, **108**(1): 27–32.

Diallo I., Lamon A., Shamsaldin A., Grimaud E., de Vathaire F. and Chavaudra J., 1996, Estimation of the radiation dose delivered to any point outside the target volume per patient treated with external beam radiotherapy. *Radiotherapy and Oncology* **38**(3): 269–271.

Hubbell J.J., 1982, Photon mass attenuation and energy-absorption coefficients from 1 keV to 20 MeV. *Int. J. Appl. Radiat. Isot.*, **33**: 1269.

ICRP. 1985, Protection of the Patient in Radiation Therapy. Publication 44. *Ann. ICRP*, **15** (2) Pergamon Press, Oxford, 1985.

Ing H., Nelson W.R. and Shore R.A., 1982, Unwanted photon and neutron radiation resulting from collimated photon beams interacting with the body of radiotherapy patients. *Med. Phys.* **9**(1): 27–33.

Johns H.E. and Cunningham J.R., 1983, *The Physics of Radiology*. C.C. Thomas, Springfield.

Kron T., Elliot A. and Metcalfe P., 1993, The penumbra of a 6-MV x-ray beam as measured by thermoluminescent dosimetry and evaluated using an inverse square root function. *Med. Phys.* **20** (5): 1429–1438.

Lope Lope R., Lozano Flores F., Gracia Sorrosal J., Font Gómez J.A. and Hernández Vitoria A., 2001, Peripheral dose in photon beams from a linear accelerator with a multileaf collimator. *Int. Conf Radiological Protection of Patients in Diagnostic and Interventional Radiology, Nuclear Medicine and Radiotherapy. Malaga, Spain,* Vienna IAEA Proceedings Series: IAEA-CN-8–97.

Miah F.K., Ahmed M.F., Begum Z., Alam B. and Chowdhury Q., 1998, Dose distribution over different parts of cancer patients during radiotherapy. *Rad. Prot. Dosim.* **77**(3): 199–203.

Roy S., Ambia A.S.M., Akhter S., Banu H. and Begum M., 2001, Dose distribution over the radiation filed and organs of the body during radiotherapy procedures. *Int. Conf Radiological Protection of Patients in Diagnostic and Interventional Radiology, Nuclear Medicine and Radiotherapy. Malaga, Spain,* Vienna IAEA Proceedings Series: IAEA-CN-85–36.

Sung S.C., 2001, Shielding for radiation scattered dose distribution to the outside fields in patients treated with high energy radiotherapy beams. *Int. Conf Radiological Protection of Patients in Diagnostic and Interventional Radiology, Nuclear Medicine and Radiotherapy. Malaga, Spain,* Vienna IAEA Proceedings Series: IAEA-CN-85–70.

Van Gasteren J.J.M., Heukelom S., van Kleffens H.J., van der Laarse R., Venselaar J.L.M. and Westermann C.F., 1991, The determination of phantom and collimator scatter components of the output of megavoltage photon beams: measurements of the collimator scatter part with a beam-coaxial narrow cylindrical phantom. *Radiotherapy and Oncology*, **20**: 250–257.

Mutations

MUTATION AND DNA REPAIR: FROM THE GREEN PAMPHLET TO 2005

JOHN W. DRAKE
Laboratory of Molecular Genetics, National Institute of Environmental Health Sciences, Research Triangle Park, North Carolina 27709-2233 USA
drake@niehs.nih.gov

Abstract. The early conceptual basis of gene structure and mutation were laid down in a handful of seminal papers, one of the best remembered being the "Green Pamphlet" of 1935 by N.V. Timofeeff-Ressovsky, K.G. Zimmer and M. Delbrück entitled "On the Nature of Gene Mutation and Gene Structure". The concepts of a molecular basis for the gene and its mutability have expanded hugely since then and have generated the entire field of DNA repair. Here, two current insights into DNA repair and mutation are displayed. Replication repair is a newly established mode of recombination repair that works through a copy-choice (template-switching) mechanism and that has been reconstructed in vitro using the enzymes of DNA replication elaborated by bacteriophage T4. In contrast, a pathological mode of primer-strand switching generates templated complex mutations, and these greatly increase in frequency when the shepherding proteins of replication repair are disabled. At the same time, many spectra are found to contain a different kind of complex mutation whose components are more widely scattered. These are argued to arise mostly through transient bouts of hypermutation and are likely to contribute to carcinogenesis and to the virulence of microbial pathogens.

Keywords: DNA repair, replication repair, templated mutations, multiple mutations

1. Introduction

Scientists who discover something new often later speak about an almost instantaneous transition from a mental vacuum of understanding to a state where it is all so obvious that remembrance of vacuums past is impossible. A reviewer of one of my papers once wrote something to the effect that the whole thing seemed perfectly obvious, even trivial, until further reflection revealed that the matter had been completely opaque the day before. Those

A.A. Cigna and M. Durante (eds.),
Radiation Risk Estimates in Normal and Emergency Situations, 271–281.

few of you who personally remember the mindset of 1935, when the Green Pamphlet (Timofeeff-Ressovsky, Zimmer, & Delbrück, 1935) described certain then very modern ideas about the nature of the gene, would understand how the idea that the gene was a molecule seemed novel. Soon that idea took hold in key minds and spread rapidly, but for half a century now has seemed ordinary rather than revolutionary. I don't know where the next such insight will surface, but it might well be when a neurobiologist explains the molecular and cellular basis of consciousness.

Although a great deal of complex experimentation was practiced in the decades of the 1930s–1950s, I have the impression that those years were dominated by a desire for the simplicity that often accompanies deep insights. The target theory created by Timofeeff-Ressovsky, Zimmer and Delbrück was fundamentally simple and could be comprehended almost immediately, whereas its modern equivalent takes into account innumerable details happily unknown in the early 1930s. The concept of DNA repair was latent in a number of papers from the 1930s and 1940s but was not foreseen by even such prescient scientists as Jim Watson and Francis Crick. Here I will lay out two lines of investigation that started with a simple idea or observation and quickly progressed to insights that were not anticipated, at least by me.

2. Replication Repair

A cell that is trying to replicate its DNA past a severe polymerase-blocking lesion on the template strand faces a good chance of death or mutation. By the middle 1970s, the *Escherichia coli* paradigm of recombination repair was becoming well established: it is an efficient cut-and-paste process that obtains the requisite genetic information from the other parental strand. However, this kind of mechanism has still not been described in eukaryotes. Higgins, Kato & Strauss (1976) proposed an elegant alternative way in which DNA damage could be circumvented (Fig. 1). A blocked primer strand could switch templates, jumping from the cognate parental strand to the other daughter strand. This switch would be facilitated by first backing up the replication fork a little way, although other routes to the same end could be imagined. The primer strand could then elongate briefly on the ectopic template, perhaps until it reached the 5′ end of the parental strand, and could then switch back to the cognate template, having bypassed the lesion. This is a copy-choice mode of recombination repair that does not involve strand cutting, although it would certainly require some shepherding. The HKS model was supported by somewhat indirect experimental results because they were obtained with a mammalian cell system that was amenable to neither genetic nor enzymatic analysis. The model has been invoked many times since then, but only recently has it received direct experimental support. In my laboratory in the 1980s, a new epistasis group was described in bacteriophage T4 whose mutant alleles increase sensitivity to the killing action of diverse DNA-damaging agents but are distinct from alleles affecting excision repair or classical recombination repair (the *uvsWXY* system) (Wachsman & Drake, 1987).

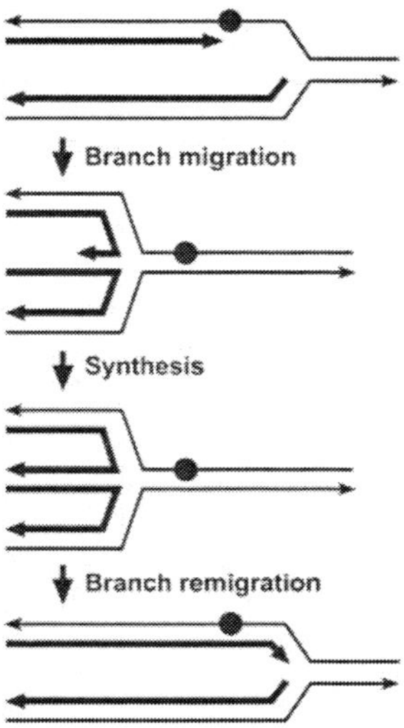

Figure 1. Replication repair. At the top, the extension of a primer strand is blocked by a lesion in the template strand. Next, the replication fork migrates backwards and the two displaced daughter strands pair. The previously blocked daughter strand can then extend for a short distance. Finally, the replication fork rearranges into a normal configuration and the primer strand is seen to have accurately bypassed the lesion.

The newly characterized mutations fell in genes encoding various proteins of DNA replication, notably gp32 (the SSB protein that binds to single-stranded DNA) and gp41 (the replicative DNA helicase). It should be mentioned here that T4 encodes most of the genes its needs for DNA transactions, and that these genes resemble their eukaryotic and archaeal counterparts more than their eubacterial equivalents. Much more recently, Farid Kadyrov, a Tartar transplanted from Moscow Region to North Carolina, set up an assay for detecting template switching using several of the proteins of T4 DNA replication and an artificial replication fork bearing a polymerase-blocking lesion a little ahead of one of the primer strands (Fig. 2).

By following the size of the labeled primer strand, he found that the blocked primer terminus quickly extends up to the block, then switches to the other daughter strand with the assistance of gp32 and gp41 and extends as far as its 5′ end, but in this first system the primer did not then switch back to its cognate template (Kadyrov & Drake, 2003). The key observation was that the proteins encoded by alleles defective in replication repair (*32mms* and *41uvs79*) could not facilitate the switch. In contrast, when the T4-encoded recombinase UvsX and the T4-encoded DNA helicase Dda replaced gp32 and gp41, both switches were catalyzed (Kadyrov & Drake, 2004). Genetic tests had placed *32mms* and *41uvs79* in one epistasis group and *uvsX* in another, so that two pathways of replication repair must operate during T4 DNA replication, a surprising result that we still do not understand. Furthermore, because the classical UvsWXY system catalyses both ordinary genetic recombination and also recombination repair (that is, survival after DNA damage even under conditions of single infection), it is possible that all recombination repair in T4 is replication repair.

Figure 2. An artificial replication fork. Strand #3 is the elongating primer and is ^{32}P-end labeled. The dot in strand #4 is a polymerase-blocking lesion resembling an abasic site and the following "g" is another blocking lesion because dCTP was not included in the reaction mixture. Strand #2 is the other daughter strand that is used as an ectopic template.

This contrasts sharply with the situation in *E. coli*, where at least a lot of recombination repair occurs by a cut-and-paste rather than a copy-choice mechanism. Because the former mechanism could not be detected in yeast (Resnick, Boyce & Cox, 1981), and because proteins involved in T4 DNA transactions are more closely related to those in archaea and eukaryotes, it is possible that the eubacteria perform most of their recombination repair differently from organisms in the other two domains. It is also possible that *E. coli* does indeed carry out some replication repair, because inactivating the *E. coli recA* system still allows numerous lesions to be survived in phage λ (Defais *et al*., 1989), and because of ingenious genetic evidence that supports the operation of replication repair (Ozgenc, Szekeres & Lawrence, 2005).

Students of the mutation process soon discover that there is a class of mutations, widely seen in diverse spectra but usually in small numbers that seem to bear some relation to template switching. Many of these mutations are complex, that is, they consist of several base-pair substitutions or small indels that arise together and are located close to one another. Frequently the same cluster appears several times. When one searches the nearby sequences, it is often possible to discover a sequence that could have templated the complex mutation. Certain kinds of complex mutations, first described by Lynn Ripley (1982), arise in the context of imperfect reverse repeats, usually with a spacer sequence between the two repeats. She called these imperfect reverse repeats quasipalindromes, which have the general sequence structure.

$$\mathtt{XXXXABCDEFGXXXXXXX\underline{G}\underline{F}\underline{E}W\underline{C}\underline{B}\underline{A}XXXX}$$

where $\mathtt{X}$ is any base, pairs such as $\mathtt{A}$ and $\mathtt{\underline{A}}$ are complementary bases such as A and T, and the pair $\mathtt{D}$ and $\mathtt{W}$ are non-complementary bases such as A and G, or a base with no counterpart or with two or more bases as the counterpart. Note first that the complement of the right copy of the inverse repeat is the sequence `ABCWEFG` which, if used to replace `ABCDEFG`, causes the mutation `D` → `W`. If the two components of the inverse repeat contain several non-complementary bases, then a complex rather than a single change will result.

As Ripley pointed out in her canonical paper on the subject (1982), one can imagine two different ectopic templates that could produce such mutations. One such template is the primer strand itself: if it copies the first component of an imperfect reverse repeat and the spacer sequence, then switches back on itself to copy the complement of that same first component, then reverts to using the parental strand as its template, some change(s) will be introduced into, in this case, the second copy of the inverse repeat. This scheme is often drawn as a hairpin-like structure and is easy to visualize mentally. The other ectopic template is the other parental strand, and this is less easy to visualize.

Because templated mutations are infrequent, there have been few focused studies concerning the mechanisms that generate them. In studies with phage T4 and its cousin RB69, we have been using the T4 *rI* gene as a mutator reporter because it has an appropriate size (291 translated bases) and many of its missense mutations produce a phenotype that is easily recognized (Bebenek *et al.*, 2001). It happens that about 0.2 of the spontaneous mutations of this gene are complex mutations of the form GCG → CTA at residues 146-148. The relatively high frequency of these mutations made it possible to ask some interesting questions about their origins. This mutation can be modelled by the switchback scheme mentioned above, or when the ectopic template is the other parental strand as shown in Fig. 3.

Our interest in templated mutations followed naturally from our work on replication repair, which also depends on template switches. We considered two hypotheses, each of which is to be contrasted with the no-effect null hypothesis. In our first hypothesis, complex mutations are errors of template choice within the general scheme of replication repair. Under this hypothesis, mutationally disabling replication repair would be antimutagenic for templated mutations, and our hotspot would be depleted. In our second hypothesis, spontaneous melting of primer termini would suffice to initiate most templated mutations, and the role of the proteins that conduct replication repair would be to shepherd the primer terminus to the correct ectopic template. Under this hypothesis, mutationally disabling replication repair would be mutagenic for templated mutations. The experiments were simple albeit labor-intensive: T4 *rI* mutational spectra were determined in the wild type and in *32mms*, *41uvs79*, and *uvsX* T4 genetic backgrounds.

A complex mutation arising at T4 *rI* 146-148: GCG → CTA

The quasipalindrome:

AAATAAAA**GCG**TAGAATCGTCTGAACAATTCTA**TAG**TTTTATGA

Two parental strands have a quasipalindrome near a replication fork. The primer end
of one daughter strand melts and finds its complement on the other parental strand:

5′-TTTATTTTCGCATCTTAGCAGACTTGTTAAGATATCAAAATACT-3′
TTTTATTT-5′

5′-TTTATTTT →

3′-AAATAAAAGCGTAGAATCGTCTGAACAATTCTATAGTTTTATGA-5′

The primer strand then extends by, say, 10 bases:

5′-TTTATTTTCGCATCTTAGCAGACTTGTTAAGATATCAAAATACT-3′
AATTCTATAGTTTTATTT-5′

The extended primer strand melts away and reanneals with its cognate complement:

5′-TTTATTTTGATATCTTAA
3′-AAATAAAAGCGTAGAATCGTCTGAACAATTCTATAGTTTTATGA-5′

The terminal mismatch is excised and the primer strand is extended:

5′-TTTATTTTGATATCTTAGCAGACTTGTTAAGATATCAAAATACT-3′
3′-AAATAAAAGCGTAGAATCGTCTGAACAATTCTATAGTTTTATGA-5′

The next round of replication converts the former daughter strand into a homoduplex:

5′-TTTATTTTGATATCTTAGCAGACTTGTTAAGATATCAAAATACT-3′
3′-AAATAAAA**CTA**TAGAATCGTCTGAACAATTCTA**TAG**TTTTATGA-5′

Figure 3. The mutational consequences of an imperfect 15-base reverse repeat in the T4 *rI* gene associated with a hotspot of complex mutations. The two components of the repeat are each underlined except for the central 3-base imperfection. The left portion of the repeat is bases 140-154 and the right portion is bases 165-179.

Table 1. Rates of templated mutation are increased in phage T4 by mutations that disable replication repair.

T4 genotype	Mutation rate (GCG → CTA)	Relative rate
Wild type	2.6×10^{-7}	1
32mms	2.1×10^{-5}	82
41uvs79	4.6×10^{-6}	18
uvsX	1.8×10^{-6}	7

Some of the results are shown in Table 1, from which it is clear that all three mutations increase the rate of templated mutation.

Mutations modeled on reverse repeats with unchanged spacers:

A T A
GCACG**T**TT**A**TA**T**GCTTTTTGAAAATAAAAGCGTAGAATCGTC
T

ATATAAA
GCACG**TTTATAT**GCTTTTTGAAAATAAAAGCGTAGAATCGTC
T

Mutations modeled on reverse repeats with inverted spacers:

A T TCAG G
AGCGTAGAAT**C**GT**CTGA**AC**A**ATTCTA**T**AGT

TGTTCAGACG
CGTAGAAT**CGTCTGAACA**ATTCTATA

A mutation modeled on a distant direct repeat

GAGA
ACGCGGAG**GAGA**GCACAATAC......GGCGCGGAG ATG
GCACAATC

Figure 4. Other complex mutations. The mutational changes are shown above the local wild-type *rI* sequence. The two components of the repeat are underlined where they are fully complementary. The two sequences in the top panel show a complex mutation first only where it differs from the wild type and then as they presumably arose in a single block requiring only two switches. Donor sequences are in bold face.

These results support our second hypothesis, which is that a key role of the proteins gp32, gp41 and UvsX is to shepherd the primer strand to the correct ectopic template. The same mutations also increase the mutation rate to other kinds of templated mutations in the T4 *rI* gene, some of which are shown in Fig. 4. A signature that distinguishes template switches to the same primer strand versus to the other parental strand is the impact on the spacer sequence between the two components of a reverse repeat. When (and only when) the ectopic template is the other parental strand, the spacer sequence can be inverted. The two mutations in the bottom panel of Fig. 4 are of this type.

3. Other Nonrandom Clusters of Mutations

Complex mutations are always closely clustered, usually within no more than about 20 bases. However, it has recently become clear that another kind of cluster is often seen in mutational spectra, again usually in small numbers of mutants (Drake et al., 2005). These mutants contain two or more mutations that are usually well separated, sometimes by hundreds or thousands of bases. If the mutations in a collection of mutants are distributed at random, then the average number of mutations per mutant can be obtained from the fraction of mutation-free individuals, which will correspond to the first term of the Poisson distribution, $e^{-a} = 1 - F$, where F is the mutant frequency. The expected fraction of mutants with exactly i mutations is $a^i e^{-a}/i!$ If one sequences M mutants, the population size that yielded those M mutants is M/F, and the expected number of doubles (mutants with two mutations) is $E_2 = Ma^2e^{-a}/2F$. The expected number of mutants with three mutations is $E_3 = E_2a/3$, and so on.

When a set of mutational spectra assembled for other purposes was surveyed for "multiples", many spectra were found to contain them in substantial excess over the expectations of a random distribution of mutations among mutants. Some examples are given in Table 2, from which we can see certain patterns. First, multiples are found among mutational spectra obtained from all of the model organisms usually used by students of the mutation process, including RNA and DNA viruses, the usual bacterium, the usual yeast, and a number of mammalian cells in culture or mammalian tissues. Second, excess numbers of multiples are also found in spectra based on a DNA polymerase copying a template strand in vitro.

One can imagine two reasons for these clusters. One is mutator mutants that have accumulated in the population. However, at least for microbes, mutators are at a selective disadvantage under most laboratory conditions because many of their offspring are mutationally disabled. In *E. coli* and *S. typhimurium*, for instance, the frequency of mutators is $\leq 10^{-5}$ in laboratory cultures. Now, even strong mutator mutations increase the value of F by at most about 100-fold, because F is the mutation frequency across all of a reporter gene that typically contains only a few sites whose mutation frequency is strongly enhanced plus many sites in which it is only weakly enhanced.

Table 2. Too many mutants with multiple mutations.

System	Reporter	F	M	E_2	$D+$
Tobacco mosaic virus	*MP*	4.3×10^{-2}	17	0.36	3+3
Bacteriophage T4/RB69	*rI*	2.0×10^{-2}	72	0.72	3
Herpes simplex virus	*supF*	4.9×10^{-4}	80	0.020	7+1
Escherichia coli	*lacI*^{d}	1.3×10^{-7}	368	0.000024	2
Saccharomyces cerevisiae	*SUP4*-o	1.9×10^{-6}	297	0.00028	2
Chinese hamster cell line	*gpt*	1.2×10^{-4}	18	0.0011	2
Human cell line	*hprt*	9.0×10^{-6}	200	0.0009	6+1
Mouse tissue	*lacI*	2.3×10^{-5}	435	0.0050	7+1
Human tissue	*HPRT*	1.9×10^{-4}	82	0.0078	5+1
T4 DNA polymerase	*lacZα*	1.1×10^{-2}	121	0.65	2
E. coli Pol I(KF)	*lacZα*	4.7×10^{-3}	118	0.28	3
HIV-1 RT	*lacZα*	1.3×10^{-1}	97	6.2	19+1
Rat Pol β	*lacZα*	1.1×10^{-1}	296	15.7	≤16

"Reporter" is the gene in which mutations were scored. "D +" is the number of observed doubles + higher multiples. Other details of the systems and references are provided elsewhere (Drake, 2005).

Thus, the frequency of doubles produced by mutator mutants will be $\leq (100F)^2(10^{-5}) = F/10$ and the fraction of doubles produced by mutators among all observed doubles D will be $\leq FM/10D$. This fraction is <0.1 for all of the entries in Table 2 for which $D \gg E_2$. Therefore, mutator mutations are unlikely to contribute very much to the observed excess of doubles (and higher-order multiples). Instead, the multiples are likely to arise during transient intervals of phenotypic hypermutation. For instance, errors of transcription and translation, and perhaps post-translational folding or processing, will sometimes produce faulty copies of proteins involved in DNA transactions. Such proteins, for instance, would include DNA polymerases, and it is well known that defects in either the polymerase or proofreading-exonuclease activities of replicative DNA polymerases are often mutators. Such a mutator polymerase could then synthesize a portion of the chromosome before turning over, producing a cluster of mutations.

We have explored two ways to decompose such a spectrum into two subpopulations that would generate the observed distributions of singles and multiples. Let the subpopulations be S_1, f_1 and S_2, f_2 where S_i is the fraction of the subpopulation and f_i is its mutation frequency ($S_1 > S_2$, $f_1 < f_2$). If there are triples as well as doubles, and if S_1 does not contribute significantly to the multiples, then a little manipulation of the Poisson distribution gives us $f_2 = 3M_3 / M_2$, $S_2 = 2FM_2 / M_s f_2^2 e^{-f_2}$, $S_1 = 1 - S_2$, and $f_1 = -\ln[(1 - F - S_2 e^{-f_2})/S_1]$. Alternatively, one may employ an iterative optimization program to find the best fit for our four parameters. Both of these methods were applied to a tobacco mosaic virus mutation spectrum consisting of 11 singles, 3 doubles and 3 triples (Malpica et al., 2002). The algebraic method gave $S_1 = 0.97$, $f_1 = 0.011$ and $S_2 = 0.03$, $f_2 = 3$, all being biologically reasonable values with no significant contribution (0.001 mutants) from class 1 to the multiples. However, these values predict too few singles (6.3 instead of 11) and too many multiples with 4 or more mutations (4.7 instead of 0) and the P value for this fit is a modest 0.40. The optimization method gave $S_1 = 0.96$, $f_1 = 0.013$ and $S_2 = 0.04$, $f_2 = 1.2$ and predicted 11 singles, 3.8 doubles, 1.6 triples, 0.6 multiples with 4 or more mutations, and an impressive goodness-of-fit $P = 0.98$.

Why should we be interested in these curiosities? First, the probability of accumulating two or more mutations in a lineage can be sharply increased by hypermutation-induced clusters, especially when lateral transfer (sex and crossing over) is infrequent. Such clusters may then contribute to adaptive evolution. This is especially the case when the multiples are advantageous but the singles are neutral and overwhelmingly the case when the singles are deleterious. An example of interest to the health of humans and economic plants and animals is the adaptations that pathogenic microbes must achieve as they move from individual to individual and especially from species to species. We already know that freshly isolated human pathogens carry mutator mutations much more often than do laboratory-grown cultures. However, most such isolates do not carry mutator mutations. Instead, they may have adapted with the assistance of transient hypermutation. A second reason to be interested in these curiosities is carcinogenesis. Generating a malignant tumor has long been known to require multiple genetic (and perhaps epigenetic) events, more than could be expected from the usual rate of somatic mutation. Many tumors turn out to contain mutator mutations, but many others do not. Thus, carcinogenesis is also likely to require the assistance of transient hypermutation.

It would seem that we have come a long way from postulating, in or around 1935, that the gene was a molecule that could be probed in the same way that atomic and subatomic structures were probed using target theory. The Green Pamphlet has given way to a vast literature on the nature and behavior of these molecules, and the processes that maintain genetic stability in the face of such an onslaught of entropy-enhancing processes that every human cell must deal with about 10,000 DNA damages per day.

Acknowledgements

This research was supported by the Intramural Research Program of the National Institutes of Health, National Institute of Environmental Health Sciences.

References

Bebenek, A., Dressman, H.K., Carver, G.C., Ng, S., Petrov, V., Yang, G., Konigsberg, W.E., Karam, J.D., Drake J.W., 2001, Interacting fidelity defects in the replicative DNA polymerase of bacteriophage RB69. *J. Biol. Chem.* **276**: 10387–1039.

Defais, M., Lesca, C., Monsarrat, B., Hanawalt, P., 1989, Translesion synthesis is the main component of SOS repair in bacteriophage lambda DNA. *J. Bacteriol.* **171**: 4938–4944.

Drake, J.W., Bebenek, A., Kissling, G.E., Peddada, S., 2005, Mutational clusters from transient hypermutability. *Proc. Natl. Acad. Sci. USA* **102**: 12849–12854.

Higgins, N.P., Kato, K., Strauss, B., 1976, A model for replication repair in mammalian cells. *J. Mol. Biol.* **101**: 417–425.

Kadyrov, F.A., Drake, J.W., 2003, Properties of bacteriophage T4 proteins deficient in replication repair. *J. Biol. Chem.* **278**: 25247–25255.

Kadyrov, F.A., Drake, J.W., 2004, UvsX recombinase and Dda helicase rescue stalled bacteriophage T4 DNA replication forks *in vitro*. *J. Biol. Chem.* **279**: 35735–35740.

Malpica, J.M., Fraile, A., Moreno, I., Obies, C.I., Drake, J.W., García-Arenal, F., 2002, The rate and character of spontaneous mutation in an RNA virus. *Genetics* **162**: 1505–1511.

Resnick, M.A., Boyce, J., Cox, B., 1981, Postreplication repair in *Saccharomyces cerevisiae*. *J. Bacteriol.* 146, 285.

Ripley, L.S., 1982, Model for the participation of quasi-palindromic DNA sequences in frameshift mutation. *Proc. Natl. Acad. Sci. USA* **79**: 4128–4132.

Ozgenc, A.I., Szekeres, E.S., Lawrence, C.W., 2005, In vivo evidence for a *recA*-independent recombination process in *Escherichia coli* that permits completion of replication of DNA containing UV damage in both strands. *J. Bacteriol.* **187**: 1974–1984.

Timofeeff-Ressovsky, N.W., Zimmer, K.G., Delbrück, M., 1935, Über die Natur der Genmutation und der Genstruktur. *Nachrichten von der Gessellschaft der Wissenschaften zu Göttingen. Biologie*. Neue Folge. **1**: 189–245.

Wachsman, J.T., Drake, J.W., 1987, A new epistasis group for the repair of DNA damage in bacteriophage T4: replication repair. *Genetics* **115**: 405–417.

CANALIZATION AND EVOLVABILITY: TEMPERING THE EFFECTS OF MUTATION IN A CHANGING ENVIRONMENT

KEVIN MCMANUS[#], CHRIS A. SCANNELL[#]
SUZANNAH RUTHERFORD[*], CHARLES C. CAREY
Division of Basic Sciences, Fred Hutchinson Cancer Research Center, 1100 Fairview Ave. N., Seattle, WA 98109-1024 USA

*to whom correspondence should be addressed.

Abstract. The biological consequences of mutation depend on the translation of genotype into phenotype. The genotype-phenotype map is controlled by developmental and physiological processes, including genetic buffering by the Hsp90 protein chaperone. Hsp90 supports over 150 signal transduction proteins, maintaining the strength of signaling through developmental pathways. During stress, Hsp90 buffering is reduced, and the expressed mutation rate is suddenly increased causing dramatic morphological changes in previously invariable quantitative and qualitative traits. Our recent work suggests that traits controlled by Hsp90 be protected by thresholds, a natural consequence of non-linear biological responses to genetic and environmental factors. A threshold model makes several predictions for the effect of Hsp90 on individual development and the behavior of populations. These include: the existence of upper and lower thresholds, the independence and cumulative effects of buffered genetic variants, the dependence on proximity to the threshold of genetic, environmental or stochastic effects, and the saturation of biological responses and insensitivity (canalization) over a wide range of genetic or environmental effects. We show here how Hsp90-buffered variation meets the predictions of threshold trait models, and suggest that the distinction by Hsp90 between continuous and discrete traits may be more fundamental than the classical division of quantitative and qualitative traits.

Keywords: canalization; evolvability; Hsp90; chaperone; genetic buffering; threshold trait

# These authors contributed equally.

A.A. Cigna and M. Durante (eds.),
Radiation Risk Estimates in Normal and Emergency Situations, 283–290.

1. Genotype is not Phenotype

Genotype, the level of inheritance, is not phenotype, the level at which most natural selection occurs. At one extreme, if the mapping from genotype to phenotype were one to one, then every genetic change would have a phenotypic effect and every nucleotide a consequence for fitness. At the other extreme, identical genotypes would produce infinitely different phenotypes. At either extreme, evolution could not occur. In reality, genes and the environment interact, allowing genetically identical individuals to display multiple phenotypes. Through selection this tendency can become quite stereotyped. For example, an adaptive developmental switch based on the amount of leaves versus catkin flowers in their diet determines whether *Nemoria arizonaria* caterpillars will develop into dramatically different twig or catkin morphs (Greene, 1989). On the other hand, unrelated individuals, differing by hundreds of thousands if not millions of nucleotides in the vast majority of cases develop into phenotypically normal human beings (Przeworski *et al.,* 2000). Between these extremes, phenotype is derived through complex interactions of genes with each other and with the environment.

An evolutionary result of the nonlinear and highly complex relationship between genotype and phenotype is the accumulation of new genetic mutations that are able to persist as long as their potential consequences remain continuously hidden. Genetic buffering is described in the classical quantitative genetics of dominance, epistasis and genotype-by-environment interactions, and is one mechanism proposed to maintain variation by silencing potentially negative mutations (Lynch & Walsh, 1998). What we discuss here is not mutation or population genetics *per se* but the developmental control of mutations and genetic variation, how DNA changes are propagated through successive levels of biological organization to become phenotypic differences upon which natural selection can act, and where and how developmental differences can be buffered to remain neutral. How is genetic buffering achieved within the context of individual development? Can genetic buffering be controlled? In considering the biological control of variation, we return several times to the explanatory utility of a threshold model in understanding how traits are determined and how variation can be normally hidden and conditionally exposed.

1.1 TRAIT THRESHOLDS

Sewall Wright first described threshold trait models to explain the sudden appearance of extra digits in breeding experiments with his beloved guinea pigs (Wright, 1934). In nature, phenotypic switches such as the developmental switch between catkin and twig morphs of *Nemoria* are common and often have obvious adaptive value. For example, the catkin morph provides a stunning mimicry of the catkins upon which they feed. However, when catkins are scarce, a threshold triggered at a discrete point in a continuously varying quantity of dietary tannin switches development to the alternate twig morph. The relationship between tannin and phenotype is steep at the threshold, preventing the occurrence of intermediate and maladapted forms. This switch paradigm describes many molecular signaling processes underlying developmental decisions. For particular systems the mechanisms of switch like responses are beginning to be understood in molecular detail (Ferrell, 1996, 1998; Gardner *et al.*, 2000).

In considering the consequences of a developmental switch for a population, these trait thresholds form barriers to abnormal development of individuals at the extremes, which are acted on by stabilizing selection. For an optimum and normally-invariant phenotype, the threshold model hypothesizes a continuous distribution in the strength of developmental signaling responsible for the trait (classically described as genetic and environmental liability), that is under strong stabilizing selection. We find in a population of wild yeast that variation in signaling responsible for osmotic stress resistance has the bell-shaped distributions expected of stabilizing selection. The extreme individuals (very high or very low signaling relative to most individuals in the population) are also less fit. Genotypes surpassing the threshold by chance are rare, until environmental pressure or genetic liability is increased and the threshold for abnormalities is more frequently surpassed. Hsp90 genetic buffering fits the threshold model, which is consistent with established Hsp90 biochemistry and genetics as well as all known features relevant to the 'Hsp90 capacitor' hypothesis for rapid morphological evolution (Rutherford & Lindquist, 1998).

1.2 PROTEIN CHAPERONES AND THE FOLDING LANDSCAPE

To understand the role of different protein chaperones in modifying the effects of mutation, and their potential for regulating trait thresholds, imagine the potential fitness consequences of a single mutation. Many levels of mutation buffering occur in development. At the simplest level of buffering, there is no effect of mutation if the mutated nucleotide is a synonymous substitution. To affect fitness, an alteration at the level of DNA must normally result in an altered protein, and the alteration must have functional consequences for the activity or regulation of that protein. The consequences of the changed protein function must propagate onward through development and physiology into phenotype. In diploids even very strong loss-of-function mutations often are not dominant in their effects, at least for laboratory raised animals, therefore many mutations even if they abolish protein function may be buffered with no consequent observable phenotype.

But now let us consider the consequences of a protein level change in a gene for which the loss of function phenotype should be obvious. From the time a newly synthesized polypeptide emerges from the ribosome until late stages of folding, proteins are protected from non-productive aggregation by specific chaperones that recognize particular features of non-native protein. Hsp70 chaperones recognize linear chains of amino acids; Hsp60/GroE type chaperones recognize "molten globule" intermediates. Hsp90s recognize nearly mature proteins that likely expose hydrophobic surface character. In general, molecular chaperones enable the proper folding of protein sequences and can increase the number of sequences that fold into productive proteins thereby smoothing the adaptive landscape of protein sequence space (Rutherford, 2003). However, in contrast with the other major chaperones, which are required by nearly every protein during folding, in unstressed cells Hsp90 function is restricted to a specific set of "client proteins," inactive signaling molecules poised in anticipation of ligand or association into an active signaling complex in the cell.

It is known that for many difficult-to-fold proteins over-expression of GroE increases the yield of functional bacterial expression systems. Decreased fitness due to mutation load, the accumulation of deleterious mutations, is buffered by over-expression of the GroE chaperone in *E. coli* in a laboratory environment (Maisnier *et al.,* 2005; Fares *et al.,* 2002). This is also found in natural environments. Endosymbionts often experience extreme mutation load in the course of their natural life cycle, as populations are usually established from a small number of individuals and isolated in a single host. *Buchneria* massively up regulates the GroE operon to as much as 10% of total cellular protein in order to buffer the effects of this mutation load. Hsp70 is a general folding factor in all organisms, and in *Drosophila* is particularly important for the recovery of damaged proteins after heat shock. Additional copies of Hsp70 in *Drosophila* can promote the reactivation of heat-damaged alcohol dehydrogenase (ADH), which otherwise would not refold properly (Feder & Krebs, 1997). In yeast, a particularly unstable luciferase is stabilized by the Hsp90 chaperone, having more activity in wild-type or Hsp90 over-expressing cells than in Hsp90 mutant strains (Nathan *et al.,* 1997). At every stage in the folding pathway and for a variety of types of client proteins, the number of primary sequences that can fold productively into functional proteins is expanded by molecular chaperones.

The overall consequence of chaperone activity is the potential accumulation of polymorphisms in genes throughout the genome and throughout populations without consequent observable phenotypic change, thereby creating a more elastic relationship between genotype and phenotype. If the types of targeted molecules of this activity were critically involved in threshold-like responses that are a normal part of developmental and signaling processes, this would enable a system that can buffer a multitude of phenotypes from consequences of evolutionary pressures. Hsp90 is poised as a unique chaperone to study buffering between evolutionary pressures and developmental and signaling pathways because its client proteins are enriched for signaling proteins, and signaling inherently exhibits a threshold like response.

2. Hsp90 Buffered Morphologies

Hsp90 affects a wide range of morphologies. In *Drosophila* it can both conceal and expose abnormal phenotypes and seems to affect specifically threshold traits (Rutherford, 2000). For example, most Hsp90 mutations can cause graded continuous changes in the activity of signaling pathways that can result in discrete phenotypic changes. One of the best-studied cases is the *Sevenless* (*Sev*) Ras/mitogen-activated protein kinase (MAPK) pathway. When the pathway is over-stimulated by a constitutively activated upstream kinase (an Hsp90 target protein), differentiation of ectopic photoreceptors produces a rough eye phenotype. A drastic reduction in signaling, achieved through either Hsp90 mutation or heterozygous loss-of-function mutations in other pathway components, shifts the signaling activity below a threshold to where eyes become normal (van der Straten *et al.*, 1997). The upper level of signaling defines an upper threshold for normal *Sev* pathway activity. However, when the pathway is under-activated by a weakly functioning component in Hsp90 heterozygous mutants, there is a loss of photoreceptors and the expression of an abnormal phenotype (Simon et al., 1991; Cutforth & Rubin, 1994). As the excessive signaling defines an upper threshold, these lower amounts of signaling define the lower threshold for normal *Sev* pathway activity.

More recent findings have confirmed the threshold model with abnormal wing and eye phenotypes (Rutherford & Lindquist, 1998). Hsp90 mutants were crossed with standard laboratory strains producing flies with observable defects. A deformed eye and thickened wing-vein phenotype were chosen and then bred into high and low expression lines. The lines were exposed to different ranges of temperatures that were shown to increase the penetrance of the trait. The thickened wing-vein phenotype was produced in high-expression lines raised at 18°C but not observed at replicate cultures raised at 25°C. Conversely, the deformed eye phenotype was enhanced at 30°C but not observed at 18°C. Together these results indicate the existence of an environmental threshold that reduces Hsp90 signaling when crossed and helps to reveal previously cryptic genetic variation.

The threshold model can also be used to explain abnormal scutellor (SC) bristle numbers in Raleigh Inbred (RI) *Drosophila* lines (Rutherford, in review). When Hsp90 is impaired in most RI lines, abnormal SC bristle numbers increase. Lab and wild RI lines can also be isolated with a naturally high frequency of abnormal SC bristles. In agreement with an upper threshold, the abnormal variation is reduced when Hsp90 is impaired. The observation of both upper and lower thresholds suggests that natural variation, in the form of essentially neutral segregating alleles in outbreed populations, can either reduce or enhance signaling through Hsp90 signaling pathways. When Hsp90's buffering capacity is compromised, reduced activity of its client proteins shifts signaling distributions to lower values that can have phenotypic consequences. Whether resulting phenotypic thresholds are crossed depends on the genetic and environmental liability as well as the proximity to trait thresholds.

We propose that most populations are closer to low thresholds because of the accumulation of more-common loss-of-function or conditionally loss of function mutations. Reducing Hsp90 pushes individuals in these populations below critical trait thresholds with the consequence that novel as well as abnormal phenotypes may result. Some populations, like the one with high SC bristle numbers, however, are already at the other extreme of a threshold because reducing Hsp90 signaling causes them to return from an abnormal to a normal phenotype with consequently less variability. In the SC bristle case, the variation does not necessarily have to be a rare gain-of-function mutation but could be a loss-of-function mutation in an inhibitory pathway.

The evolutionary role of Hsp90 in uncovering morphological variation is challenged by the idea that variation controlled by the protein has necessarily pleiotropic and unconditionally deleterious side-effects (Wagner et al., 1999; Dickinson & Seger, 1999). However, this has been overturned by research showing that large morphological changes buffered by Hsp90 can occur independently of strongly deleterious fitness effects (Rutherford, in review). The same lines originally created for studying temperature thresholds were used to measure the fitness costs of the deformed eye (*dfe*) trait. Lines with high *dfe* penetrance carried multiple polymorphisms leading to the eye deformity whereas related lines with low *dfe* penetrance carried few. First, the competitive ability, daily and lifetime fecundity, and egg to adult viability were measured between three high and three low lines and no significant difference in these fitness measures was found. Second over 1,432 recombinant isogenic lines containing small portions of the wildtype (*Samarkand*) background were created to specifically test the fitness effects of the alleles responsible for the *dfe* trait. Logically it seemed the recombinant lines with greater penetrance would carry more *dfe* polymorphisms. There was, however, no correlation between fitness and *dfe* penetrance for any line. Likewise, when quantitative trait loci (QTL) were mapped to 5-10 cM resolution, there were no consistent associations between trait QTL and effects on fitness. These observations indicate how the almost certainly polygenic structure of Hsp90 buffered abnormalities could be used to avoid potentially deleterious side effects. Alleles with negative fitness costs could be lost early in selection and be replaced with less deleterious alleles for that trait. This would allow diverse features controlled by Hsp90 to be exposed without serious fitness costs, easily producing viable and fertile selection lines.

3. Canalization by Hsp90

Not only are traits controlled by Hsp90 typically threshold traits, they are also highly invariant or canalized. Traits like SC bristle number or eye morphology have discrete phenotypic states without continuous variation between them. This previously unrecognized property of Hsp90 buffering suggests the difference between discrete and continuous may be more organic than the traditional division of qualitative and quantitative traits. We suggest here that the specificity of Hsp90 for normally invariant traits is a consequence of the thresholds involved in their expression. Thresholds are created by non-linear and cooperative (sigmoidal) responses, which are common in biology, with their inflection points defining the thresholds. Steep cooperativity and positive feedback both ensure that developmental processes will be invariable over a large range of signaling inputs and suddenly become variable and change over a very narrow range. For instance, the non-linear kinetics of MAPK signaling has been well documented (Ferrule 1996; Nijhout *et al.*, 2003; Gardner *et al.*, 2000). This is also the basis for binary switches (Gardner *et al.*, 2000) and all-or-none phenotypic responses.

The Hsp90 capacitor hypothesis suggests that Hsp90 promotes evolutionary change as a response to environmental stress. Importantly, its function is stress-sensitive and reversible. Upon exposure to an environmental stress such as increased temperature, Hsp90 can reveal developmental variation across an entire population. Rare individuals with potentially abnormal morphological traits can be immediately exposed to selection upon stress, and phenotypes can be quickly fixed by selection of polygenetic variation (genetic assimilation (Falconer & Mackay, 1996)). On the other hand, potentially adaptive mutations are traditionally thought to spread through the population one individual at a time and consequently become fixed much more slowly. Also, while most populations have a large reservoir of quantitative trait variation with a high evolvability, Hsp90 seems to mainly protect canalized traits with little capacity to evolve. Thus, when Hsp90 function is compromised, threshold traits of normally invariant phenotypes would be the most drastically affected. Finally, Hsp90 is a chaperone for over 150 targets and is involved in many integral cell processes (www.picard.ch/downloads/downloads.htm). Its targets are key proteins in cell cycle, protein synthesis, gene regulatory chromatin remodeling, growth control, and morphogenesis networks. With such a range of influence, Hsp90 buffering can affect virtually any morphological feature in a large range of organisms, extending from adult flies to *arabidopsis* seedlings (Rutherford & Lindquist, 1998; Queitsch *et al.*, 2002). Hsp90's dual roles as a biochemical buffer of misfolded proteins and a phenotypic buffer of diverse morphology places it in a singular position to promote the diversification and survival of stressed populations in novel environmental conditions.

References

Cutforth T. and G.M. Rubin, 1994, Mutations in Hsp83 and cdc37 impair signaling by the sevenless receptor tyrosine kinase in Drosophila, *Cell* **77**(7): 1027–1036.

Dickinson W.J. and J. Seger, 1999, Cause and effect in evolution, *Nature* **399**(6731): 30.

Falconer D.S. and T.F.C. Mackay, 1996, *Introduction to quantitative genetics* (Longman, Essex, England.

Fares M.A., M.X. Ruiz-Gonzalez, A. Moya, S.F. Elena and E. Barrio, 2002, Endosymbiotic bacteria: groEL buffers against deleterious mutations, *Nature* **417**(6887): 398.

Feder M.E. and R.A. Krebs, 1997, Ecological and evolutionary physiology of heat shock proteins and the stress response in Drosophila: complementary insights from genetic engineering and natural variation, *Exs* **83**: 155–173.

Ferrell J.E., Jr., 1996, Tripping the switch fantastic: how a protein kinase cascade can convert graded inputs into switch-like outputs, *Trends Biochem Sci* **21**(12): 460–466.

Ferrell J.E., Jr., 1998, How regulated protein translocation can produce switch-like responses, *Trends Biochem Sci* **23**(12): 461–465.

Gardner T.S., C.R. Cantor and J.J. Collins, 2000, Construction of a genetic toggle switch in Escherichia coli, *Nature* **403**(6767): 339–342.

EFFECTS OF CHRONIC LOW-LEVEL IRRADIATION ON RADIOSENSITIVITY OF MAMMALS: MODELING AND EXPERIMENTAL STUDIES

OLGA SMIRNOVA*
Research and Technical Center of Radiation-Chemical Safety and Hygiene, Shchukinskaya st., 40, Moscow 123182, Russia
E-mail: nestr@theor.jinr.ru

MORIO YONEZAWA
Research Institute for Advanced Science and Technology, Osaka Prefecture University, 1-2 Gakuen-cho, Sakai, Osaka 599-8570, Japan

*To whom correspondence should be addressed.

Abstract. Effects of chronic low-level irradiation on radiosensitivity of mammals (mice) are studied experimentally and by making use of the methods of mathematical modelling. Our own and reference experiments show that chronic low-level short-term and long-term exposures induce, respectively, elevated radiosensitivity and lowered radiosensitivity (radioresistance) in mice. The manifestations of these radiosensitisation and radioprotection effects are, respectively increased and decreased mortality of pre-irradiated specimens (in comparison with previously unexposed ones) after challenge with acute irradiation. The reason of the animals' death in the experiments was the hematopoietic syndrome among acute radiation syndromes. Therefore the theoretical investigation of the influence of pre-irradiation on radiosensitivity of mice is conducted by making use of the biologically motivated mathematical models which describe the dynamics of hematopoietic system in mice exposed to challenge acute exposure following the chronic one. Modelling results indicate that the radiosensitization effect of chronic low-level short-term (less than 1 month) pre-irradiation on mice is due to increased radiosensitivity of lymphopoietic, granulocytopoietic, and erythropoietic systems accompanied by increased or close to the normal level radiosensitivity of thrombocytopoietic system. In turn, radioprotection effect of chronic

A.A. Cigna and M. Durante (eds.),
Radiation Risk Estimates in Normal and Emergency Situations, 291–299.

low-level long-term (more than 1 month) pre-irradiation on mice is caused by decreased radiosensitivity (radioresistance) of the granulocytopoietic system appeared at the level of one functional cell pool, blood granulocytes. It is important that modelling estimations of the duration of low-level chronic preexposures, which result in radioprotection and radiosensitisation effects on mice, are in agreement with the relevant experimental data. The developed models can be useful in planning of prospective experiments in this field and in estimating the risks of chronic low-level irradiation.

Keywords: priming low-level chronic irradiation; challenge acute exposure; acquired radioresistance; mice; mathematical modelling; mechanisms, adaptive response

1. Introduction

Investigation of the effects of chronic low-level irradiation on mammals is a challenging problem of radiation biology and ecology. Difficulties in this field are, in particular, due to non-linear dose-effect relationships observed in some experiments. Their results conflict with a radiobiological paradigm according to which the radiation hazard increases with increasing doses. An objective of our work is twofold. First, it involves experimental studies of the effects of chronic pre-irradiation on changes of mouse radiosensitivity which manifests itself in changes of mortality of preirradiated specimens after challenge acute irradiation in comparison with that for previously unexposed ones. Second, our work seeks to investigate theoretically the biological mechanisms, which are responsible for the modifications of radiosensitivity in pre-irradiated mice. Our theoretical consideration is based on experimental observations according to which the injury of the hematopoietic system is prevalent after acute irradiation with both mid-lethal and sub-lethal doses. Therefore it is reasonable to suppose that the change of radiosensitivity of pre-irradiated mice is due to the change of radiosensitivity of the hematopoietic system in them and to choose the blood-forming system as an object of our theoretical studies. As a tool of our theoretical investigations, we use the mathematical models of the major hematopoietic lines.

2. Materials and Methods

2.1 EXPERIMENTAL EQUIPMENT AND PROCEDURES

Mature mice of NA-2 strain and their first progeny were used (Megumi *et al.*, 1968). The animals were housed in cages at 24 ± 1°*C* and 60 ± 10% relative humidity. The animals were maintained on a 7 a.m. - 7 p.m. light/dark cycle. The experiment was planned to examine chronic effects of the radionuclides produced by explosion of the atomic bomb. The two radionuclides, ^{137}Cs and ^{90}Sr, were selected as the most toxic products with very long half-lives. Female mature mice were administered 4.10^{-9} Ci/mL of ^{137}Cs and 1.10^{-9} Ci/mL of ^{90}Sr in drinking water after the mating and during lactation. Their first progeny was administered ^{137}Cs and ^{90}Sr in milk during 4 weeks after their birth. Then, during the next 4-6 weeks, they were administered 4.10^{-9} Ci/mL of ^{137}Cs and 1.10^{-9} Ci/mL of ^{90}Sr in drinking water. Thus, the young mice were subjected to chronic internal irradiation by administration of ^{137}Cs and ^{90}Sr in the natal period and for 8-10 weeks after their birth. When eight-ten weeks old, these mice were exposed to challenge whole-body *X*-irradiation with sub-lethal doses of 7 Gy and 8 Gy, after which mortality was checked daily for 30 days. Experiments with pre-irradiated groups of mice, as well as with previously non-irradiated (control) groups, were run concurrently because of the variation of radiosensitivity of the experimental animals from unidentified causes. The experimental mice were not *Pseudomonas*-free animals. Therefore there were early (before day 10 after challenge irradiation) deaths in irradiated animals, presumably caused by bacterial infections.

2.2 MATHEMATICAL MODELS

In our approach (Smirnova, 1989; Zukhbaya and Smirnova, 1991; Kovalev and Smirnova, 1996; Smirnova and Yonezawa, 2003, 2004) hematopoietic system is regarded as a complex of four major hematopoietic lines: thrombocytopoiesis, lymphopoiesis, erythropoiesis, and granulocytopoiesis. The models of these systems consider the principal stages of development of hematopoietic cells and take into account the specific features of thrombocytopoiesis and granulocytopoiesis: the variable average ploidy of megakaryocytes and the existence of a bone marrow depot of granulocytes.

In constructing the model of a hematopoietic line, we take into account the feedback loop mechanism of sustaining the homeostasis in this system which is accomplished by the specific inhibitor of cell reproduction, the chalone. The latter is a product of the vital activity and decay of the cells of the hematopoietic line (Ketlinsky *et al.*, 1992).

The models of the major hematopoietic lines consider three cell compartments according to the degree of the maturity and differentiation of the cells: (X_1) the bone-marrow precursor cells (from stem cells in the respective microenvironment to morphologically identifiable dividing cells); (X_2) the nondividing maturing bone-marrow cells; (X_3) the mature blood cells. The model of the granulocytopoietic system considers one more cell compartment: tissue granulocytes (X_4).

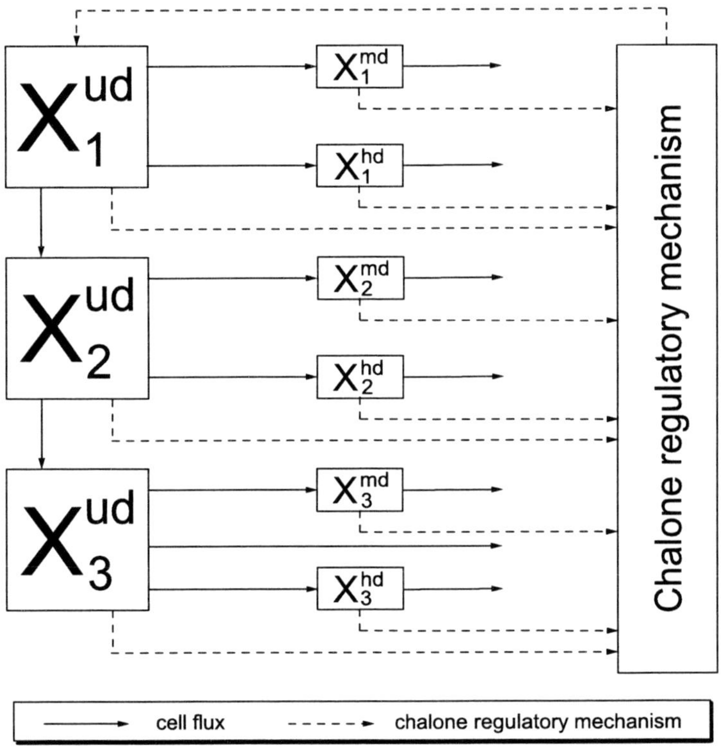

Figure 1. Block diagram of the lymphopoiesis model.

Effects of ionizing radiation on the radiosensitive cell compartments are described by making use of the one-target-one-hit theory of cell damage (Kudryashov and Berenfeld, 1982). According to this theory the damage rate of the radiosensitive cells is proportional to the dose rate N. With regard for experimental data (Belousova *et al.*, 1979), the cells of each radiosensitive compartment are divided into three groups with respect to their reaction to irradiation: (X_i^{ud}) undamaged cells; (X_i^{md}) moderately damaged cells that die within 1-2 days (mitotic death); (X_i) heavily damaged cells that die within 4-7 hours following irradiation (interphase death).

An illustration of the main ideas and relations forming the basis of our modelling approach is the block diagram of the model of one of the major hematopoietic lines, lymphopoiesis (Figure 1). Detailed descriptions of the mathematical models of the major hematopoietic lines are given in papers (Smirnova, 1989; Zukhbaya and Smirnova, 1991; Kovalev and Smirnova, 1996; Smirnova and Yonezawa, 2003, 2004).

3. Results and Discussion

The developed models are used to simulate the dynamics of lymphopoietic, erythropoietic, thrombocytopoietic, and granulocytopoietic systems in mice exposed to challenge acute irradiation following the chronic one and in mice exposed to the same challenge irradiation only. The dose *D* of the challenge acute exposure is chosen from the range of doses causing the hematopoietic syndrome. The dose rate *N* and the duration *T* of the priming chronic irradiation are varied.

Our analysis of the modelling results is based on conventional radiobiological concepts (Bond *et al.*, 1965). According to these concepts, radiosensitivity of an organism's cell system can be characterized by a severity of postirradiation injury of this system, namely, by a depth of postirradiation depletion of its functional cell pool. Proceeding from this, we take, as the index of radiosensitivity of a hematopoietic line in preirradiated animals, the minimal level up to which the concentration of functional cells in this system is reduced after challenge irradiation. As the standard (control) value of the radiosensitivity index of the hematopoietic line, we take the minimal level up to which the concentration of functional cells of this system in previously unexposed animals is reduced after the same challenge irradiation. The smaller index of radiosensitivity of the hematopoietic line in preirradiated animals (in comparison with the standard one) corresponds to increased radiosensitivity of this system in preirradiated animals, the larger index (in comparison with the standard one) corresponds to decreased radiosensitivity of this system.

In the framework of the models we calculate minimal levels, up to which the concentrations of functional cells of the major hematopoietic lines (lymphocytes in blood, erythrocytes in blood, thrombocytes in blood, granulocytes in blood, and granulocytes in tissues) are decreased in mice exposed to the challenge acute irradiation with a dose *D* following the chronic preirradiation with a dose rate *N* and duration *T*. We also calculate the minimal levels up to which the concentrations of functional cells of the major hematopoietic lines are decreased in mice exposed only to challenge irradiation at the same dose *D*.

The modelling results enable one to reveal the following. Preliminary chronic irradiation, in chosen ranges of N and T, leads to increase of radiosensitivity of lymphopoietic and erythropoietic systems. The effect of chronic preirradiation on thrombocytopoietic system can manifest itself in increased or close to the normal radiosensitivity of this system. As for the granulocytopoietic system, chronic irradiation can cause both radiosensitization effect (increase of radiosensitivity) and radioprotection effect (decrease of radiosensitivity). On the level of the second pool of functional cells, tissue granulocytes, only the radiosensitisation effect of chronic preirradiation takes place.

The modelling results show the following. Preliminary chronic irradiation with the duration T exceeding 1 month and with the dose rate N not exceeding 0.03 Gy/day induces the radioresistance in granulocytopoietic system. Besides, such preexposures practically do not affect the radiosensitivity of thrombocytopoietic and erythropoietic systems and cause a weak radiosensitisation effect on the lymphopoietic system. Taking into account that the functional cells of the lymphopoietic and granulocytopoietic systems take part in immune protection of the mammalian organism against infections, one can expect that after challenge acute irradiation the mortality will be lower in mice preliminary irradiated in regimes stated above than in earlier unexposed ones.

To verify this modelling prediction we compare the modelling results on the radiosensitivity of the above-indicated hematopoietic lines under exposures in question and the respective experimental data on the radiosensitivity of mice. With this aim, we represent the experimental data on mouse radiosensitivity by making use of the dimensionless index, increment of survival rate. It equals the difference in survival rate of pre-irradiated and unexposed mice after challenge irradiation. Obviously, the positive value of the increment of survival rate points to the state of radioresistance of preirradiated specimens at the moment of challenge exposure and the negative one testifies to the state of radiosensitisation. In turn, we represent the modelling results on the radiosensitivity of the above-indicated critical body systems by making use of the dimensionless index, the increment of radioresistance (Smirnova and Yonezawa, 2003, 2004). It equals the ratio of the difference in the minimal levels of the concentration of the functional cells in a critical system of preirradiated and unexposed specimens after identical challenge irradiation to the minimal level of the functional cell concentration in unexposed specimens after the same challenge exposure. Obviously, the positive value of the increment of radioresistance shows that the critical body system is in the radioresistance state at the moment of challenge irradiation and the negative value of the increment of radioresistance corresponds to the radiosensitisation state.

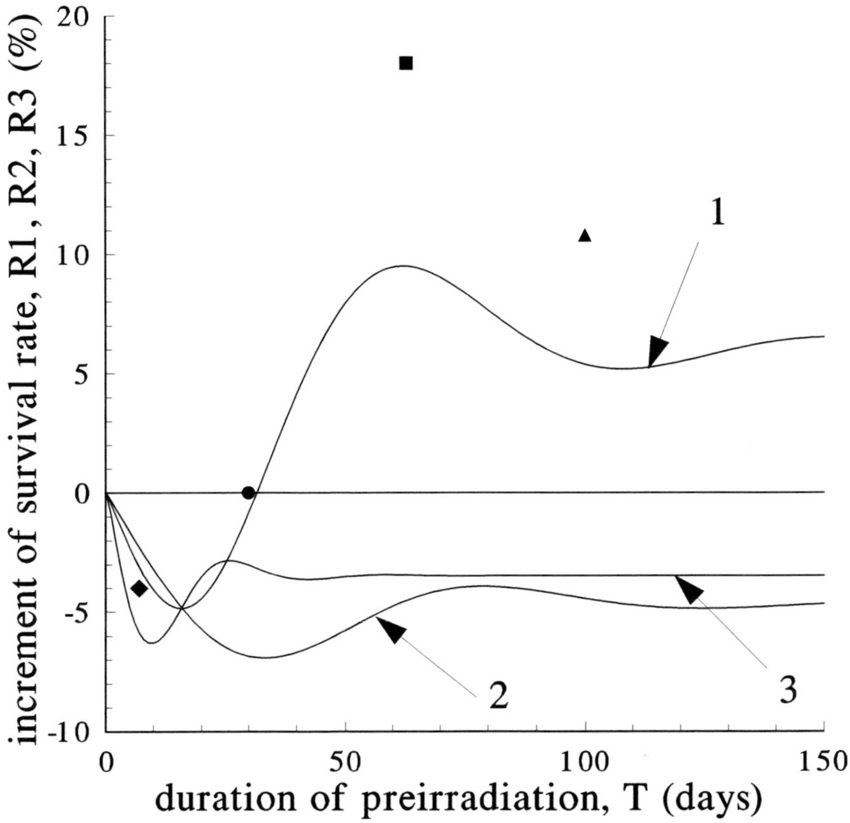

Figure 2. The increment of radioresistance calculated in the framework of the model of the granulocytopoietic system for the pool of blood granulocytes (function R1(T), curve 1) and for the pool of tissue granulocytes (function R2(T), curve 2) and the increment of radioresistance calculated in the framework of the model of the lymphopoietic system for the pool of blood lymphocytes (function R3(T), curve 3) at the priming dose rate of 0.03 Gy/day and the challenge dose of 4 Gy. The experimental values of the increment of survival rate obtained for female NA-2 mice at the challenge doses of 8 Gy (box) (Megumi *et al.*, 1968), for SHK mice at the challenge dose of 8 Gy (triangle) (Fomenko *et al.*, 1991), and for (CBA.C57/BL6)F1 mice at the challenge dose of 9 Gy (diamond and circle) (Konradov *et al.*, 1993).

The modelling results on the increment of radioresistance, which are calculated for pools of blood granulocytes, tissue granulocytes, and blood lymphocytes at various durations of priming chronic irradiation T and at fixed values of priming dose rate and challenge dose, are shown in Figure 2 as the functions $R1(T)$, $R2(T)$, and $R3(T)$, respectively. The qualitative comparison of these modelling results and experimental data (Megumi *et al.*, 1968; Fomenko *et al.*, 1991; Konradov *et al.*, 1993) shows the following. At the chosen regimes of pre-irradiation, the radioresistance of an organism as a whole at the moment of the challenge exposure is determined mainly by the radioresistance of granulocytopoietic system manifested on the level of the pool of blood granulocytes. In particular, the radioprotection and radiosensitisation effects of pre-irradiation on this cell system correlate, respectively, with the radioprotection and radiosensitisation effects of the pre-exposure on the mouse organism as a whole. Moreover, the duration of the chronic pre-irradiation, at which the radiosensitisation effect on mice changes into the radioprotection one, practically coincides with the duration of the chronic pre-irradiation, at which the radiosensitisation effect on the above-mentioned cell system changes into the radioprotection one. In Figure 2, it is the time interval of 30 days, which the sign of the effect is reversed at. The juxtaposition of the modelling curves and the experimental points in Figure 2 also shows that the chosen duration of preliminary chronic exposure to internal irradiation (8–10 weeks) proves to be optimum for elucidating the radioprotection effect of pre-exposure on mortality of mice after challenge irradiation. At this duration of low-level pre-irradiation, radioresistance of granulocytopoietic system displayed at the level of the pool of blood granulocytes is close to maximum.

4. Conclusion

Effects of chronic pre-irradiation on radiosensitivity of mice are studied experimentally and by making use of mathematical modelling methods. The results of these joint investigations show that the radioprotection effect of chronic low-level long-term pre-irradiation on mice is a consequence of the acquired radioresistance of the granulocytopoietic system. In turn, the radioresistance of the granulocytopoietic system is due to its adaptation to low-level long-term pre-irradiation which results in the relaxation of the concentration of blood granulocytes to a new stationary level increased in comparison with the normal one. It is important to emphasise that the evaluations of the duration of pre-irradiation, which results in radioresistance of mice, carried out on the basis of the modelling and experimental investigations, practically coincide. All this demonstrates the effectiveness of employment of biologically motivated mathematical models, parallel with relevant experiments, in elucidation of the mechanisms of radioprotection effect of low level preliminary exposures on mammals and in planning prospective experiments.

References

Belousova, O.I., Gorizontov, P.D. and Fedorov, M.I., 1979, *Radiation and haemopoietic system*, Atomizdat, Moscow (in Russian).

Bond, V.P., Fliedner, T.M. and Archambeau, J.O., 1965, *Mammalian Radiation Lethality: A Disturbance in Cellular Kinetics*, Academic Press, New York.

Fomenko, L.A., Kozhanovskaya, Y.K. and Gaziev, A.I., 1991, Formation of micronuclei in bone-marrow cells of chronically irradiated mice and subsequent acute exposure to gamma-radiation, *Radiobiologiya* **31**: 709–715 (in Russian).

Ketlinsky, S.A., Simbircev, A.S. and Vorob'ev, A.A., 1992, *Endogenous immunomodulators*, Hippokrat, Sankt-Peterburg (in Russian).

Konradov, A.A., Lyubimova, N.V., and Pelevina, I.I., 1993, Modification of radiosensitivity of animals after exposure within the zone of Chernobyl disaster, *Radiatsionnaya Biologiya. Radioekologiya* **33**: 499–507 (in Russian).

Kovalev, E.E. and Smirnova, O.A., 1996, *Estimation of radiation risk based on the concept of individual variability of radiosensitivity*, AFRRI Contract Report 96–1, Defense Nuclear Agency Contract DNA001–93-C-0152, Armed Forces Radiobiology Research Institute, NWED QAXM, Bethesda.

Kudryashov, Ju.B. and Berenfeld, B.S., 1982, *Bases of Radiation Biophysics*, Moscow State University Press, Moscow (in Russian).

Megumi, T., Yonezawa, M. and Nishio, K., 1968, Changes of resistance against X-rays induced by continuous administration of 137Cs and 90Sr in mice, *Annual Report of the Radiation Center of Osaka Prefecture* No 9, pp. 94–97.

Smirnova, O.A., 1989, The model of homeostasis of hematopoiesis system under chronic irradiation, in: *Modeling of Population Dynamics*, Gorkiy University Press, Gorkiy, pp. 39–45 (in Russian).

Smirnova, O.A., and Yonezawa, M., 2003, Radioprotection effect of low level preirradiation on mammals: modeling and experimental investigations, *Health Phys.* **85**: 150–158.

Smirnova, O.A. and Yonezawa, M., 2004, Radioresistance in mammals induced by low-level chronic irradiation: modeling and experimental investigations, *Health Phys.* **87**: 366–374.

Zukhbaya, T.M. and Smirnova, O.A., 1991, An experimental and mathematical analysis of lymphopoiesis dynamics under continuous irradiation, *Health Phys.* **61**: 87–95.

PROLONGED ENVIRONMENTAL STRESS INDUCES MUTATIONS AND PROVIDES NONSPECIFIC ADAPTATION OF DROSOPHILA POPULATIONS

I.B. MOSSE *, I.V. GLUSHKOVA, I.P. ANOSHENKO, V. AKSYUTIKT
Institute of Genetics and Cytology at National Academy of Sciences, Minsk, Belarus, Akademicheskaya str.27, E-mail: i.mosse@igc.bas-net.by

*To whom correspondence should be addressed.

Abstract. Adaptation of natural drosophila populations from the area radiocontaminated due to the Chernobyl accident (Vetka district of Gomel region) and from the control area (Berezinsky National Reserve) to irradiation was studied. Additional exposure to 40 Gy γ-rays resulted in different responses of population samples - much more dominant and recessive lethal mutations arose in the population from the control area than from that with "radiation history". Viability of flies from Vetka district was much higher after additional irradiation in comparison with flies from Berezinsky Reserve. These facts prove that Vetka population is more resistant to radiation than Berezinsky one. When the population samples were kept under laboratory conditions without irradiation for 8 generations it was revealed that the mutation level in both populations increased at keeping under such conditions. Adaptation of Vetka population to irradiation remained, besides, the control population also became more resistant to ionizing radiation. It means that keeping of natural drosophila populations under laboratory conditions is a strong stress (limited space, overpopulation, other temperature and light conditions), which increases a mutation process and induces non-specific adaptation. In order to study the process of insect adaptation in detail we investigated experimental drosophila populations of 4 kinds: control, supplied with melanin, irradiated in each generation and irradiated with melanin. The samples of these populations were exposed to 30 Gy x-rays in the 55th generation. Dominant and recessive lethal mutation frequencies, induced by additional radiation exposure, were the lowest in the irradiated population - this population was best adapted to ionizing radiation impact. The population irradiated with melanin was less resistant. This may be explained by a radioprotective action of melanin — more radiosensitive genotypes were not eliminated by

A.A. Cigna and M. Durante (eds.),
Radiation Risk Estimates in Normal and Emergency Situations, 301–308.

natural selection and remained in the population preventing its adaptation. Thus melanin which is an effective radioprotector against chronic irradiation is able to prevent adaptation process.

Keywords: radiation; Chernobyl accident; natural population; drosophila; mutation; environmental stress; adaptation; melanin

1. Introduction

Study on response of natural populations of living organisms to an increase in radiation background is a very important and complicated problem.

Earlier we studied genetic processes in natural populations of insects (drosophila, potato bittle and so on) living in areas radiocontaminated due to the Chernobyl disaster (Mosse and Makeeva, 1994). We revealed some morphogenetic changes in these populations.

Since 12-13 years passed after this catastrophe, natural insect populations had to change their genetic structure because of chronic radiation effect. It's well known that prolonged action of mutagenic factors results in adaptation of populations of living organisms to harmful agents (Ayala, 1966; Marques, 1973). Such adaptation is associated with death of more sensitive individuals and with selection and breeding of more resistant ones (Cordeiro *et al.*, 1973; Notel, 1976). It was important to investigate if natural insect populations from radiocontaminated regions became more adapted to irradiation than population from "clean" area. It was also very interesting to check if such adaptation will reverse after radiation ceasing.

The second task of our research was to study melanin influence on an adaptation process in order to investigate mechanisms of population adaptation. We revealed earlier that pigment melanin decreased very effectively mutation frequencies induced by ionizing radiation in animals (drosophila, mice) and cultured human cells (Mosse *et al.*, 1997; Mosse *et al.*, 2000). Besides, melanin reduced significantly "mutation load" accumulated in experimental drosophila populations due to irradiation within more than 100 generations (Mosse and Lyach, 1994; Mosse *et al.*, 1996). Melanin is the only radioprotector, which is effective against chronic irradiation (Mosse *et al.*, 2001; Mosse *et al.*, 2002).

2. Material and methods

2.1 NATURAL POPULATIONS

We studied natural populations of *Drosophila melanogaster* from radiocontaminated area (Vetka district of Gomel region with 24 Ci/km^2 of ^{137}Cs and 0,5 Ci/km^2 of ^{90}Sr) and from Berezinsky Natural Reserve as a control area.

Males from these two populations were exposed to 40 Gy γ-rays on Laboratory Cs-machine for Microbiological Installation. The dose rate was 29 Gy/min. We analysed mutations of two kinds — dominant lethal and recessive lethal mutations after additional 40 Gy irradiation in populations from different areas. In order to estimate the dominant lethal mutation frequency and viability of flies, we mated irradiated males (55 flies from each area) individually with virgin females and placed them in bottles with fresh food daily for 3 days. We estimated the numbers of laid eggs, undeveloped eggs, alive larvae and the number of imago. Dominant lethal mutations (DLM) were estimated as a proportion of the number of undeveloped eggs to that of laid eggs.

DLM (%) = undeveloped egg number / laid egg number x 100%.

Viability was estimated as a proportion of the imago number to that of laid eggs.

Viability (%) = imago number / laid egg number x 100%.

The frequency of sex-linked recessive lethal mutations (RLM) was estimated by the standard method of Muller (Ayala, and Kiger, 1984). According to this method wild-type males were mated to several virgin females of a special test line Muller-5 with marker genes “yellow” and “white apricot”. If recessive lethal mutation in X-chromosome arises, wild-type males are absent in the second generation of such mates.

Statistic processing of the experimental data was confirmed by the methods of Student and of Fisher.

2.2 EXPERIMENTAL POPULATIONS

Drosophila melanogaster populations of four types have been investigated: (1) control; (2) regularly supplied with melanin together with food at the concentration of 0,5 mg/ml; (3) irradiated at 15 Gy dose per generation (6 Gy at the 3-day larvae stage and 9 Gy at the imago stage); (4) irradiated and supplied with melanin.

Experimental populations were developed on the basis of samples including 200 individuals (100 males and 100 females) taken from a panmictic population. The population was bred from eight isogenic lines at an equal ratio and kept under equilibrium conditions within 5 generations.

All the populations were kept in glass jars (one liter) and were irradiated in these jars through the necks covered with cheesecloth. The populations of different types were simultaneously irradiated. X-irradiation was conducted under the following conditions: 200 kV; 5 mA; focal distance 30 cm; dose rate 1,4 Gy/min. Melanin was obtained from 1-dioxyphenylalanine.

All adult flies in each generation were placed in new bottles with fresh nutrient medium for 1 day of egg laying and then the flies were removed.

We estimated frequencies of dominant lethal and recessive lethal mutations within 55 generations in all populations every 5 generations. The samples of each population were exposed to 30 Gy x-rays in the 55-th generation. Frequencies of DLM and RLM, induced by such additional irradiation, were analysed.

3. Results and Discussion

3.1 NATURAL POPULATION STUDY

The data, obtained after irradiation of the samples from natural drosophila populations, were presented in Table. The DLM frequency in the control population (Berezinsky Reserve) after exposure was equal to 63,1±0,9% and such parameter in Vetka district was 42,8±0,9%.

Irradiation death of flies from various populations at late ontogenetic stages differs even more sharply - viability of the population from the control area was 15,5% as a result of irradiation and that of the population from district with high radiation background was 42,3%.

We obtained similar data with RLM method (table). The mutation level, induced by irradiation in the Berezinsky population, was 12,6±1,1% and the frequency of RLM, induced in Vetka population, was 6,6±0,7%. Thus, flies from radiocontaminated area were shown to be much more adapted to irradiation than insects from the control region.

It was interesting to investigate if such adaptation remains after irradiation ceasing. For this population samples were kept under laboratory condition without irradiation for 8 generations Figure 1 shows that the DML frequencies in the control population significantly increased at keeping under such conditions. Acute 30 Gy irradiation was used after 8 generations. Adaptation of Vetka population to irradiation remained.

Besides the control population also became more resistant to ionizing radiation as well as Vetka population.

Table 1. Response of drosophila populations from control and radiocontaminated areas to 40 Gy exposure.

Number	Berezinsky Reserve	Vetka district
Laid eggs	2831	3192
Undeveloped eggs	1786	1365
DLM (%)	63,1±0,9	42,8±0,9*
Laid eggs	2831	3192
Imago	438	1349
Viability (%)	15,5±0,7	42,3±0,9 *
Analyzed chromosomes	831	1429
Mutations	105	95
RLM (%)	12,6±1,1	6,6±0,7*

* p< 0,01

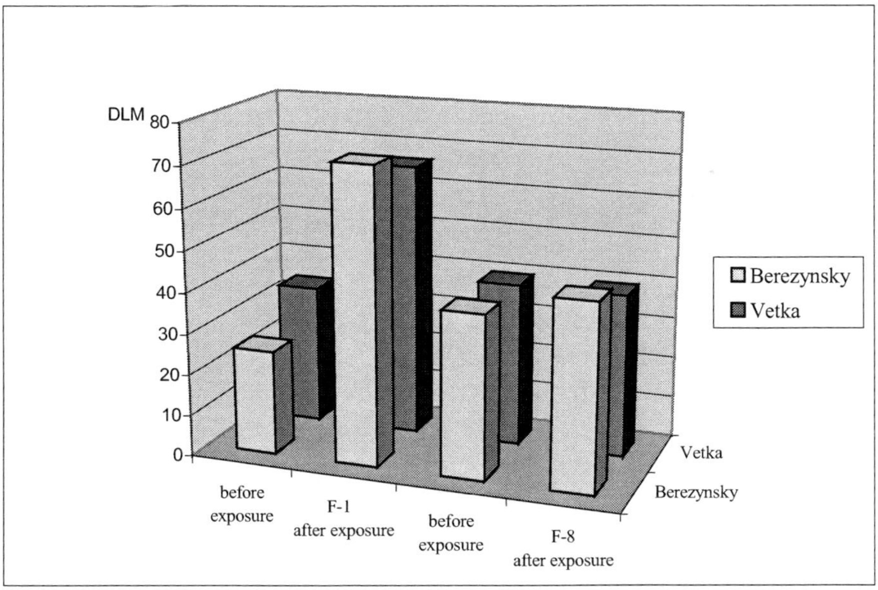

Figure 1. Influence of additional irradiation on the DLM frequency in the natural populations in the 1-st and the 8-st generation after catching.. Berezynsky – population from Berezynsky reserve (control area), Vetka – population from Vetka district.

It means that keeping of natural drosophila populations under laboratory conditions is a strong stress (limited space, overpopulation, other than in nature temperature and light conditions), which increases mutation process and induces non-specific adaptation. Earlier we revealed that adaptation of the natural drosophila population from radiocontaminated areas was non-specific –populations with "radiation history" became more resistant not only to high radiation doses but also to chemical mutagen ethylmethansulfonate (Glushkova *et al.*, 2002). So, prolonged stress can also be a cause of mutation level increase and non-specific adaptation formation.

These facts should be taken into account in studying dynamics of the mutation level during radionuclides removal in animals, which were caught in radiocontaminated regions and placed under vivarium conditions.

3.2 EXPERIMENTAL POPULATION STUDY

In order to study the process of insect adaptation to ionizing radiation in detail we investigated experimental drosophila populations, irradiated with or without radioprotector melanin.

Analysis of DLM and RLM in the 1-st generation after irradiation and then every 5 generations (in the 5-th, 10-th, 15-th and so on) revealed that mutation levels in the constantly irradiated population decreased step by step from the 1-st to 20-25th generations and the RLM level reached the control one. Mutation frequencies of s in the population, supplied with melanin and irradiated, were significantly lower than in the population, irradiated without melanin, and decreased also step by step almost until the control level.

These facts may be explained by an adaptation of exposed populations to radiation. For verification of this suggestion we took samples from populations of all the 4 types (control, supplied with melanin, irradiated and irradiated with melanin) in the 55-th generation and exposed them to 30 Gy x-rays. Results of 30 Gy x-rays. Results of these experiments are presented in Figure 2, which shows that additional mutation frequencies, induced by 30 Gy exposure were the lowest in the population with "radiation history" in comparison with these parameters in other populations. This fact proves that the irradiated population was best adapted to ionizing radiation.

It was surprising that the population, irradiated with melanin, was less resistant than the others. This may be explained by a radioprotective action of this pigment - more radiosensitive genotypes were not eliminated by natural selection and remained in the population owing to melanin

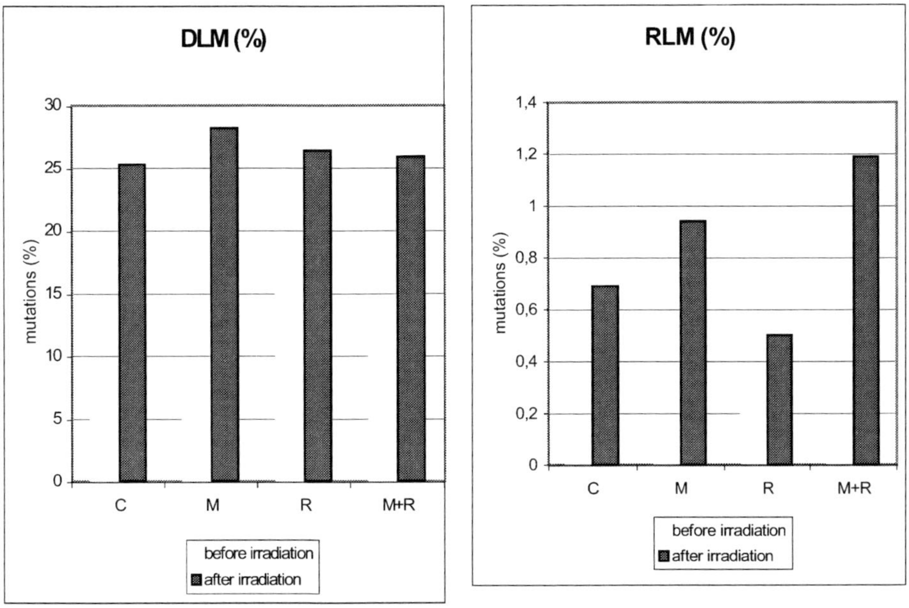

Figure 2. Effects of 30 Gy exposure on the mutation process in experimental drosophila populations of different kinds: C - control; M - melanin; R - radiation; M+R - melanin + radiation. DLM - dominant lethal mutations, RLM - recessive lethal mutations.

protection. Thus population adaptation to radiation results from elimination (death) of radiosensitive genotypes, that's why radioprotector effective against chronic irradiation is able to prevent this process.

4. Conclusions

1. Natural insect populations from radiocontaminated areas are more resistant to additional irradiation than control populations.
2. Keeping of natural populations under laboratory or vivarium conditions is a strong stress (limited space, overpopulation, other than in nature temperature and light conditions), which increases mutation process and induces non-specific adaptation.
3. Pigment melanin is an effective radioprotector against chronic irradiation and is able to prevent process of population adaptation.

References

Ayala, F.I., and Kiger, I.A., 1984, *Modern Genetics*, Univ. of California, Davis.

Ayala, F.I., 1966, Evolution of fitness. Improvement in the productivity and size of irradiated populations of Dr. serrata and Dr. Birchii, *J. Genetics*.**53**:833–895.

Cordeiro A.R., Marques E.K., Veiga-Neto A.J., 1973, Radioresistance of a natural population of Dros. willistoni, living in a radioactive environment, *J. Mutat. Res.* **19**:325–329.

Glushkova T.I., Mosse I.B., Anoshenko I.P., 2002, Genetic monitoring of natural drosophila populations from radiocontaminated regions of Belarus, *J. Radiation biology. Radioecology*. **42**(2):124–129.

Marques E.K., 1973, The development of radioresistance in irradiated Dros. nebulosa populations, *J. Mutat. Res.* **17**:59–72.

Mosse I., Dubovic B., Plotnikova S., Kostrova L., Molophei V., 2001, Melanin is effective radioprotector against chronic irradiation and low radiation doses, in:, *Radiation Protection and Health,* Proc. of IRPA. Regional Congress on Radiation Protection in Central Europe, Dubrovnik, Croatia, pp.1–6.

Mosse I., Molophei V., Plotnikova S., Kostrova L., 2002, Genetic effects of low radiation doses and their modification by different chemicals, in: *Proc. of the European Nuclear Conference,* Foratom, Lille, France, CD/scientific/paperSS/E1/abs.52/ pp.1–5.

Mosse I.B. and Lyakh I.P., 1994, Influence of melanin on mutation load in Drosophila populations after long - term irradiation, *J. Rad. Res.* **139**:356–358.

Mosse I.B., Dubovic B.V., Vilczok T., Plotnikova S., Kostrova L., Subbot S.T., 1996, Melanin decreases remote consequences of long-term irradiation, in: *Proc. of 9-th Int. Congr, on Rad. Protection*, Vienna, Austria, **4**: 168–170.

Mosse I.B., Kostrova L.N., Subbot S.T., Maksimenya I., Molophei V.P., 2000, Melanin decreases clastogenic effects of ionizing radiation in human and mouse somatic cells and modifies the radioadaptive response. *J. Radiation and Envir.Biophysics.* **39**(1):47–52.

Mosse I.B., Plotnikova S., Kostrova L., 1997, Population-genetic aspects of the low doses radiological risk and melanin influence on genetic radiosensitivity, in: *Proc. of Int. Conf. on Low doses of Ionizing Radiation. Biological Effects and Regulatory Control*, Seville, Spain, pp. 104–108.

Mosse, I.B. and Makeeva, E.N., 1994, Genetic monitoring of natural Drosophila populations from Belarus regions with a different level of radiation background, *Proc. of Int. Conf., "Nuclear Power and Industry",* Obninsk, Russia: 279–281.

Notel H., 1976, Investigations on radiosensitive and radioresistant populations of Drosophila melanogaster, *J. Mutat.Res.* **36:**245–248.

MODULATION OF MUTATION RATES AND ADAPTATION OF BACTERIA

IVAN MATIC
INSERM U571 Faculté de Médecine,
Université Paris 5
156 rue Vaugirard, 75015 Paris, France
E-mail: matic@necker.fr

Abstract. The level of genetic variability that maximises the fitness of population varies with the degree of its adaptation to the environment. It is low when environment is stable and high when environment is unstable and hostile. Since environmental conditions are changing, the adaptation is never permanent. Consequently, because the genetic variability in bacteria is first generated by mutagenesis, it could be expected that populations with high mutation rates would have better chance for successful evolution. Indeed, bacteria with elevated mutation rates are frequently found among natural isolates. Experimental observations and theoretical calculations suggest that there must be positive selection for higher mutation rate in spite of the fact that majority of newly generated mutations are deleterious or lethal. Mutator alleles rise to a high frequency through their association with the favourable mutations they generate that counterbalance the load of deleterious mutations. However, when adaptation is achieved, the load of deleterious mutations counterselects high-mutation rates. Therefore, evolution of bacterial populations may happen through alternating periods of high and low mutation rates that provide a remarkable potential for the fine tuning of the rates of generation of genetic variability in the function on the adaptation to environmental conditions.

Keywords: bacteria, mutagenesis, adaptation, evolution

A.A. Cigna and M. Durante (eds.),
Radiation Risk Estimates in Normal and Emergency Situations, 309–317.

1. Introduction

Because newly arisen mutations are mainly deleterious or lethal, it is believed that the mutation rate of living organisms evolved to be as low as possible and that it results from a balance between the effects of deleterious mutations and the metabolic costs of further reducing mutation rates (Drake *et al.,* 1998). Any variant that has increased mutation rates is expected to have reduced fitness due to increased production of deleterious mutations. However, contrary to these intuitive predictions, strains having high mutation rates (mutators) are not rare in natural bacterial populations. They have been found in populations of *E. coli* (Matic *et al.,* 1997), *Salmonella enterica* (Le Clerc *et al.,* 1996), *Neisseria meningitidis* (Richardson *et al.,* 2002), *Haemophilus influenzae* (Watson *et al.,* 2004), *Staphylococcus aureus* (Prunier *et al.,* 2003), *Helicobacter pylori* (Bjorkholm *et al.,* 2001), *Streptococcus pneumoniae* (Del Campo *et al.,* 2005) and *Pseudomonas aeruginosa* (Oliver *et al.,* 2000), with frequencies ranging from 0.1 to above 60%. Experimental (Le Clerc *et al.,* 1998; Mao *et al.,* 1997) and theoretical (Boe *et al.,* 2000) studies indicated that the frequency of mutators observed among natural isolates is much higher than expected from mutation/selection equilibrium, which suggests that there are situations in nature where being mutator confers a selective advantage.

2. Mismatch-repair Deficient Mutators

Mutator phenotypes in general result from alterations in genes coding for DNA repair enzymes and for proteins that assure accuracy of DNA replications. These mutant genes are called mutator alleles. The majority of strong mutators found in the laboratory (*E. coli* (Sniegowski *et al.,* 1997); *S. enterica* serovar Typhimurium (Le Clerc *et al.,* 1998)) and in nature (*E. coli* (Matic *et al.,* 1997); *S. enterica* (Le Clerc *et al.,* 1997); *N. meningitidis* (Richardson *et al.,* 2002); *P. aeruginosa* Del Campo *et al.,* 2005)) have a defective mismatch-repair system (Schofield & Hsieh, 2003) due to the inactivation of *mutS* and *mutL* genes. This repair system controls the fidelity of DNA replication by eliminating biosynthetic errors, and participates in the processing of some DNA lesions and in transcription-coupled repair. In addition, the mismatch-repair system is involved in the maintenance of chromosomal structural integrity and in the control of horizontal gene transfer by preventing recombination between non-identical DNA sequences. The mismatch-repair system involves several proteins, of which two, MutS and MutL, have been highly conserved during evolution. MutS protein recognises seven of eight possible base pair mismatches, C-C mismatches (the least frequent replication error) being refractory. In addition, MutS protein binds unpaired bases, allowing for repair of frameshift errors. The efficiency with which different mismatches are repaired is determined by the different affinities of MutS protein for various mismatches. MutL plays the role of "molecular matchmaker" between MutS-mismatch complex and other proteins involved in the repair process. The inactivation of *mutS* or *mutL* genes results in strong mutator phenotype, with a 10^2-fold increased rate of transition (G:C->A:T and A:T->G:C) and 10^3-fold increased rate of frameshift mutations. In addition, *mutS* or *mutL* knockout mutants have a strong hyper-recombination effect, resulting in 10^1-10^3-fold increase in the rate of chromosomal rearrangements. Molecular characterisation of *E. coli* and *P. aeruginosa* natural *mutS* and *mutL* mutants has revealed that these genes were inactivated by a variety of mechanisms: frameshifts, insertions, premature stop codons and deletions (Oliver *et al.,* 2002; Li *et al.,* 2003).

Any bacterial population is expected to harbour a sub-population of mismatch repair mutants due to spontaneous mutations in mismatch repair genes. The frequency of mismatch repair deficient mutators in cultures of *E. coli* K-12 that are not subjected to any selective pressure was estimated to be less than $3 \cdot 10^{-5}$ (Mao *et al.,* 1997; Boe *et al.,* 2000). How can mismatch repair deficient mutators outgrow non-mutator cells when they produce so many mutations, the vast majority of which are expected to be deleterious? There are two possibilities:

(i) mismatch-repair mutators have higher rates of replication than non-mutators due to the absence of the metabolic load imposed by DNA repair enzymes, or

(ii) (ii) cells with a high mutation rate have a selective advantage because they adapt faster than non-mutators.

These two hypotheses have been experimentally and theoretically tested and the first one was rejected. If there is a selective advantage due to decreased metabolic load, then this advantage should be independent of the initial ratio of mutator to non-mutator cells. However, this is not the case. For example, when *E. coli* K-12 *mutS* mutator population size was about 10^3-fold smaller than that of the non-mutator one, the mutator cells were out-competed in *in vivo* competition experiments (Giraud *et al.,* 2001). However, the mutator outgrows the non-mutator strain when the ratio of mutator: non-mutator population size is above certain threshold (Giraud *et al.,* 2001; Chao, *et al.,* 1983; Labat *et al.,* 2005), which corroborates the hypothesis that mutators are selected because they produce more adaptive mutations. This selective advantage occurs despite the fact that mismatch repair defective mutators start off with a small selective disadvantage (about 1%) relative to non-mutators (Boe *et al.,* 2000; Tröbner & Piechocki, 1984a).

3. Selection of High Mutation Rates

Natural selection favours mutator cells by favouring beneficial mutations generated in these cells due to its mutator phenotype. However, it also indirectly favours the mutator allele due to its physical linkage with beneficial mutations. The linkage between beneficial mutations and mutator alleles is particularly strong in bacteria because the rate of gene exchange in these asexual organisms is, in general, very low. Mutators can increase in frequency even when their initial population size is very small when several beneficial mutations are required for adaptation (Mao *et al.,* 1997; Giraud *et al.,* 2002). For example, if the probability of generating each beneficial mutation is 100-fold higher in mutators than in non-mutators, then the probability that three beneficial mutations will be generated in mutators is 10^6-higher than in non-mutators. The hitchhiking of mutator alleles with beneficial mutations is possible only when fitness gain counterbalances the fitness loss due to generation of deleterious mutations (Taddei *et al.,* 1997).

Selection of mutator alleles also depends on many other phenomena. For example, the increase in frequency of mutators depends on total population size (Tenaillon *et al.,* 1999), on mutator strength (Taddei *et al.,* 1997), and on the rate of gene exchange (Tenaillon *et al.,* 2000). Furthermore, it depends on the stability of the environment. For example, mutator alleles are particularly advantageous upon a shift in environmental conditions (Tanaka *et al.,* 2003; Travis & Travis, 2002). The dynamics of selection of mutators depends also on environmental spatial heterogeneity, which will allow or prevent the competition between the cells carrying different adaptive mutations. Theoretical modelling predicts that mutators will be particularly favoured in temporally and spatially heterogeneous environmental conditions (Travis & Travis, 2004).

4. Counter-selection of Mutators

Experimental and theoretical studies showed that the frequency of mutator strains in a population could rapidly increase to almost 100%. However, the majority of natural isolates are not mutators. A major factor that diminishes the fitness of mutators is the continuous production of deleterious mutations, particularly in a stable environment, once adaptation is achieved (Giraud *et al.,* 2001). A second factor is that neutral, beneficial and deleterious mutations can have very different impacts on fitness in different environments. Consequently, no single genotype is optimally adapted to all environments. For example, an adaptive mutation in one environment can be deleterious in another (a phenomenon called antagonistic pleiotropy). Therefore, migration from one environment to another might contribute to the reduction of fitness of mutators in natural populations, as observed in "*in vivo*" laboratory experiments for *E. coli* (Giraud *et al.,* 2001) and *S. enterica* serovar Typhimurium (Nilsson *et al.,* 2004) *mutS* mutators. Finally, continuous passages through strong bottlenecks result in the accumulation of deleterious mutations due to genetic drift. This phenomenon, called Muller's ratchet, is particularly deleterious for strong mutator populations (Funchain *et al.,* 2000). The loss of fitness of mutator strains should be positively correlated to the mutator strength (Taddei *et al.,* 1997).

Therefore, in the long run, fitness cost associated with high mutation rates is expected to cause the loss of adaptive mutations from bacterial populations, with consequential elimination of the mutator genome. However, some adaptive mutations generated in mutator backgrounds can be saved either through their horizontal transfer to non-mutator background or when the mutation rate of the adapted mutator strain is reduced before the load of deleterious mutations becomes too high. The reduction of mutation rate might be achieved by several mechanisms: reversion of the mutator mutation, acquisition of suppressor mutations (e.g., in population of *mutT* mutators (Tröbner & Piechocki, 1984b)), or reacquisition of a wild-type allele of antimutator gene *via* horizontal gene exchange (see below and in Denamur *et al.,* 2000).

5. Modulation of Mutation Rates and Antibiotic Resistance

Particularly interesting example of how selection modulates mutation rates is provided by the studies of the link between mutation rates and antibiotic resistance. *In vitro* and *in vivo* studies have provided clear evidence that antibiotic treatment can contribute to the selection of the mutators (Mao *et al.,* 1997; Giraud *et al.,* 2002). Mutators are favoured under such conditions because they generate antibiotic-resistance-conferring mutations at a higher rate than non-mutators do. This phenomenon is not specific to one antibiotic or for one family of antibiotics, suggesting that mutations can contribute to various modes of evolution of antibiotic resistance. Indeed, whenever mutations can confer or increase resistance to the antibiotics, or reduce the biological cost of resistance on bacterial fitness, it is more likely that they will appear in the populations of cells with higher mutation rates. Furthermore, when resistance is associated with the acquisition of several mutations, the advantage of being a mutator increases significantly (Tenaillon *et al.,* 1999). For example, the sequential appearance of several mutations responsible for the evolution of extended spectrum β-lactamases is favoured by high mutation rates (Orencia *et al.,* 2001). Therefore, the presence of mutator bacteria is a risk factor in the treatment of infectious diseases, as they might be responsible for therapeutic failures.

A positive correlation between multiple-antibiotic resistance and high mutation rates is expected to be frequent in natural populations of bacterial pathogens (Blazquez, 2003). Such a correlation was found for *P. aeruginosa* isolated from lungs of patients with cystic fibrosis, bronchiectasis and chronic obstructive pulmonary disease (Oliver *et al.*, 2000; Macia *et al.*, 2005). This is probably due to extensive antibiotic therapy, chronicity of the infection, strong compartmentalisation and low migration rates that limit competition between strains and, therefore, increase the probability of persistence of strong mutators. Indeed, the prevalence of *P. aeruginosa* mutator strains increases with time in the lungs of cystic fibrosis patients (Ciofu *et al.*, 2005). However, this correlation was not observed in *P. aeruginosa* isolated from acutely infected patients (Le Clerc *et al.*, 1996). The explanation of this discrepancy is provided by the studies of relationship between antibiotic resistance and mutation rates in *E. coli* natural populations.

The correlation between multiple-antibiotic resistance and high mutation rates was also not found in *E. coli* natural populations, despite the capacity of strong *E. coli* mutators to generate mutations conferring resistance to antibiotics *in vivo* and *in vitro* (Mao *et al.*, 1997; Giraud *et al.*, 2002; Denamur *et al.*, 2002; Tanabe *et al.*, 1999). Among *E. coli* natural isolates, the highest frequency of antibiotic resistance was found among weak mutators, while strains with high mutation rates had significantly lower antibiotic resistance (Denamur *et al.*, 2005). Such absence of the correlation between high mutation rates and antibiotic resistance can be explained by the dynamics of selection and counter-selection of mutator alleles. Selection acting on mutator alleles is modulated by many factors including changes in the environment and the opportunity for competition between strains. Mutator clones can rapidly specialise to one environment, which renders them less fit in other environments (Giraud *et al.*, 2001). The initial gain and long-term loss of fitness is proportional to the strength of the mutator allele. Computer simulations predict that weak mutators will be selected less rapidly than strong mutators but, once selected, they will persist for much longer than strong mutators (Taddei *et al.*, 1997). Therefore, in bacterial populations that cycle between different environments, e.g., *E. coli*, strong mutators (around 1% (Matic *et al.*, 1997; Baquero *et al.*, 2004)) can be rapidly selected, but cannot persist. Such selective conditions probably favour weak mutator strains (around 10-30% (Matic *et al.*, 1997; Baquero *et al.*, 2004)) that, in the long run, might have more chance to accumulate multiple antibiotic resistances than the strong mutators (Denamur *et al.*, 2005). In conclusion, in the function of spatial and temporal environmental variations, and consequently, of the nature and the strength of selective pressures, the mutation rate is constantly adjusted to a level that favours adaptation to a given environment.

References

Baquero M.R., A.I. Nilsson, M.C. Turrientes, D. Sandvang, J.C. Galan, J.L. Martinez, N. Frimodt-Moller, F. Baquero, and D.I. Andersson, 2004, Polymorphic mutation frequencies in *Escherichia coli*: emergence of weak mutators in clinical isolates, *J. Bacteriol.* **186**: 5538–5542.

Bjorkholm B., M. Sjolund, P.G. Falk, O.G. Berg, L. Engstrand, and D.I. Andersson, 2001, Mutation frequency and biological cost of antibiotic resistance in *Helicobacter pylori*, *Proc. Natl. Acad. Sci. U S A* **98**: 14607–1412.

Blazquez J., 2003, Hypermutation as a factor contributing to the acquisition of antimicrobial resistance, *Clin. Infect. Dis.* **37**: 1201–1209.

Boe L., M. Danielsen, S. Knudsen, J.B. Petersen, J. Maymann, and P.R. Jensen, 2000, The frequency of mutators in populations of *Escherichia coli*, *Mutat Res* **448**: 47–55.

Chao L., and E.C. Cox, 1983, Competition between high and low mutating strains of *Escherichia coli*, *Evolution* **37**: 125–134.

Ciofu O., B. Riis, T. Pressler, H.E. Poulsen, and N. Hoiby, 2005, Occurrence of hypermutable *Pseudomonas aeruginosa* in cystic fibrosis patients is associated with the oxidative stress caused by chronic lung inflammation, *Antimicrob. Agents Chemother.* **49**: 2276–2282.

del Campo R., M.I. Morosini, E.G. delaPedrosa, A. Fenoll, C. Munoz-Almagro, L. Maiz, F. Baquero, and R. Canton, 2005, Population structure, antimicrobial resistance, and mutation frequencies of *Streptococcus pneumoniae* isolates from cystic fibrosis patients, *J. Clin. Microbiol.* **43**: 2207–2214.

Denamur E., G. Lecointre, P. Darlu, O. Tenaillon, C. Acquaviva, C. Sayada, I. Sunjevaric, R. Rothstein, J. Elion, F. Taddei, M. Radman, and I. Matic, 2000, Evolutionary implications of the frequent horizontal transfer of mismatch repair genes, *Cell* **103**: 711–721.

Denamur E., S. Bonacorsi, A. Giraud, P. Duriez, F. Hilali, A. Amorin, E. Bingen, A. Andremont, B. Picard, F. Taddei, and I. Matic, 2002, High frequency of mutators strains among human uropathogenic *Escherichia coli* isolates, *J. Bacteriol.* **184**: 605–609.

Denamur E., O. Tenaillon, C. Deschamps, D. Skurnik, E. Ronco, J.L. Gaillard, B. Picard, C. Branger, and I. Matic, 2005, Intermediate mutation frequencies favor evolution of multidrug resistance, *Genetics* **171**: 825–827.

Drake J.W., B. Charlesworth, D. Charlesworth, and J.F. Crow, 1998, Rates of spontaneous mutation, *Genetics* **148**: 1667–1686.

Funchain P., A. Yeung, J.L. Stewart, R. Lin, M.M. Slupska, and J.H. Miller, 2000, The consequences of growth of a mutator strain of *Escherichia coli* as measured by loss of function among multiple gene targets and loss of fitness, *Genetics* **154**: 959–970.

Giraud A., I. Matic, O. Tenaillon, A. Clara, M. Radman, M. Fons, and F. Taddei, 2001, Costs and benefits of high mutation rates: adaptive evolution of bacteria in the mouse gut., *Science* **291**: 2606–2608.

Giraud A., I. Matic, M. Radman, M. Fons, and F. Taddei, 2002, Mutator bacteria as a risk factor in treatment of infectious diseases, *Antimicrob. Agents Chemother.* **46**: 863–865.

Gutierrez O., C. Juan, J.L. Perez, and A. Oliver, 2004, Lack of association between hypermutation and antibiotic resistance development in *Pseudomonas aeruginosa* isolates from intensive care unit patients, *Antimicrob. Agents Chemother.* **48**: 3573–3575.

Labat F., O. Pradillon, L. Garry, M. Peuchmaur, B. Fantin, and E. Denamur, 2002, Mutator phenotype confers advantage in *Escherichia coli* chronic urinary tract infection pathogenesis, *FEMS Immunol. Med. Microbiol.* **44:** 317–321.

LeClerc J.E., B. Li, W.L. Payne, and T.A. Cebula, 1996, High mutation frequencies among *Escherichia coli* and *Salmonella* pathogens, *Science* **274**: 1208–1211.

LeClerc J.E., W.L. Payne, E. Kupchella, and T.A. Cebula, 1998, Detection of mutator subpopulations in *Salmonella typhimurium* LT2 by reversion of his alleles, *Mutat Res* **400**: 89–97.

Li B., H.C. Tsui, J.E. LeClerc, M. Dey, M.E. Winkler, and T.A. Cebula, 2003, Molecular analysis of *mutS* expression and mutation in natural isolates of pathogenic *Escherichia coli*, *Microbiology* **149**: 1323–1331.

Macia M.D., D. Blanquer, B. Togores, J. Sauleda, J.L. Perez, and A. Oliver, 2005, Hypermutation is a key factor in development of multiple-antimicrobial resistance in *Pseudomonas aeruginosa* strains causing chronic lung infections, *Antimicrob. Agents Chemother.* **49**: 3382–3386.

Mao E.F., L. Lane, J. Lee, and J.H. Miller, 1997, Proliferation of mutators in a cell population, *J. Bacteriol.* **179**: 417–422.

Matic I., M. Radman, F. Taddei, B. Picard, C. Doit, E. Bingen, E. Denamur, and J. Elion, 1997, Highly variable mutation rates in commensal and pathogenic *Escherichia coli*, *Science* **277**: 1833–1834.

Nilsson A.I., E. Kugelberg, O.G. Berg, and D.I. Andersson, 2004, Experimental adaptation of *Salmonella typhimurium* to mice, *Genetics* **168**: 1119–1130.

Oliver A., R. Canton, P. Campo, F. Baquero, and J. Blazquez, 2000, High frequency of hypermutable *Pseudomonas aeruginosa* in cystic fibrosis lung infection, *Science* **288**: 1251–1254.

Oliver A., F. Baquero, and J. Blazquez, 2002, The mismatch repair system (*mutS*, *mutL* and *uvrD* genes) in *Pseudomonas aeruginosa*: molecular characterization of naturally occurring mutants, *Mol. Microbiol.* **43**: 1641–1650.

Orencia M.C., J.S. Yoon, J.E. Ness, W.P. Stemmer, and R.C. Stevens, 2001, Predicting the emergence of antibiotic resistance by directed evolution and structural analysis, *Nat. Struct. Biol.* **8:** 238–242.

Prunier A.L., B. Malbruny, M. Laurans, J. Brouard, J.F. Duhamel, and R. Leclercq, 2003, High rate of macrolide resistance in *Staphylococcus aureus* strains from patients with cystic fibrosis reveals high proportions of hypermutable strains, *J. Infect. Dis.* **187**: 1709–1716.

Richardson A.R., Z. Yu, T. Popovic, and I. Stojiljkovic, 2002, Mutator clones of *Neisseria meningitidis* in epidemic serogroup A disease, *Proc. Natl. Acad. Sci. U S A* **99**: 6103–6107.

Schofield M.J., and P. Hsieh, 2003, DNA mismatch repair: molecular mechanisms and biological function, *Annu. Rev. Microbiol.* **57**: 579–608.

Sniegowski P.D., P.J. Gerrish, and R.E. Lenski, 1997, Evolution of high mutation rates in experimental populations of *E. coli*, *Nature* **387**: 703–705.

Taddei F., M. Radman, J. Maynard-Smith, B. Toupance, P.H. Gouyon, and B. Godelle, 1997, Role of mutators in adaptive evolution, *Nature* **387**: 700–702.

Tanabe K., T. Kondo, Y. Onodera, and M. Furusawa, 1999, A conspicuous adaptability to antibiotics in the *Escherichia coli* mutator strain, *dnaQ49*, *FEMS Microbiol. Lett.* **176**: 191–196.

Tanaka M.M., C.T. Bergstrom, and B.R. Levin, 2003, The evolution of mutator genes in bacterial populations: the roles of environmental change and timing, *Genetics* **164**: 843–854.

Tenaillon O., B. Toupance, H. Le Nagard, F. Taddei, and B. Godelle, 1999, Mutators, population size, adaptive landscape and the adaptation of asexual populations of bacteria, *Genetics* **152**: 485–493.

Tenaillon O., H. Le Nagard, B. Godelle, and F. Taddei, 2000, Mutators and sex in bacteria: conflict between adaptive strategies, *Proc. Natl. Acad. Sci. U S A* **97**, 10465–10470.

Travis J.M., and E.R. Travis, 2002, Mutator dynamics in fluctuating environments, *Proc. R. Soc. Lond. B Biol. Sci.* **269**: 591–597.

Travis E.R., and J.M. Travis, 2004, Mutators in space: the dynamics of high-mutability clones in a two-patch model, *Genetics* **167**: 513–522.

Tröbner W., and R. Piechocki, 1984a, Competition between isogenic *mutS* and mut^+ populations of *Escherichia coli K12* in continously growing cultures, *Mol. Gen. Genet.* **198**: 175–176.

Tröbner W., and R. Piechocki, 1984b, Selection against hypermutability in *Escherichia coli* during long term evolution, *Mol. Gen. Genet.* **198**: 177–178.

Watson M.E.J., J.L. Burns, and A.L. Smith, Hypermutable *Haemophilus influenzae* with mutations in *mutS* are found in cystic fibrosis sputum.*Microbiology* **150**: 2947–2958.

NUCLEAR CYTOPLASMIC INTERACTION HYPOTHESIS AND THE ROLE OF TRANSLOCATIONS IN *NICOTIANA* ALLOPOLYPLOIDS

ANDREW R. LEITCH*, KAR YOONG LIM,

School of Biological and Chemical Sciences, Queen Mary, University of London, E1 4NS, UK

KAMILA SKALICKA, ALES KOVARIK

Institute of Biophysics, Academy of Sciences of the Czech Republic, CZ-61265 Brno, Czech Republic

*To whom correspondence should be addressed.

Abstract. The nuclear cytoplasmic interaction hypothesis (NCI) states that in a newly formed allopolyploid genetic instabilities are induced giving rise to altered paternal genome structure and chromosomal translocations. The hypothesis predicts that plants emerging from a "bottleneck of sterility" are stable, with increased fertility, fixed for particular translocations that are "species-specific", and have a degraded paternal genome. We investigate this hypothesis in the allopolyploids *Nicotiana tabacum* (tobacco), *N. rustica* and *N. arentsii*. Each of these natural allopolyploids have a similar chromosome complement, 2n = 4x = 48. We review the cytological data available for these species. From those studies using genomic *in situ* hybridisation (GISH) we found evidence in support of NCI only in *N. tabacum*. To our surprise there is also supporting evidence in the form of structurally similar translocations in a synthetic tobacco line that is only three generations old. These data suggest that the mechanisms of genetic change act early and fast. However in the synthetic material no translocation resolvable by GISH had gone to fixation. Nevertheless the presence of translocations does support the argument that in natural tobacco at least the genomic restructuring that occurred after polyploidy may have facilitated the establishment and stabilisation of the polyploid genome.

Keywords: allopolyploidy, genomic shock, translocations, NCI.

A.A. Cigna and M. Durante (eds.),
Radiation Risk Estimates in Normal and Emergency Situations, 319–326.

1. Introduction

Polyploidy, in which the entire chromosome complement multiplies (3x, 4x…) has recently been the subject of many reviews (Adams & Wendel 2005; Levy & Feldman 2002; Levy & Feldman 2004; Liu & Wendel 2003; Osborn *et al.,* 2003; Soltis & Soltis 1999; Soltis *et al.,* 2004a; Soltis *et al.,* 2004b; Soltis & Soltis 2000). Interest stems from the abundance of plants that are polyploid, indeed perhaps as high as 30%-70% of all angiosperms are recognisably polyploid based on chromosome counts. Indeed recent sequencing studies suggest that all angiosperms have polyploidy as some point in their ancestry (Lagercrantz & Lydiate 1996). Given the high number of agricultural crops that are polyploid (e.g. cotton, wheat, maize, banana), a genetic understanding of polyploidy is essential for plant breeding. Furthermore polyploids can have a profound impact on ecosystems (Ainouche *et al.,* 2004; Brochmann *et al.,* 2004; Leitch & Bennett 1997). Allopolyploidy occurs when two related species hybridise and the genome multiplies, e.g. two species with genome designations AA and BB gives rise to an allopolyploid with genome designation AABB. In a single step the allopolyploid can be genetically isolated from its parents. This event is perhaps the most rapid and best-understood form of sympatric speciation. The genetic consequence of the union of two genomes that may have diverged from each other and their common ancestor many thousands or millions of years ago can potentially be profound. The outcome however is not easily predicted. Indeed for some classes of DNA, e.g. amplified length polymorphisms (AFLPs), there can be significant changes observed immediately after the formation of a polyploid (e.g. in *Aegilops-Triticum* synthetic polyploids) (Ozkan *et al.,* 2001) while in others no changes are recorded (e.g. in *Gossypium* synthetic polyploids) (Liu *et al.,* 2001).

This paper considers the nuclear cytoplasmic inheritance hypothesis (NCI) proposed by Gill (1991) and extended by Song (1995). There are two components to this hypothesis. (i) The paternal genome in the newly formed allopolyploid is in the alien environment of the maternal cytoplasm. This renders the paternal genome unstable and vulnerable to genetic change. For example, analysis of restriction fragment length polymorphisms (RFLPs) in synthetic *Brassica napus* revealed that the paternal genome was more unstable than the maternal genome (Song *et al.,* 1995). Furthermore in synthetic tobacco, investigated tandem and dispersed repeats are lost specifically from the paternal genome (Skalicka *et al.,* 2005). (ii) The newly formed polyploid must pass through a bottleneck of sterility that is overcome by “species-specific translocations”. These translocations are thought to enhance fertility and cytoplasmic compatibility in the newly formed allopolyploids (Gill 1991). Jiang & Gill (1994a) supported this argument in their analyses of two tetraploid wheats that have "species-specific" translocations and Leitch and Bennett (1997) noted that two cultivars of tobacco shared the same translocations (Leitch & Bennett 1997). Lim *et al.* (2004) extended the latter analysis to a larger range of tobacco cultivars and observed "species-specific" translocations common to all tobaccos investigated (Lim *et al.,* 2004).

Here we investigate the NCI hypothesis by extending the analysis of Lim *et al.* (2004) on tobacco. Tobacco is an allotetraploid (2n = 4x = 48) that formed up to 0.2 million years ago (Clarkson *et al.,* 2005). DNA sequencing, morphology and chromosomal analyses reveal that the closest living parents are *N. sylvestris* (2n = 2x = 24, section *Sylvestres*) the maternal S-genome donor (Bland *et al.,* 1985; Chase *et al.,* 2003; Goodspeed 1954; Kenton *et al.,* 1993; Lim *et al.* 2000; Mosconne *et al.* 1996) and *N. tomentosiformis* (2n = 2x = 24, section *Tomentosae*) the paternal, T-genome donor (Chase *et al.,* 2003; Murad *et al.* 2004). DNA sequence also reveals that tobacco is monophyletic (Chase *et al.,* 2003; Clarkson *et al.,* 2004). Here we compare the genome structure of natural tobacco with that of synthetic tobacco plants. The most successful attempt to reconstruct tobacco is the Th37 line made by Burk (1973). This was made by crossing *N. sylvestris* (maternal parent) with *N. tomentosiformis* (paternal parent) and the hybrid chromosomally doubled by tissue culture. A single plant was recovered that was used to generate breeding material, generations 3 and 4 were studied (Skalicka *et al.,* 2003; Skalicka *et al.,* 2005) and reviewed here.

2. Chromosome Translocations in *Nicotiana* Polyploids

Lim *et al.* (2004) reported that a "species-specific" translocation might occur on chromosome T9/s and cultivar-specific translocations on T9/s, S2/t, and S11/t (Fig. 1D). These translocations are resolved using genomic *in situ* hybridisation (GISH). In this method total genomic DNA from each of the progenitor diploid species are labelled in different ways and hybridised simultaneously to chromosomes of tobacco *in situ*. The two labels are then detected using different fluorochromes that fluoresce at different wavelengths. The method clearly resolves the S- and the T-genome of tobacco and reveals in the different colours chromosomes carrying translocations. The "species-specific" translocation does lend support to the hypothesis that translocations have been important in the early evolution of tobacco.

The GISH data also lends support to the NCI hypothesis that the paternal genome is the most unstable since the *N. tomentosiformis* total genomic DNA probe produces a less clearly resolved signal that the *N. sylvestris* total genomic DNA probe (compare the clarity of the yellow and red chromosomes in Fig. 1D). These data are also consistent with the observed losses of tandem and dispersed repeats specifically from the *N. tomentosiformis*-derived genome in natural tobacco (Lim *et al.* , 2004) and its synthetic hybrids (Skalicka *et al.,* 2005). These changes have all occurred with a time frame of less than 0.2 M years, the maximum expected age of tobacco (Clarkson *et al.,* 2005).

However GISH has also been applied to two other *Nicotiana* allopolyploids (a) *N. arentsii* (2n = 4x = 48, WWUU), with diploid progenitors most closely related to *N. wigandioides* (2n = 2x = 24) the W genome donor and *N. undulata* (2n = 2x = 24) the U genome donor and; (b) *N. rustica* (2n = 4x = 48, UUPP) with diploid progenitors most closely related to *N. undulata* (U genome donor) and *N. paniculata* (2n = 2x = 24) the P genome donor. In both of these allopolyploids translocations were not resolved using GISH (Lim *et al.,* 2004; Lim *et al.,* 2005).

An analysis of the Th37 lines revealed the surprising observation that it shares translocations observed in normal tobacco. Two plants carried a translocation that looked very similar to S2/t found in all cultivars and T9/s found in all tobaccos studied to date (Skalicka *et al.,* 2005). These can be seen by comparing the karyotypes of tobacco cv. 095-55 (Lim *et al.,* 2004) with that from Th37 plant number 1 (Skalicka *et al.,* 2005) shown in Fig. 1. Thus translocations similar to those reported as "cultivar-specific" and "species-specific" are also seen in early generations of synthetic tobacco. Perhaps "cultivar-specific" translocations are a response to selection pressure derived through breeding and "species-specific" translocations a response to recover fertility in the early generations of this synthetic material as predicted from the NCI hypothesis. Nevertheless if this is so it is not an ubiquitous response applicable to all species even in *Nicotiana* or indeed in all of the synthetic tobaccos.

3. Conclusion

It has been proposed that the process of allopolyploidy perturbs the genome such that the "genomic shock" results in allopolyploid-induced genetic changes. Genomic shock may be responsible for the activation of transposons (McClintock 1984), retrotransposons (Kashkush *et al.,* 2003; Melayah *et al.,* 2004), genomic translocations (Gill 1991), the loss and gain of DNA sequences and epigenetic reprogramming (Kashkush *et al.,* 2002; Levy & Feldman 2004; Ozkan *et al.,* 2001).

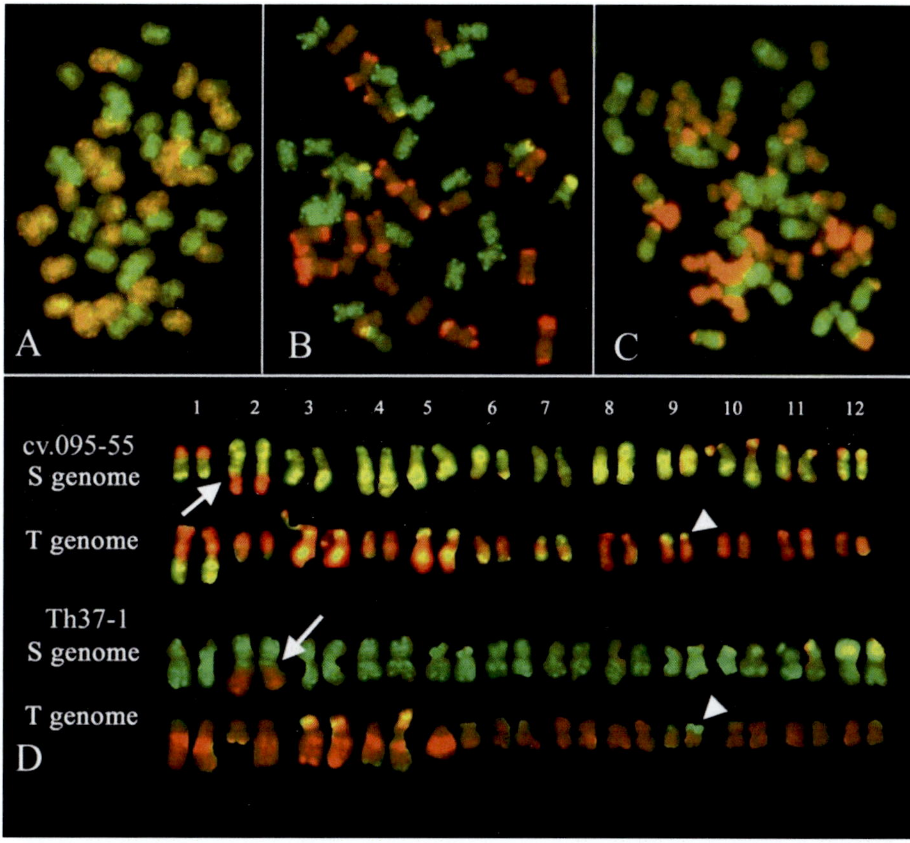

Figure 1. A.-D. GISH to metaphase chromosome spreads of (A) *N. arentsii* – U genome in yellow, W genome in orange, (B) *N. rustica* – U genome in green, P genome in red (C) *N. tabacum* cv. 095-55, S genome in green, T genome in orange. (D) Karyotypes of 095-55 and of synthetic tobacco Th37.1. The chromosomes are arranged with the S genome above (yellow chromosomes) and the T genome below (red or orange chromosomes). Note the translocation on the S genome chromosome number 2 (called S2/t, arrows) and on the T genome chromosome number 9 (T9/s, arrowheads) in both the natural and synthetic tobacco.

In tobacco it may also lead to genomic restructuring including chromosomal translocations and perturbations to the paternally derived genome (Skalicka *et al.,* 2005) as predicted by the NCI hypothesis. However these phenomena were not observed in the allopolyploids *N. arentsii* or *N. rustica* indicating that they are a phenomenon of tobacco and not a generality in *Nicotiana*. Perhaps the translocations in tobacco formed as a random polyploid-induced process in the *de novo* polyploid. Subsequently, in a pool of altered karyotypes, only those that gave rise to fertile offspring survived and the fittest combinations proliferated. Probably fixation of genome restructuring occurs in the first few generations, thereafter change might be much slower. None of the changes we observed in the synthetic tobaccos have gone to fixation, and the material remains genetically unstable. However, individual lineages might be stabilized since plants were mostly homozygous for "species-specific" translocations. But two out of three lines do already closely resemble tobacco in terms of their translocations and this points to a common outcome affecting the synthetic tobacco as may have occurred to tobacco in nature. The remaining lineage showed a perfectly "additive" karyotype without translocations as in *N. rustica* and *N. arentsii* indicating that intergenomic translocation events are not an essential condition of successful establishment of a fertile *N. tomentosiformis* x *N. sylvestris* hybrid. It will be worthwhile following subsequent generations of these plants and monitor fitness with regard to the occurrence of translocations.

References

Adams K.L., Wendel J.F., 2005, Polyploidy and genome evolution in plants. *Current Opinion in Plant Biology* **8**:135–141.

Ainouche M.L., Baumel A., Salmon A., 2004, *Spartina anglica* C.E. Hubbard: a natural model system for studying early evolutionary changes that affect allopolyploid genomes. *Biological Journal of the Linnean Society*.

Bland M.M., Matzinger D.F., Levins C.S., 1985, Comparison of the mitochondrial genome of *Nicotiana tabacum* with its progenitor species. *Theoretical and Applied Genetics* **69**:535–541.

Brochmann C., Brysting A.K., Alsos I.G., Borgen L., Grundt H.H., Scheen A.-C., Elven R., 2004, Polyploidy in arctic plants. *Biological Journal of the Linnean Society* **82**: 521–536.

Chase M.W., Knapp S., Cox A.V., Clarkson J.J., Butsko Y., Joseph J., Savolainen V., Parokonny A.S., 2003, Molecular systematics, GISH and the origin of hybrid taxa in Nicotiana (Solanaceae). *Annals of Botany* **92**:107–127.

Clarkson J.J., K.Y.L., A.K., Chase M., Knapp S., A.R.L., 2005, Long-term genome diploidization in allopolyploid *Nicotiana* section *Repandae* (Solanaceae). *New Phytologist* in press.

Clarkson J.J., Knapp S., Garcia V.F., Olmstead R.G., Leitch A.R., Chase M.W., 2004, Phylogenetic relationships in Nicotiana (Solanaceae), inferred from multiple plastid DNA regions. *Molecular Phylogenetics and Evolution* **33**:75–90.

Gill B.S. (1991, Nucleocytoplasmic interaction , NCI) hypothesis of genome evolution and speciation in polyploid plants. In *Proceedings of the Kihara Memorial International Symposium on cytoplasmic engineering in wheat* , Sasakuma, T., Kinoshita, T., ed.), : 48–53, Yokohama, Japan.

Goodspeed T.H., 1954, *The genus Nicotiana*, Chronica Botanica Company, Massachusetts, USA.

Jiang J., Gill B.S., 1994, Different species-specific chromosome translocations in *Triticum timopheevii* and *T. turgidum* suort the diphyletic origin of polyploid wheats. *Chromosome Research* **2**:59–64.

Kashkush K., Feldman M., Levy A.A., 2002, Gene loss, silencing and activation in a newly synthesized wheat allotetraploid. *Genetics* **160**:1651–1659.

Kashkush K., Feldman M., Levy A.A., 2003, Transcriptional activation of retrotransposons alters the expression of adjacent genes in wheat. *Nature Genetics* **33**:102–106.

Kenton A., Parokonny A.S., Gleba Y.Y., Bennett M.D., 1993, Characterization of the *Nicotiana tabacum* L. genome by molecular cytogenetics. *Molecular and General Genetics* **240**:159–169.

Lagercrantz U., Lydiate D.J., 1996, Comparative genome mapping in *Brassica. Genetics* **144**:1903–1910.

Leitch I.J., Bennett M.D., 1997, Polyploidy in angiosperms. *Trends in Plant Science* **2**:470–476.

Levy A.A., Feldman M., 2002, The impact of polyploidy on grass genome evolution. *Plant Physiology* **130**:1587–1593.

Levy A.A., Feldman M., 2004, Genetic and epigenetic reprogramming of the wheat genome upon allopolyploidization. *Biological Journal of the Linnean Society*, **82**: 607–613.

Lim K.Y., Matyasek R., Kovarik A., Leitch A.R., 2004, Genome evolution in allotetraploid *Nicotiana. Biological Journal of the Linnean Society* **82**:599–606.

Lim K.Y., Matyasek R., Lichtenstein C.P., Leitch A.R., 2000, Molecular cytogenetic analyses and phylogenetic studies in the *Nicotiana* section Tomentosae. *Chromosoma* **109**:245–258.

Lim K.Y., R.M., A.K., J.F., A.R.L., 2005, Molecular cytogenetics and tandem repeat sequence evolution in the allopolyploid *Nicotiana rustica* compared with diploid progenitors *N. paniculata* and *N. undulata. Cytogenetics and Genome research* **109**:298–309.

Liu B., Brubaker C.L., Mergeai G., Cronn R.C., Wendel J.F., 2001, Polyploid formation in cotton is not accompanied by rapid genomic changes. *Genome* **44**:321–330.

Liu B., Wendel J.F., 2003, Epigenetic phenomena and the evolution of plant allopolyploids. *Molecular Phylogenetics and Evolution* **29**:365–379.

McClintock B., 1984, The significance of responses of the genome to challenge. *Science* **226**:792–801.

Melayah D., Lim K.Y., Bonnivard E., Chalhoub B., Dorlhac de Borne F., Mhiri C., Leitch A.R., Grandbastien M.-A., 2004, Distribution of the Tntl retrotransposon family in the amphidiploid tobacco , *Nicotiana tabacum*, and its wild *Nicotiana* relatives. *Biological Journal of the Linnean Society* **82**:in press.

Mosconne E.A., Matzke M.A., Matzke A.J.M., 1996, The use of combined FISH/GISH in conjunction with DAPI counterstaining to identify chromosomes containing transgene inserts in amphidiploid tobacco. *Chromosoma* **105**:231–236.

Murad L., Bielawski J.P., Matyasek R., Kovarik A., Nichols R.A., Leitch A.R., Lichtenstein C.P., 2004, The origin and evolution of geminivirus-related DNA sequences in *Nicotiana. Heredity* in press.

Osborn T.C., Pires J.C., Birchler J.A., Auger D.L., Chen Z.J., Lee H.S., Comai L., Madlung A., Doerge R.W., Colot V., Martienssen R.A., 2003, Understanding mechanisms of novel gene expression in polyploids. *Trends in Genetics* **19**:141–147.

Ozkan H., Levy A.A., Feldman M., 2001, Allopolyploidy-induced rapid genome evolution in the wheat (*Aegilops-Triticum*) group. *Plant Cell* **13**:1735–1747.

Skalicka K., Lim K.Y., Matyasek R., Koukalova B., Leitch A.R., Kovarik A., 2003, Rapid evolution of parental rDNA in a synthetic tobacco allotetraploid line. *American Journal of Botany* **90**:988–996.

Skalicka K., Lim Y., Matyasek R., Matzke M., Leitch A., Kovarik A., 2005, Preferential elimination of repeated DNA sequences from the paternal, *N. tomentosiformis* genome donor of a synthetic, allotetraploid tobacco. *New Phytologist* **166**:291–303.

Soltis D.E., Soltis P.S., 1999, Polyploidy: recurrent formation and genome evolution. *Trends in Ecology & Evolution* **14**:348–352.

Soltis D.E., Soltis P.S., Pires J.C., Kovarik A., Tate J.A., Mavrodiev E., 2004a, Recent and recurrent polyploidy in *Tragopogon* (Asteraceae): cytogenetic, genomic, genomic and genetic comparisons. *Biological Journal of the Linnean Society.*, **82**: 485–501.

Soltis D.E., Soltis P.S., Tate J.A., 2004b, Advances in the study of polyploidy since Plant speciation. *New Phytologist* **161**:173–191.

Soltis P.S., Soltis D.E., 2000, The role of genetic and genomic attributes in the success of polyploids. *Proceedings of the National Academy of Sciences of the United States of America* **97**:7051–7057.

Song K.M., Lu P., Tang K.L., Osborn T.C., 1995, Rapid genome change in synthetic polyploids of *Brassica* and its implications for polyploid evolution. *Proceedings of the National Academy of Sciences of the United States of America* **92**:7719–7723.

MOLECULAR DYNAMICS STUDY OF RADIOSENSITIVE MUTANT ALLELE OF PROTEIN KINASE ycdc28-srm [G20S] USING hcdk2 AS MODEL

DMITRY KRETOV, NATALIA KOLTOVAYA
KHOLMIRZO KHOLMURODOV*
Joint Institute for Nuclear Research,
Laboratory of Radiation Biology,
M. Reg., Dubna 141980; Russia

*To whom correspondence should be addressed.

Abstract. The cyclin-dependent kinases play an essential role in the timing of cell division and repair, furthermore a high incidence of genetic alterations of CDKs or deregulation of CDK inhibitors have been observed in several cancers. These arguments make of CDC28 of *Saccharomyces cerevisiae* yeast a very attractive model to study CDK regulation mechanisms. We observed certain gene mutations, including *cdc28-srm* [G20S], affect cell cycle progression, maintenance of different genetic structures (Devin, 1990), checkpoint-control (Li, 1997) and increase cell radiosensitivity (Koltovaya, 1998). A *cdc28-srm* mutation is not a temperature-sensitive mutation and differs from known *cdc28-ts* mutations, since it shows the evident phenotypic manifestations at 30°C. The mutation is on the third glycine site in the conserved sequence GxGxxG of the G-rich loop, whose position is opposite to the activation T-loop. Despite its established importance, the role of the G-loop is still unclear. The crystal structure of the human CDK2 served as model for the catalytic core of other CDKs, including CDC28. Nanoseconds long molecular dynamics trajectories of the CDK2/ATP complex were analysed. The MD simulations of corresponded substitution CDK2-G16S in conserved G-loop shows this amino acid importance and the induced conformational change in CDK2 structure, resulting in the ATP removal from G-loop and in new amino acids rearrangement in the T-loop.

Keywords: molecular dynamics simulations, protein kinase, G-loop

A.A. Cigna and M. Durante (eds.),
Radiation Risk Estimates in Normal and Emergency Situations, 327–339.

1. Introduction

The cyclin dependent kinases (CDKs) belongs to the serine/threonine-specific protein kinases subfamily. The enzymes catalyze the transfer of γ-phosphate in adenosine triphosphate (ATP) to a protein substrate. CDKs are crucial regulators in timing and co-ordination of eukaryotic cell cycle events. Transient activation of these kinases at specific cell stages is believed to trigger the principal cell cycle transitions, including the DNA replication and the entry into mitosis. In yeast, transition events are controlled by a single CDK (CDC28 in *Saccharomyces cerevisiae* (Mendenhall, 1998)]) and several cyclins, while in human, cell cycle progression is governed by several CDKs and cyclins. In particular, CDK4-cyclin D is required to pass through G1, CDK2-cyclin E for the G1 to S phase transition, CDK2-cyclin A to progress through the S phase and CDC2-cyclin B to reach the M phase. The CDK2 and CDC2/CDK1 proteins have been extensively studied. Both are closely related to yeast CDC28/CDK1 (the identity in amino-acid sequence is 62% for CDK2 and 60% for CDC2).

The central role that CDKs play in cell division timing, in cell cycle regulation and repair together with the high incidence of genetic alteration of CDKs or deregulation of CDK inhibitors observed in several cancers made CDC28 a very attractive model for structural and functional CDK studies. We observed that certain gene mutations in *Saccharomyces cerevisiae*, including *cdc28-srm*, affect cell cycle progression and different genetic structures maintenance (Devin, 1990). The *cdc28-srm* resulted into decreased mitotic stability of natural chromosomes excess and recombinant circular plasmids containing centromeres, as well as decreased rates of spontaneous mitochondrial rho^- mutations (Devin, 1990), increased cellular radiosensitivity (Koltovaya, 1998; Koltovaya, 1995) and conferred checkpoint defects (Li, 1997). A *cdc28-srm* is not a temperature-sensitive mutation and differs from known *cdc28-ts* mutations, since it shows the evident phenotypic manifestations at 30°C.

Crystallographic studies on several eukaryotic protein kinases showed they present the same fold and tertiary structure. The crystal structure of the human CDK2 (De Bondt, 1993; Jeffrey, 1995) was used as model for the catalytic core of other CDKs, including CDC28. The CDK2 structure is bilobed with an N-terminus, mainly consisting of β-sheet, and a C-terminus, mostly composed by α-helix structures. ATP binds the cleft between the two lobes. CDKs are inactive as monomers. Cell cycle dependent oscillations in CDK activity are induced by complex mechanisms, as binding to positive regulatory subunits (cyclins) and phosphorylation at positive (pT160 in T-loop) and negative (pT14, pY15 in G- loop) regulatory sites.

The aim of the present work was to analyse the structure of the yeast pleiotropic mutant allele *cdc28-srm* by molecular dynamics simulation studies using human complex CDK2/ATP as model.

The observation of a key mutation in an extended protein structure is significant, since it allows to investigate the mechanism of the protein kinase regulation activity. In our case a mutation in a highly conservative sequence in G-rich loop was obtained. Despite its established importance, the role of G-loop is still unclear. The consequences of the conserved glycine substitution in the G-rich loop on the structure of the protein kinase were simulated using human CDK2 as model. The substitution of G20S in yeast CDC28 corresponds to G16S in human CDK2.

2. Materials and Methods

2.1 MD SIMULATIONS OF CONFORMATIONAL CHANGES OF PROTEIN

For the MD simulations, the SANDER modules of the program package AMBER 8.0 (Case, 2003) and of the modified version of AMBER 7.0, for a special-purpose computer MDGRAPE-2 (Narumi, 2000), were used. The starting geometries for the simulations were prepared using X-ray structures from the Prookhaven Protein Data Bank (http://www.pdb.org). The all-atom force field (Cornell, 1995) was used in the MD simulations. A system was solvated with TIP3P molecules (Jorgensen, 1983) generated in a spherical (non-periodic) water bath. The system temperature was kept constant by the Berendsen algorithm with 0.2 ps coupling time (Berendsen, 1984). Only bond lengths involving hydrogen atoms were constrained using the SHAKE method (Ryckaert, 1997). The integration time step in the MD simulations was 1 fs. The simulation procedures were the same in all the calculations (Kholmurodov, 2005). Firstly, a potential energy minimisation was performed for each system on an initial state. Then, the MD simulation was performed on the energy-minimised states. The temperatures of the considered systems were gradually heated to 300 K and then kept at 300 K for the next 1 million time steps (Kholmurodov, 2004; Kholmurodov, 2003). The trajectories at 300 K for 1-ns were compared and studied in detail. The simulations and images of simulated proteins results were analysed by RasMol (Sayle, 1995) and MOLMOL (Koradi, 1996) packages.

3. Results

3.1 THE CDK2/ATP STRUCTURAL CONFORMATIONS

First, the inactive complex CDK2/ATP was analysed. Analysis of the CDK2/ATP binary complex (De Bondt, 1993) indicates that ATP interacts with several residues lining the cleft between the two lobes. The adenine base is positioned in a hydrophobic pocket between the βsheet of the small lobe and the L7 loop between β-5 and α-2. The ATP phosphates are held in position by ionic and hydrogen-bonding interactions with several residues, including K33, D145, and the backbone amides of the G-rich loop between β-1 and β-2.

Sequencing analysis of *cdc28-srm* reveals a single nucleotide substitution of glycine with serine in position 20 (G20S). This occurs at the third glycine in the conserved sequence GxGxxG of G-rich loop. Since glycine is the smallest amino acid, serine is larger. Both the amino acids belong to the neutral polar class. They are highly solvable in water, their polar R-groups may indeed form hydrogen bonds with water (for glycine and serine polar R-groups – H- and HO-CH_2-, respectively).

Simulated wild-type CDK2-G16/ATP structure (Fig. 1) was compared to the CDK2-S16/ATP structure, after conformational changes evaluation. The resulting wild-type CDK2-G16/ATP and CDK2-S16/ATP structural conformations are shown in Fig. 2. The picture displays the initial (left) and the final 1-ns (right) states. For CDK2-G16 and CDK2-S16 structures several animation movies were performed to display the real-time dynamical motions. Positional changes between the ATP, residue 16 and T-loop (the latter covers a left bottom α-helix shown in Fig. 1) were analysed.

Comparing initial and final states of wild-type CDK2-G16/ATP structure, no big difference was visually observed. Moreover, the position of the amino acid residue 16 did not change within 1 million time MD steps in comparison with the initial configuration. So, for the wild-type protein the original state is kept in a relatively stable conformation.

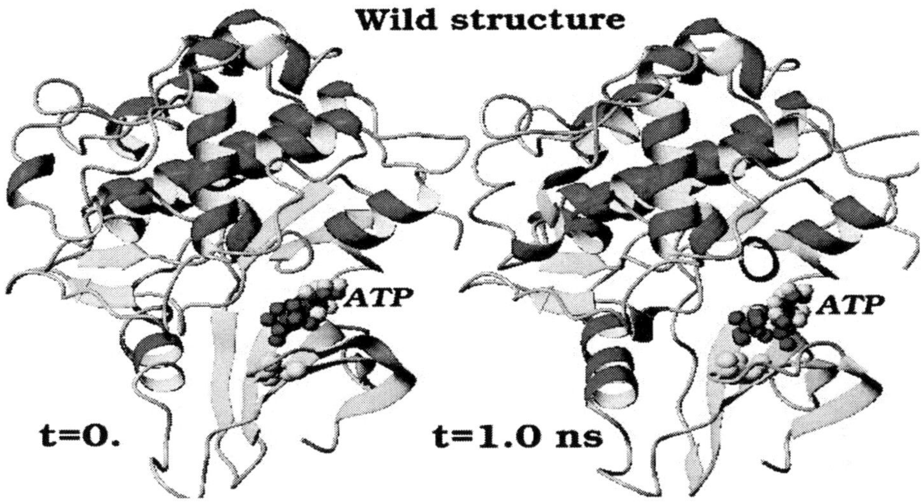

Figure 1. The initial and final (1-ns state) structures of the CDK2/ATP of the wild-type complex. The ATP molecule and residue 16 of the G-loop are represented by balls model.

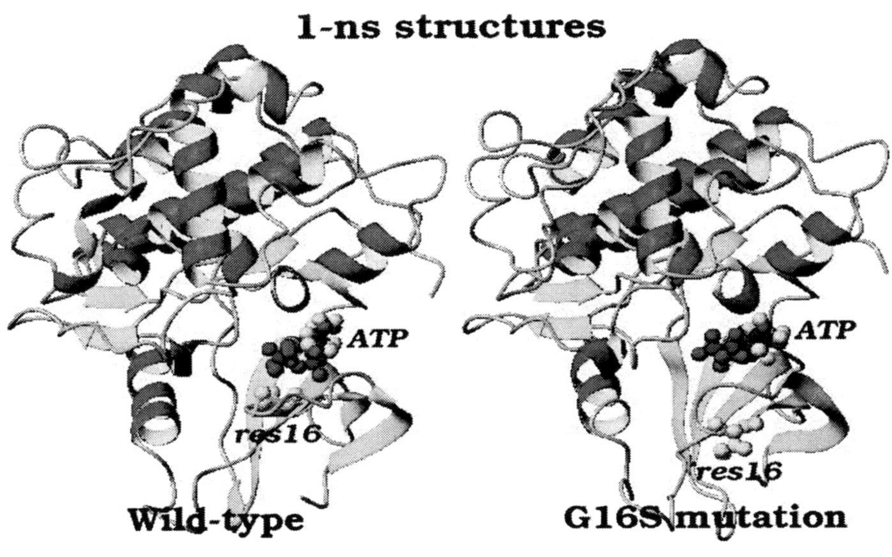

Figure 2. Comparison between final (1-ns state) structures of wild-type (left), and S16 mutant (right) protein. The ATP molecule and the residue 16 of the G-loop are represented by balls model.

Regarding the CDK2-S16 variant (Fig. 2), a completely different dynamic and conformational shape was found. First, the amino acid residue 16 moves dramatically far from the ATP location site. In comparison to the wild-type structure, the distance between S16 and ATP position, as snapshots show, increases of about 2-2.5 times in average. At the same time, such movement results in relative shift of the T160 residue and the whole T- and G-loops positions.

The increase of the ATP-S16 distance induces changes in hydrogen bonds formation involving the ATP and G-loop. On the other hand, the conformational change mutation induced (CDK2') results in an interhelical protein movement, covering a phosphorylation point (viz. T160).

3.2 A HYDROGEN BOND NETWORK

The T160, ATP and S16 positions (an "activation triangle") in the final (1-ns) state are represented in Fig.3, aiming to estimate (although indirectly) the possibility of the hydrogen bond formation in the ATP and G-loop region. Generally a hydrogen bond is regarded as being effective when the distance between the hydrogen atom of the proton donor and the proton acceptor is less then 2.6, 2.4 Å. The evaluation of the distances between the ATP-res16 and ATP-T160 relative to the "activation triangle" are shown in Figs. 4-5, respectively.

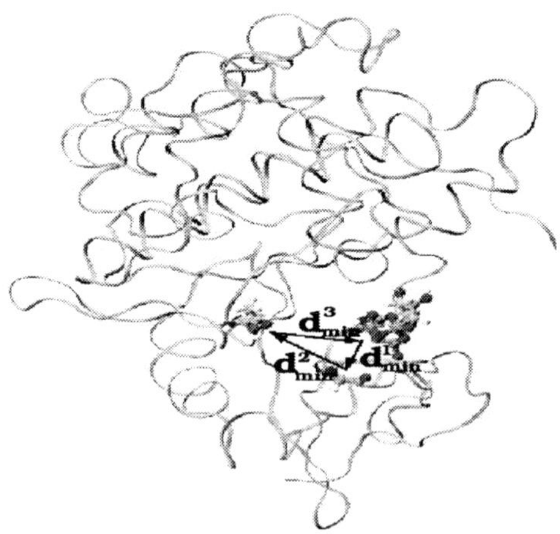

Figure 3. The relative positions of the T160, ATP and res16 (an "activation triangle") are shown. The ATP molecule, residues T160 and 16 are represented by balls model.

The ATP-res16 distance for the wild type and mutant structures shows a completely different behaviour (Fig. 4). The ATP-res16 distance in the wild-type structure evidently lies within ~2.5 Å during the all 1-ns dynamical changes. At the same time, ATP-res16 distance lies within of 7.5 Å in the mutated structure. The G16S mutation causes an increase (three fold higher) in the ATP-res16 distance. Thus, the all hydrogen bond network in the ATP-res16 binding site is obviously corrupt in the CDK2 mutated structure.

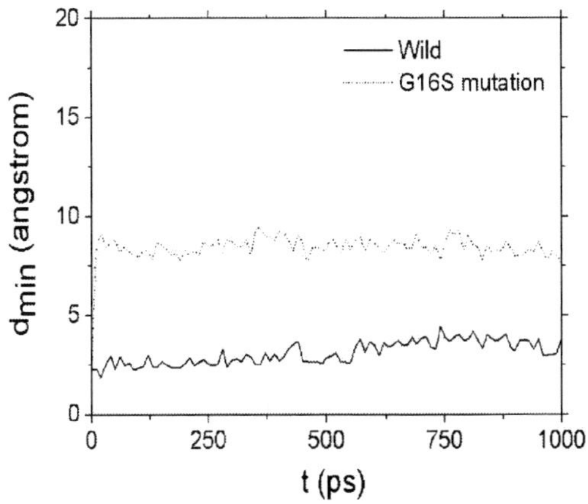

Figure 4. The time dependence of the ATP-res16 distance are shown for the CDK2-G16/ATP (solid line) and CDK2-S16/ATP (dot line) in accordance to the “activation triangle”.

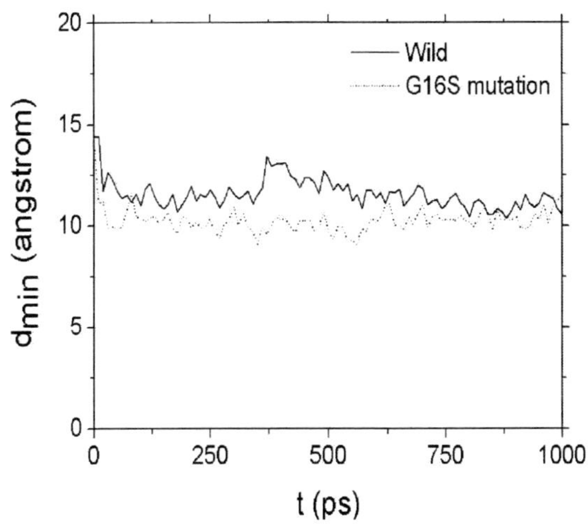

Figure 5. The time dependence of the ATP-T160 distance are shown for the CDK2-G16/ATP (solid line) and CDK2-S16/ATP (dot line) in accordance to the “activation triangle”.

The time dependence dynamics for the ATP-T160 distance are presented in Fig. 5. The ATP-T160 distance for the wild-type protein is comparably shorter then that for the CDK2-S16/ATP variant. The ATP-T160 distance trend agrees with the previous results

3.3 ROTATIONAL CHANGES OF THE AMINO ACID RESIDUES AROUND OF PHOSPHORILATED REGULATORY SITE T160

The CDK2/ATP' dynamical peculiarities in the neighbour of posphorilation site (T160) were analysed in detail, showing by snapshots and animation movies all the amino acid positions in the T-loop.

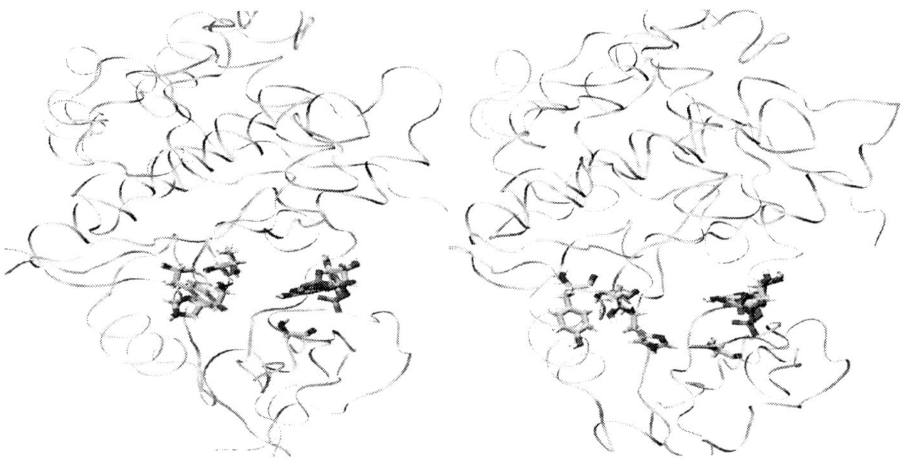

Figure 6. The initial (left) and final (right) configurations of the wild-type protein CDK2-G16. The pictures display the orientational changes of two neighbouring amino acids around of phosphorylation site (viz. the residue T160). It is seen that the Y159 and H161 exchange their relative orientations with respect to the residue T160.

From these data, a rotation of the two amino acids neighbouring T160 (viz. residues 159 and 161) was observed. The residue 159 (Y159) is tyrosine with a non-charged polar R-group, and the residue 161 (H161) is histydine with a polar charged R-group. A summary of these founding is presented by the following two snapshots set (Figs. 6 and 7).

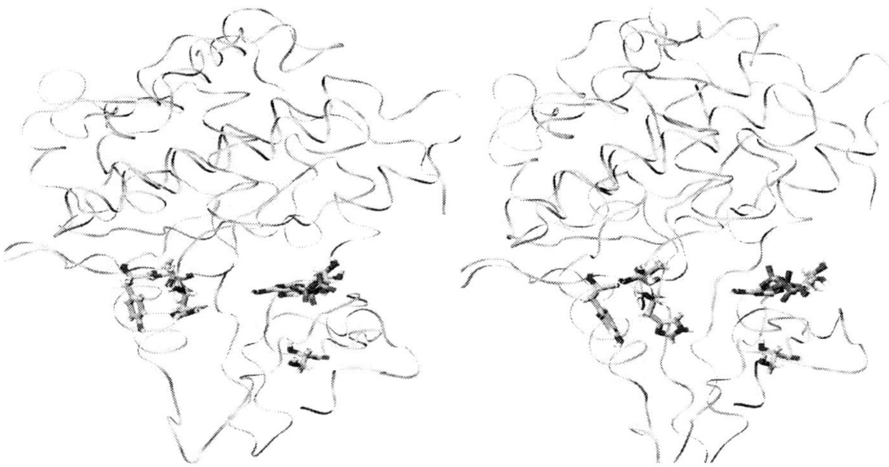

Figure 7. The initial (left) and final (right) configurations of the protein CDK2-S16. The snapshots display the orientational changes of two neighbouring amino acids (Y159 and H161) around of a phosphorylation site (T160). The positions of the ATP molecule and the res16 are also drawn.

The two carbon rings of Y159 and H161 residues occupy well-separated positions (Fig. 6). The carbon rings during their dynamical motions (at around of 0.5 ns from the start) abruptly exchange their orientation positions. The rings begin to "screen" the amino acid residue T160. Such "screening" does not change until the end of the dynamics (right snapshot: final 1-ns state). On the contrary, the carbon rings overlapping do not emerge for the S16 variant (Fig. 7). During the all simulation period, there has to be no "screening" of the site T160 to see.

Thus, the mutation induced changes for the protein structure affects the overlapping of the carbon rings at a key phosphorylation point T160. The "screening" phenomena described above were connected to the positional changes of the activation T-loop induced by the mutation at the residue 16.

From the "activation triangle" described above, the T160-res16 distance (see Fig. 8) for the wild-type CDK2-G16/ATP, CDK2-S16/ATP and wild-type monomeric (without ATP) CDK2 proteins was estimated. The T160-res16 distance in the mutant CDK2-S16/ATP structure is significantly larger then those in the wild-type CDK2-G16/ATP ones (Fig. 8). At the same time, the T160-res16 distance for the wild-type CDK2/ATP and wild-type monomeric CDK2 structures follows almost the same law. So, the above "screening" phenomena could not be originated from the presence of ATP molecule or T160-ATP exchanges.

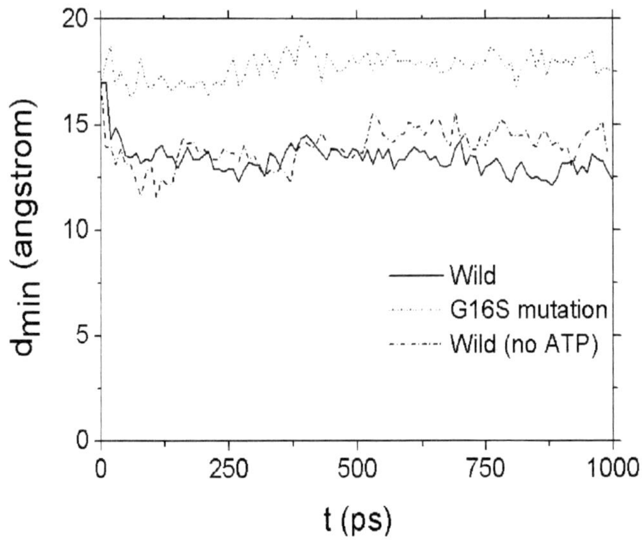

Figure 8. The time dependence of the T160-res16 distance are shown for the CDK2-G16/ATP wild-type (solid line), a wild-type CDK2 (dashed line) and CDK2-S16/ATP (dot line) in accordance to the "activation triangle".

4. Discussion

The G-loop enables protein kinase to adopt a wide range of backbone conformations. The substitution of the glycine residues in the G-loop, particularly the first and the second glycine (GxGxxG), with either alanine or serine results in a dramatic decrease in cAPK activity, showing the high domain significance. The functional role of the G-loop has been described in detail for cAPK (Hemmer, 1997; Tsigelny, 1999; Aimes, 2000; Johnson, 2001), but its involvement in CDK regulation has not been yet discussed. The G-loop catalytic function – that is, correct ATP binding and alignment – is believed to be the same as in cAPK, but exhibiting a new inhibitory function for CDK (Bartova, 2004). The mutation *cdc28-srm*, having a pleotropic manifestation in yeast cells, was analysed. This mutation is localised on the glycine-rich loop (G-loop) and corresponds to the substitution of the third glycine with serine. MD simulation analysis shows a dramatic increase in CDK2-S16/ATP the distance between ATP and the residue 16 in the G-loop. The shift is equal 5 Å.

The crystal structure of inactive monomeric CDK2 shows the T-loop (residues 147-153) would block access of substrates to the active site and that ATP would bind with the wrong geometry for efficient catalysis. Binding to cyclin A simultaneously, the T-loop of CDK2 moves away from the substrate binding cleft and repositions of the G-loop (residues 11-18), so that they can interact properly with the ATP phosphates. This complex has a low but detectable activity. Less dramatic changes occur in the CDK2/cyclin A complex structure, following activation by phosphorylation of T160 (Bartova, 2004).

It is known that the G-loop is changed during the early stage of inactive (CDK2/ATP), partly active (CDK2/cyclin A/ATP), and fully active (pT160-CDK2/cyclin A/ATP) CDK2 simulations in comparison with its conformation, as found in the crystal structures (Bartova, 2004; Halmes, 2001; Cook, 2002). The G-loop moves away from the ATP phosphate moiety binding site after the interaction of CDK2 with cyclin A and again after CDK2/cyclin A/ATP complex phosphorylation at the T160 site. The shift of the G-loop is equal to 3.5 Å (CDK2/cyclin A/ATP) and 8.6 Å (pT160-CDK2/cyclin A/ATP) in comparison with the G-loop position found in the CDK2/ATP system. It was interesting to observe this shift in the mutant allele CDK2-G16S/ATP as well.

We found unexpected consequence of substitution G16S - new orientation of two neighbouring amino acids of T160 in the T-loop. It may influence the interactions between protein kinase and cyclins. In further studies we intent to examine several other complexes of the CDK2, including cyclin A and substrate.

Acknowledgements

This work was performed under the computer hardware and software support by the Computational Astrophysics Laboratory (CAL) of the Institute of Physical and Chemical Research (RIKEN), Japan. The authors are grateful to the Director of CAL RIKEN Professor Toshikazu Ebisuzaki, for providing special-purpose computers MDGRAPE-2 for MD simulations of proteins with a total processing power of 25 teraflops.

References

Aimes R.T., Hemmer W., Taylor S.S., 2000, Serine-53 at the tip of the glycine-rich loop of cAMP-dependent protein kinase: Role in catalysis, P-site specificity, and interaction with inhibitors, *Biochemistry*, **39**: 8325–8332.

Berendsen H.J.C., Postma J.P.M., van Gunsteren W.F., DiNola A., Haak J.R., 1984 – Molecular dynamics with coupling to external bath, *J. Chem. Phys.*, **81**(8): 3684–3690.

Bartova I., Otyepka M., Kriz Z., Koca J., 2004, Activation and inhibition of cyclin-dependent kinase-2 by phosphorylation; a molecular dynamics study reveals the functional importance of the glycine-rich loop. *Protein Science*, **13**: 1449–1457.

Case D.A., Pearlman D.A., Caldwell J.W., Cheatham III T.E., Ross W.S., Simmerling C.L., Darden T.A., Merz K.M., Stanton R.V., Cheng A.L., Vincent J.J., Crowley M., Ferguson D.M., Radmer R.J., Seibel G.L., Singh U.C., Weiner P.K., Kollman P.A., 2003, AMBER 8.0. University of California.

Cook A., Lowe E.D., Chrysina E.D., Skamnaki V.T., Oikonomakos N.G., Johnson L.N., 2002, Structural studies on phospho-CDK2/Cyclin A bound to nitrate, a transition state analogue: Implications for the protein kinase mechanism. *Biochemistry*, **41**: 7301–7311.

Cornell W.D., Cieplak P., Bayly C.I., Gould I.R., Merz Jr. K.M., Ferguson D.M., Spellmeyer D.C., Fox T., Caldwell J.W., Kollman P.A., 1995, *J. Am. Chem. Soc.*, **117**: 5179–5197.

De Bondt H.L., Rosenblah J., Jancarik J., Jones H.D., Morgan D.O., Kim S.-H., 1993, Crystal structure of cyclin-dependent kinase 2, *Nature,* **363**: 595–602.

Devin A.B., Prosvirova T., Peshekhonov V.T., Chepurnaya O.V., Smirnova M.E., Koltovaya N.A., et al., 1990, The Start gene CDC28 and genetic stability of yeast, *Yeast,* **6**: 231–243.

Hemmer W., Mc Glone M., Tsigelny I., Taylor S.S., 1997, Role of the glycine triad in the ATP-binding site of cAMP-dependent protein kinase, *J. Biol. Chem.*, **272**: 16946–16954.

Halmes J.K., Solomon M.J., 2001, The role of Thr160 phosphorylation of CDK2 in substrate recognition. *Eur. J. Biochem*, **268**: 4647–4652.

Jeffrey P.D., Russo A.A., Polyak K., Gibbs E., Hurwitz J., Massague J., Paoletich N.P., 1995, Mechanism of CDK activation revealed by the structure of a cyclin A – CDK2 complex, *Nature*, **373**: 313–320.

Johnson D.A., Akamine P., Radzio-Andzelm E., Madhusudan N., Taylor S.S., 2001, Dynamics of cAMP-dependent protein kinase, *Chem. Rev.*, **101**: 2243–2270.

Jorgensen W.L., Chandrasekhar J., Madura J.D., 1983 – Comparison of simple potential functions for simulating liquids, *J. Chem. Phys.*, **79**(2): 926–935.

Kholmurodov Kh., *PEPAN (Physics of Particles and Nuclei),* v. 36, No. 2, pp. 1–16 (2005); Hirano Y., Ebisuzaki E., *Chem-Bio Informatics J.*, v. 3, No. 2, pp. 86–95 (2003).

Kholmurodov Kh., Ebisuzaki T., 2004, *ICMS-CSW2004,* v. C4, pp. 9–11.

Kholmurodov Kh. T. et al., 2003, Methods of Molecular Dynamics for Simulation of Physical and Biological Processes, *PEPAN (Physics of Elementary Particles and Atomic Nuclei),* **34**(2): 474–501.

Koltovaya N.A., Kadyshevskaya K.Yu., Devin A.B., 1995, DNA repair and cell cycle regulation: CDC28 gene and new mitotic checkpoint genes, *Yeast*, **11**: 73.

Koltovaya N.A., Arman I.P., Devin A.B., 1998, Mutations of the CDC28 gene and the radiation sensitivity of Saccharomyces cerevisiae, *Yeast,* **14**: 133–146.

Koradi R., Billeter M., Wuthrich K., 1996, MOLMOL: a program for display and analysis of macromolecular structure, *J. Mol. Graphics,* **4**: 51–55.

Li X, Cai M., 1997, Inactivation of the cyclin-dependent kinase Cdc28 abrogates cell cycle arrest induced by DNA damage and disassembly of mitotic spindles in Saccharomyces cerevisiae, *Mol Cell Biol,* **17**: 2723–2734.

Mendenhall M.D., Hodge A.E., 1998, Regulation of Cdc28 cyclin-dependent protein kinase activity during the cell cycle of the yeast, *Microbiol. Mol. Biol. Rev.,* **62**: 1191–1243.

Okimoto N., Yamanaka K., Suenaga A., Hirano Y., Futatsugi N., Narumi T., Yasuoka K., Susukita R., Koishi T., Furusawa H., Kawai A., Hata M., Hoshino T., Ebisuzaki T., 2003, *Chem-Bio Informatics J.*, **3**(1): 1–11; Narumi T., Susukita R., Furusawa H., Yasuoka K., Kawai A., Koishi T., Ebisuzaki T., 2000, MDM version of AMBER.

Ryckaert J.P., Ciccotti G., Berendsen H.J.C., 1997 – Numerical integration of the cartesian equations of a system with constraints, *J. Comput. Phys.*, v. 23: 327–341 (1997).

Sayle R.A., Milner-White E.J., 1995, RasMol: Biomolecular graphics for all., *Trends in Biochem. Sci.*, **20**: 374–376.

Tsigelny I., Greenberg J.P., Cox S., Nichols W.L., Taylor S.S., Ten Eyck L.F., 1999, 600 ps molecular dynamics reveals stable substructures and flexible hinge points in cAMP dependent protein kinase, *Biopolymers,* **50**: 513–524.

GENOME RECONSTITUTION IN THE EXTREMELY RADIATION RESISTANT BACTERIUM *Deinococcus radiodurans*

JOHN R. BATTISTA*
Department of Biological Sciences, Louisiana State University, 202 Life Sciences Building, Baton Rouge, Louisiana, 70803, USA,

MICHAEL M. COX
Department of Biochemistry, University of Wisconsin-Madison, 433 Babcock Dr, Madison, Wisconsin, 53706-1544, USA

*To whom correspondence should be addressed.

Abstract. The bacterium *Deinococcus radiodurans* is highly resistant to the effects of ionizing radiation, (IR). Cultures of *Deinococcus* survive exposure to IR doses of 5,000 Gray (Gy) with no lethality. At 5,000 Gy, several hundred double strand breaks are introduced into the genome. Recovery from this DNA damage occurs in two phases, one independent of RecA protein and one that requires RecA. A number of proteins are induced in *Deinococcus* in response to exposure to ionizing radiation, and many of those most highly expressed are novel proteins. A variety of *Deinococcus* proteins involved in genome reconstitution are now under biochemical investigation. Many of these proteins exhibit unusual properties, but a complete explanation for the radiation resistance of *Deinococcus* is not yet available.

Keywords: *Deinococcus radiodurans*, RecA, SSB, DdrA, DNA damage, DNA repair, genome, ionizing radiation, resistance

A.A. Cigna and M. Durante (eds.),
Radiation Risk Estimates in Normal and Emergency Situations, 341–359.

1. Deinococcus Radiodurans

Deinococcus radiodurans is part of a small, but growing family of bacteria that include some of the most radiation-resistant organisms known. It was first described in 1956 (Anderson, Nordon *et al.*, 1956), as a survivor of a food irradiation trial. Subsequent work has identified scores of additional radiation-resistant microbial species, but *Deinococcus radiodurans* has emerged as the primary model system for exploring this extraordinary phenotype.

D. radiodurans R1 (Battista and Rainey, 2001) is a red pigmented, non-spore-forming, non-motile, spherical bacterium that forms pairs and tetrads in liquid culture. The species is chemo-organotrophic with respiratory metabolism, growing optimally at 30°C in rich media with aeration. *D. radiodurans* is most closely related to the genera in family *Thermaceae*, including the genera *Thermus*, *Meiothermus*, *Marinithermus*, *Vulcanithermus*, *Oceanithermus*, and *Truepera* (Rainey, Nobre *et al.*, 1997; Albuquerque, Simoes *et al.*, 2005). These genera form the phylum Deinococcus/Thermus within the Bacteria. The genera *Deinococcus* and *Trupera* are comparable in terms of their radioresistance, whereas other members of the phylum are considered radiosensitive. The radiation-resistance of *D. radiodurans* is thought to have evolved as an adaptation to desiccation (Mattimore and Battista, 1996). As is the case with IR exposure, desiccation will produce single and double strand breaks in the genome, accumulating over time. The adaptations evident in this bacterium allow for rapid restoration of the genome when circumstances are favorable for repair and growth.

2. Additional Radiation-resistant Bacterial Species

Highly radiation-resistant species occur in at least seven bacterial genera, including *Deinococcus, Rubrobacter, Hymenobacter, Kocuria, Acinetobacter, Methylobacterium*, and *Chrococcidopsis*. The bacterial phyla represented include α-Proteobacteria, Actinobacteria, γ-Proteobacteria, Flexibacter-Cytophaga-Bacterioides, Cyanobacteria, and Deinococcus-Thermus. There is no evolutionary pattern to the occurrence of these species, some of which are more radiation-resistant than is *Deinococcus*. They could have evolved independently. Alternatively, the repair systems and strategies underlying radiation–resistance could be a vestige of an ancient system that arose in early life forms to deal with a challenging environment. Radiation-resistant species are also found among the archaea (Cox and Battista, 2005).

A recent study of soil samples from arid Sonoran desert and nonarid Louisiana forest environments reinforces the connection between desiccation and radiation-resistance, and also expands the known list of radiation-resistant bacterial species (Rainey, Ray *et al.*, 2005). Species isolated from the nonarid soil sample could not survive at IR doses exceeding 13,000 Gy. In contrast, the arid sample contained numerous strains that could tolerate exposure to 30,000Gy, including isolates from the genera *Deinococcus*, *Geodermatophilus*, and *Hymenobacter* (Rainey, Ray *et al.*, 2005). Nine new species of *Deinococcus* were identified in this sampling.

3. The *Deinococcus* Genome

The *D. radiodurans* chromosome is 3.28 x 10^6 base pairs with 66.6 mol % GC content. The genome is segmented; consisting of a 2.64 Mbp chromosome, a 0.41 Mbp chromosome, designated chromosome I and II, respectively, a 0.18 Mbp megaplasmid, and a 0.045 Mbp plasmid (Lin, Qi *et al.*, 1999; White, Eisen *et al.*, 1999). There is reported to be four identical copies of the chromosome per stationary phase cell, and as many as ten per cell in exponentially growing cells (Hansen, 1978; Harsojo, Kitayama *et al.*, 1981).

The *D. radiodurans* R1 genome has been sequenced and annotated (White, Eisen *et al.*, 1999; Makarova, Aravind *et al.*, 2001). In addition the genomes of the related species *Deinococcus geothermalis*, and *Thermus thermophilus* strains HB8 and HB27 have been sequenced. A comparison of proteins encoded by HB27 with those of *D. radiodurans* did not lead to any insights into why *D. radiodurans* is more radioresistant than the *Thermus* species (Henne, Bruggemann *et al.*, 2004).

4. The *D. radiodurans* Genome and DNA Repair

D. radiodurans is distinguished by its capacity to tolerate ionizing radiation. The γ radiation survival curves of actively growing cultures of *D. radiodurans* R1 exhibit a shoulder of resistance to at least 5000Gy, (Harris, 2004, Tanaka, 2004), followed by a loss of viability. The ionizing radiation resistance of a culture increases as cells enter stationary phase, with the shoulder dose extending to approximately 15,000 Gy (Daly, Ling *et al.*, 1994; Daly, Ouyang *et al.*, 1994; Daly and Minton, 1995). The D_{37} dose for *D. radiodurans* R1 is approximately 6500Gy, at least 50 fold higher than the D_{37} dose of *E. coli* cultures irradiated under the same conditions (Daly, Gaidamakova *et al.*, 2004). Ionizing radiation-induced DNA damage is extensive in irradiated *D. radiodurans* cultures, indicating that the radioresistance of this species is not solely the result of a passive process that prevents the introduction of DNA damage. For example, the energy deposited by 6500Gy γ radiation should introduce approximately 200 DNA double strand breaks, assuming one double strand break forms per 10Gy per 5×10^9 dalton dsDNA) per genome copy in *D. radiodurans*, and assessments of the number of double strand breaks formed post-irradiation closely approximates this number (Gerard, Jolivet *et al.*, 2001; Daly, Gaidamakova *et al.*, 2004).

The *D. radiodurans* genome encodes a large portion of the ensemble of DNA repair proteins found in *E. coli*. With the exception of alkylation transfer and photoreactivation, all of the major prokaryotic DNA repair pathways are represented. Since *E. coli* is much more radiosensitive relative to *D. radiodurans*, this observation suggests that *D. radiodurans* possesses unique mechanisms for dealing with ionizing radiation-induced DNA damage. Two possibilities are likely: a) *D. radiodurans* encodes previously unidentified DNA repair proteins or, b) the species has evolved mechanisms that facilitate the action of the encoded DNA repair proteins. These possibilities are not mutually exclusive and both predict that *D. radiodurans* encodes unprecedented mechanisms that allow the cell to cope with the lethal effects of ionizing radiation.

Approximately 53% of the 3187 open reading frames believed to encode proteins in *D. radiodurans* R1 were assigned a function based on similarity to other gene products found in the protein databases. The remainder encodes proteins of unknown function.

The Battista laboratory has used microarray analysis to track induction of genes following γ irradiation (Tanaka, Earl *et al.*, 2004). Upon treatment with 3000Gy, approximately 64 proteins are induced, some quite dramatically; but less than half of the affected loci are induced 5 fold or more (Tanaka, Earl *et al.*, 2004). Among these are proteins that are clearly homologues of DNA repair enzymes in *E. coli* and other organisms, as well as proteins with no function that can be assigned as a result of homology to known proteins. The complete list includes 14 proteins with no evident relationship to any protein currently in sequence databases. Seven additional proteins are induced that have orthologs in other organisms, but which have no assigned function. There are also many other proteins that are not highly induced, and yet are likely to play a significant role in recombinational DNA repair processes, (see Table 1).

5. Potential Role of the RecF Pathway for Recombinational DNA Repair

The *D. radiodurans* homologues of proteins involved in major DNA recombination-dependent repair processes in *E. coli* are summarized in Table 1. The list includes proteins involved in replication, recombinational DNA repair, DNA excision repair, mismatch repair, and replication restart.

Table 1. Comparison of *E. coli* and *D. radiodurans* repair proteins (Keck) For each protein, the region of homology is indicated as residue numbers "##-###", followed by the total number of residues in the protein "/###". The gene for SSB is not included because of apparent complexities in the database (see text). No homology found for DnaC, DnaT, PriB, PriC, RecB, RecC, RecE, RecT, SbcB.

Protein	Region of homology	% identity	% similarity	comments
DnaA (DR0002)	25-442/471 *E. coli* 9-416/454 *D. rad*	40	55	
DnaB (DR0549)	25-464/471 *E. coli* 6-444/448 *D. rad*	47	67	
DnaG (DR0601)	22-434/581 *E. coli* 1-403/571 *D. rad*	34	51	
Ligase (Dnlj) (DR2069)	11-669/671 *E. coli* 29-681/708 *D. rad*	42	57	
Pola (DR1707	5-924/928 *E. coli* 43-952/965 *D. rad*	35	49	
PriA (DR2606)	202-729/731 *E. coli* 404-922/924 *D. rad*	26	42	
RecA (DR2340)	4-324/354 *E. coli* 16-336/364 *D. rad*	57	72	
RecD (DR1902)	30-598/609 *E. coli* 218-705/716 *D. rad*	27	40	N-terminal extension in Dr
RecF (DR1089)	3-330/358 *E. coli* 6-326/360 *D. rad*	28	43	
RecG (DR1916)	6-693/694 *E. coli* 107-777/785 *D. rad*	39	53	N-terminal extension in Dr
RecJ (DR1126)	68-570/579 E. coli 3-461/685 *D. rad*	34	51	C-terminal extension in Dr
RecN (DR1477)	2-553/553 *E. coli* 34-564/564 *D. rad*	31	49	
RecO (DR0819)	7-157/242 *E. coli* 10-159/224 *D. rad*	18	34	Low homology; required PSI-BLAST
RecQ (DR1289)	9-600/608 *E. coli* 8-605/824 *D. rad* 557-605/608 *E. coli* 680-728/824 *D. rad* 549-606/608 *E. coli* 768-825/824 *D. rad*	46 36 33	63 64 59	HRDC domain repeated 3 times in Dr
RecR (DR0198)	1-199/202 *E. coli* 1-196/20 *D. rad*	42	55	C-terminal extension in Dr
RuvA (DR1274)	1-199/203 *E. coli* 1-197/201 *D. rad*	33	49	
RuvB (DR0596)	13-332/337 *E. coli* 2-321/333 *D. rad*	56	75	A second ortholog with weaker similarity is present (DR 1898)
RuvC (DR0440)	4-168/174 *E. coli* 3-166/179 *D. rad*	33	51	
SbcC (DR1922)	27-1032/1049 *E. coli* 22-896/909 *D. rad*	21	35	
SbcD (DR1921)	1-293/400 *E. coli* 24-319/417 *D. rad*	28	46	C-terminal changes

The absentees are intriguing. In *E. coli*, there are three broad pathways of recombination, called the RecBCD, RecE, and RecF pathways, in each case named after one of the defining gene products. The RecF pathway in *E. coli* is somewhat cryptic when defined in terms of conjugational recombination, and was initially detected only in strains lacking *recBC* function (Clark, 1971; Horii and Clark, 1973; Clark and Sandler, 1994). However, the RecF pathway proteins have important functions in the repair of stalled replication forks (Cox, 1997; Roca and Cox, 1997).

Based on the genome sequence of *D. radiodurans*, the RecF pathway is likely to play a more prominent role in this bacterium than it does in *E. coli*, Table 1. First, *D. radiodurans* lacks orthologous genes for the *E. coli*, Ec) RecB and RecC proteins, whose nuclease/helicase activities underpin the major DSB repair pathway in *E. coli* (Kowalczykowski and Eggleston, 1994; Kuzminov, 1999; Smith, 2001). Presumably, the RecBCD recombination pathway is therefore absent. Second, *D. radiodurans* lacks a *recE* ortholog, making the RecE pathway unlikely. In contrast, *D. radiodurans* does possess orthologs of every RecF recombination pathway gene, *recA*, *recF, recJ, recN, recO, recQ, recR, ssb, ruvA, ruvB, and ruvC*, Limited genetic studies have confirmed the importance of two RecF pathway genes in *D. radiodurans*. Deletion of the *recN* gene results in increased radiosensitivity in the mutant bacteria (Funayama, Narumi *et al.*, 1999). Mutation of the *recR* gene results in a strain that is extremely sensitive to crosslinking reagents like mitomycin C, but which remains resistant to ionizing radiation (Kitayama, Narumi *et al.*, 2000). These data indicate that the RecF recombination pathway plays a central role in nucleic acid metabolism in *D. radiodurans*. Given the fact that recombination is a major pathway for the repair of double strand breaks, it is likely to play a substantial role in the reconstitution of the *D. radiodurans* genome and DNA repair.

6. Explanations for the Extreme Radioresistance of *Deinococcus radiodurans*

In the absence of a passive defense to block ionizing radiation, at least three kinds of mechanisms can be envisioned that might contribute to the radiation resistance of *D. radiodurans*. First, the genome itself may be organized or structured to facilitate DNA repair. This could involve, for example, proteins like the eukaryotic SMC, *s*tructural *m*aintenance of chromosomes (Losada and Hirano, 2001)) proteins, serving to keep DNA broken ends from diffusing away from either their cognate ends or from prospective recombination partners. Second, the bacterium could possess mechanisms to protect its DNA from nucleolytic degradation following DNA damage. Finally, *Deinococcus* could possess more efficacious DNA repair processes relative to more radiosensitive species. *D. radiodurans* may express novel proteins that facilitate well characterized repair processes or that catalyze special DNA repair functions. Of course, these mechanisms are not mutually exclusive, and an understanding of radiation resistance in *Deinococcus radiodurans* may require an understanding of all three possibilities.

Very little is known about genome structure in *Deinococcus radiodurans*. A recent report suggests that a ring-like nucleoid seen in transmission electron microscope images might indicate that the genomic DNA is arranged as a tightly structured toroid, and that this might play an important role in DNA repair and radioresistance (Levin-Zaidman, Englander *et al.*, 2003). Although some of the underlying assumptions in this work conflict with the conclusions of earlier studies, e.g., the authors' assignment of the four compartments seen in EM as divisions of a single cell as opposed to the more common conclusion that the divisions are boundaries of four individual cells, as described above), genome structure could in principle contribute to radioresistance in important ways. However, two recent studies question the relevance of a toroidal structure. A recent investigation into the fine structure of the *D. radiodurans* nucleoid using cryoelectron microscopy did not find evidence that the genome of this species is organized in manner consistent with the formation of a toroid (Eltsov and Dubochet, 2005). In addition, a survey of nucleoid structures in radiation-resistant bacteria (Zimmerman and Battista, 2005) demonstrated that most but not all deinococcal species had a ring-like appearance. Two family members, *D. radiopugnans* and *D. geothermalis* had no clearly defined structure. Thus, there is no obvious relationship between the shape of a *D. radiodurans* nucleoid and the radioresistance of the species.

The idea that DNA end-protection might be involved in radioresistance comes from our experience with *Deinococcus*. When *E.coli* is subjected to high levels of ionizing radiation, the chromosomes are rapidly degraded (Pollard and Kraus, 1973; Pollard and Randall, 1973; Pollard and Fugate, 1978; Williams, Shibata *et al.*, 1981). The induction of the SOS response can ameliorate this degradation, but this appears to be a function of the induction of repair processes rather than inhibition of nuclease action (Krasin and Hutchinson, 1981). The degradation of DNA appears to be much more controlled in the case of *Deinococcus*. If cells are irradiated and then washed and suspended in 10 mM Mg SO_4, the cells can be held for 96 hours with minimal additional DNA degradation. Addition of growth media at any point will lead to reconstitution of the genome and complete cell recovery (J. Battista, unpublished results). As shown later, the capacity to recover from irradiation after long incubations in $MgSO_4$ is compromised in mutant strains lacking a functional DR0423 gene. We now assign the protein DdrA (DR0423) to a novel DNA end-protection system in *Deinococcus,* based on its detailed characterization.

At first, the idea of proteins that bind specifically to a DNA end may not sound very exciting, but consider this in the context of recovery from long-term desiccation, the environmental factor that guided *Deinococcus* evolution. If cells are held in an arid environment for days or weeks or even years, DNA damage will accumulate. DNA repair processes are notorious energy sinks, and a nutrient-poor environment may preclude timely repair. Proteins that bind to and protect broken DNA ends could help maintain the genome until conditions appeared that would allow cell growth and DNA repair.

A variety of additional enzymatic mechanisms can be envisioned based on systems present in other organisms, although some are rare in bacteria. The most reliably accurate form of double strand break repair is recombinational DNA repair (Paques and Haber, 1999; van den Bosch, Lohman *et al.*, 2002). Given the lack of mutation seen in *Deinococcus* after high levels of radiation, and the demonstrable importance of the RecA protein in recovery from radiation damage (Gutman, Carroll *et al.*, 1994; Minton, 1996), it is reasonable to assume that recombinational DNA repair plays a major role in overall radioresistance. There are two additional mechanisms for DSB repair that have been well-characterized in yeast and mammals (Paques and Haber, 1999; van den Bosch, Lohman *et al.*, 2002). The first is single-strand annealing, SSA), where one DNA strand is degraded from an end until a region of complementarity is exposed where the two ends can be annealed. The annealing leads to a general repair of the now-joined chromosome fragments. The second mechanism is nonhomologous end-joining (NHEJ). Single-strand annealing has the potential to facilitate nonmutagenic repair. In contrast, NHEJ is inherently mutagenic in eukaryotes (Paques and Haber, 1999; van den Bosch, Lohman *et al.*, 2002), and seems an unlikely candidate for the high-fidelity repair seen in *Deinococcus*. Nonhomologous end-joining was thought to be limited to eukaryotes. However, the same process has recently been reported to exist in *Bacillus subtilis* and several other bacterial species, based on the Ku-like and ligase-like proteins YkoU and YkoV, respectively (Weller, Kysela *et al.*, 2002). There are no proteins encoded by *Deinococcus* with significant homology to YkoU and YkoV that can be identified by a simple BLAST search.

It is likely that DNA repair in *Deinococcus* occurs in two distinct stages. Daly and Minton (Daly and Minton, 1996) first documented a RecA-independent repair process in the first 1.5 hours after irradiation, based on increases in the length of cellular chromosome fragments (Daly and Minton, 1996). Observations of cell morphology (Levin-Zaidman, Englander *et al.*, 2003) are consistent with the proposal. Daly and Minton found that the RecA-independent process was nevertheless dependent on homology, and they proposed that single-strand annealing might play a role. RecA-dependent processes are required after this first stage, and complete recovery does not occur without them (Daly and Minton, 1996). The RecA-mediated recombinational DNA repair would require the presence of multiple genomes as a source for overlapping DNA fragment.

7. Proteins Likely to be Important to Genome Reconstitution *in Deinococcus radiodurans*

7.1 THE RECA PROTEIN OF *D. radiodurans*

The *D. radiodurans*, Dr) RecA protein is likely to be central to DSB repair mechanisms in the bacteria. RecA proteins are found in essentially all bacteria, where they are involved in recombinational DNA repair processes (Brendel, Brocchieri *et al.*, 1997; Roca and Cox, 1997). Homologs are found in archaea and eukaryotes (Kowalczykowski and Eggleston, 1994; Brendel, Brocchieri *et al.*, 1997; Roca and Cox, 1997). The classical RecA protein, such as *E. coli*, (Ec RecA) forms a nucleoprotein filament on ssDNA, and is a DNA-dependent ATPase. The RecA protein often has multiple functions. The Ec RecA protein promotes DNA strand exchange reactions in its recombination mode (Cox, 1999; Lusetti and Cox, 2002), and promotes the autocatalytic proteolysis of the LexA repressor, UmuD, and several other proteins in a regulatory mode (Mustard and Little, 2000). Ec RecA also participates directly in the mutagenic translesion bypass DNA synthesis promoted by DNA polymerase V (Tang, Shen *et al.*, 1999).

Bacterial RecA proteins generally exhibit a high degree of homology and are of similar molecular weights. The Dr RecA(361 amino acids, M_r 38,013) has an amino acid sequence that is 57% identical, 72% similar) to that of the Ec RecA protein,352 amino acids, M_r 37,842, The Dr RecA protein has been purified to homogeneity and characterized in some detail (Earl, Rankin *et al.*, 2002). *In vitro*, the protein forms filaments on DNA, hydrolyzes ATP and dATP in a DNA-dependent fashion and at rates comparable to the Ec RecA, and promotes DNA strand exchange (Kim, Sharma *et al.*, 2002). Thus, in most respects, the Dr RecA protein appears to be a typical member of the RecA protein class. However, there is one major and very interesting distinction. The DNA strand exchange reactions of the Ec RecA protein and all homologues examined to date are ordered such that the ssDNA is bound first, and the dsDNA is bound second. This order of DNA binding makes sense from a biological standpoint, since the RecA protein must be targeted to single strand gaps at stalled replication forks and other damaged DNA sites. In contrast, the Dr RecA protein promotes an obligate inverse DNA strand exchange reaction (Kim and Cox, 2002), binding the duplex DNA first and the homologous single stranded DNA, (ssDNA) substrate second. It is likely that this reaction pathway reflects the function of Dr RecA protein in DSB repair, but it opens up many questions concerning how the protein can be targeted to the DNA ends where it is needed.

Although the Dr RecA protein promotes an inverse DNA strand exchange, direct binding to dsDNA proceeds with a 10-15 min lag. This binding is potentiated by the presence of small amounts of ssDNA. If any proteins stimulated dsDNA binding, the decrease in the binding lag should be easy to detect, and it will be interesting to see if mediator proteins exist in *Deinococcus* with this function.

Based on the properties of an E142K mutant of the Dr RecA protein, Satoh *et al.* (Satoh, Narumi *et al.*, 2002) suggested that the effect of Dr RecA on DSB repair reflected primarily the protein's regulatory role rather than its recombination functions. In particular, the mutant protein, called RecA424, does not complement the null phenotype of a *recA* knockout in *E. coli*, but does appear to promote LexA cleavage, and strains containing the mutation seem to retain much of their resistance to γ radiation. However, the E142K mutation does retain significant DNA strand exchange activity in some assays, and much more work is needed to resolve the question of the function of Dr RecA. Assuming that *D. radiodurans* must find and splice together overlapping segments of its chromosomes to reconstruct a functional genome, a DNA pairing activity such as that provided by RecA would be expected to be at the center of such a process.

It should be noted that although *D. radiodurans* possesses a LexA homologue that functions as a stress response regulator, the Dr LexA does not regulate expression of the *recA* gene (Narumi, Satoh *et al.,* 2001). Instead, *recA* appears to be under the regulation of another regulator called IrrE (Earl, Mohundro *et al.*, 2002).

7.2 THE SSB PROTEIN OF *D. radiodurans*

The Dr SSB has been defined, purified, characterized, and its structure determined. The *ssb* gene in *Deinococcus* was originally annotated as a tripartite gene, with two frameshifts needed to get a functional reading frame for translation (White, Eisen *et al.*, 1999; Makarova, Aravind *et al.*, 2001; Lipton, Pasa-Tolic *et al.*, 2002; Liu, Zhou *et al.*, 2003). Defining the gene and cloning it required the correction of three separate sequencing errors in the database (Eggington, Haruta *et al.*, 2004). Two of the error corrections made the original reading frame contiguous, obviating the need for translational frameshifting. One error was well outside the gene originally annotated as SSB, and rendered that reading frame contiguous with another next to it. The combined gene encodes the largest bacterial SSB polypeptide identified to date. It has two OB folds rather than the one present in most bacterial SSB proteins, and functions as a dimer rather than a tetramer (Eggington, Haruta *et al.*, 2004). The gene is very similar to *ssb* genes found in *Thermus* species, to which *Deinococcus* is closely related. The protein is very active in stimulating the DNA strand exchange promoted by RecA proteins from both *E. coli* and *D. radiodurans*, being more active than the EcSSB in both cases.

The vast majority of bacterial SSBs are homotetrameric proteins with single-OB-fold-containing monomers, *e.g.* Ec SSB. However, with two OB folds per monomer, Dr SSB is a surprising exception to this general rule. The structure of Dr SSB was determined at 1.8-Å resolution, revealing a novel SSB architecture that provides a unique structural platform for interactions between molecules while still maintaining the canonical arrangement of 4 OB domains observed in other bacterial SSB structures (Bernstein, Eggington *et al.*, 2004). Dr SSB forms a dimer in solution (Eggington, Haruta *et al.*, 2004), which is preserved in the crystal structure. In contrast to Ec SSB, a novel protein-protein interaction surface is formed by the N-terminal OB fold and connector region of Dr SSB. This secondary interface buries an extensive surface, 1290 $Å^2$) in an arrangement that interlocks the N-terminal OB domains and connector β-hairpins of interacting SSB proteins. This unique interaction could stabilize the association of dimers in conditions of high local protein concentration, *i.e.* when arrayed on ssDNA) and might thus alter the biochemical properties of Dr SSB relative to other SSBs.

7.3 THE RECQ PROTEIN OF *D. radiodurans*

A central feature of all three recombination pathways in *E. coli* is generation of ssDNA that is used as a substrate in strand exchange with homologous dsDNA. The ssDNA is most commonly generated by the combined efforts of a helicase and nuclease. In the RecF pathway in *E. coli*, these functions are carried out by the RecQ helicase and the RecJ protein, a 5' to 3' exonuclease) (Kowalczykowski, Dixon *et al.*, 1994). The activities of RecQ and RecJ produce ssDNA with a 3' end, the substrate for RecA-mediated strand invasion. RecQ and RecJ have been shown to act in concert at replication forks stalled by UV-induced DNA damage in *E. coli*, degrading the nascent lagging-strand DNA to produce ssDNA (Courcelle and Hanawalt, 1999). This activity is important for restoring stalled forks and allowing replication to proceed. Whether RecQ and RecJ proteins carry out related roles in *D. radiodurans* has not been investigated.

RecQ proteins in bacteria and eukarya are highly homologous in three domains (Morozov, Mushegian *et al.*, 1997). First, all RecQ homologues have a helicase domain that contains sequence motifs necessary for using the energy of ATP binding and hydrolysis to unwind dsDNA. C-terminal to the helicase domain, RecQ proteins have a domain of unknown function called the RecQ-conserved domain. The complete three-dimensional structure of a major fragment of the E. coli RecQ protein, encompassing the helicase domain and the RecQ-conserved domain, has been elucidated by Jim Keck and his coworkers (Bernstein, Zittel *et al.*, 2003). The third region of homology forms the C-terminal domain of RecQ proteins called the Helicase and RNaseD-like C-terminal, or HRDC, domain (Morozov, Mushegian *et al.*, 1997). This domain is proposed to be important for DNA binding (Morozov, Mushegian *et al.*, 1997), and strong DNA-binding has recently been demonstrated by the Keck group (Bernstein and Keck, 2003). The NMR structure of the *S. cerevisiae* RecQ homolog,Sgs1) HRDC domain has been solved, revealing a compact structure of 5 helices (Liu, Macias *et al.*, 1999). Interestingly, the HRDC domain is repeated three times in the primary structure of the RecQ protein from *Deinococcus radiodurans*.

7.4 NEW PROTEINS WITH POTENTIAL FUNCTIONS IN THE REPAIR OF RADIATION-DAMAGED DNA IN *Deinococcus*

A set of five novel *D. radiodurans* proteins has been identified that are at once highly induced upon exposure to IR, have interesting phenotypes with respect to genome reconstitution, and are all regulated by the positive regulator IrrE, the same regulon that controls transcription of the *recA* gene. These are the proteins designated DR0003, DR0070, DR0326, DR0423, and DRA0346. When their possible association with genome reconstitution was recognized, none of these proteins had a homologue in any existing database. More focused BLAST searches have identified homologues for two of them.

A focus on these particular novel proteins has recently been reinforced. The genome of another bacterial species that is highly radiation-resistant, *Rubrobacter xylanophilus (Carreto, Moore et al., 1996; Ferreira, Nobre et al.,* 1999*)*, has been sequenced. This bacterium was isolated from hot springs with a high radon background in Portugal and Italy, and it is actually somewhat more radiation-resistant than is *Deinococcus radiodurans*. The sequencing work is not yet published. Even though *Rubrobacter* and *Deinococcus* are not closely related, they are separated by a billion years of evolution), this new genome features very clear, but distant) homologues of DR0003, DR0070, DR0326, and DR0423. These are the very first homologues found for DR0003, DR0070, and DR0326.

Stable knockouts of the *recA* gene, as well as of DR0003, DR0070, DR0326, DR0423, and DRA0346 have been generated, along with double knockouts entailing every possible combination of these six genes (Tanaka, Earl *et al.*, 2004). The double knockouts have allowed the definition of three epistasis groups. Two of these are RecA-independent, defined by DR0423 and DR0070, renamed DdrA and DdrB, respectively), and one is RecA-dependent, defined by DRA0346, named PprA. This has helped to solidify the idea of a two phase genome reconstitution pathway after IR exposure in *Deinococcus* (Daly and Minton, 1997) featuring RecA-dependent and RecA-independent parts.

7.5 THE DDRA (DR0423) PROTEIN

A more detailed BLAST search, along with one brief report in the literature (Iyer, Koonin *et al.*, 2002), has revealed that DR0423 is part of a family that includes the important eukaryotic recombination protein Rad52. Rad52 in turn is related to a group of DNA-binding proteins from cryptic phages in prokaryotes, including the Redβ, RecT, and Erf proteins (Passy, Yu *et al.*, 1999; Iyer, Koonin *et al.*, 2002).

The DR0423 protein has been renamed DdrA (DNA damage response protein A), (Harris, Tanaka *et al.*, 2004), This 23 kDa protein binds tightly to the 3' ends of single-stranded DNA, and protects those ends from the action of exonucleases (Harris, Tanaka *et al.*, 2004). *In vivo*, strains missing gene DR0423 exhibit only a modest increase in radiation-sensitivity. However, the reconstitution of the chromosomes is delayed and requires holding the cells in a rich media. If the cells are held instead in a solution of $MgSO_4$, lacking a carbon source, wild type cells recover whereas the viability of cells lacking DR0423 declines precipitously over a period of several days. Further, genome reconstitution is observed in the wild type R1 strain under these nutrient-poor conditions as seen in pulsed-field gels, but not in the strains lacking DR0423. DR0423 is induced more than any other protein (except for DR0070) following heavy irradiation, and it appears to be part of a DNA end-protection system.

7.6 THE PPRA PROTEIN, DRA0346

The PprA protein binds to duplex DNA and promotes the ligation of DNA fragments (Narumi, Satoh *et al.*, 2004). Extensive BLAST searching has revealed this protein has a distant relationship to the human recombination protein XRCC5,Harris, Battista, and Cox, unpublished data.

7.7 AN X FAMILY DNA POLYMERASE

A DNA polymerase from the X family has also been implicated in the reconstitution of the genome after IR exposure in *Deinococcus* (Lecointe, Shevelev *et al.*, 2004). X family polymerases have a wide variety of functions in eukaryotic DNA repair. The *Deinococcus* polymerase is the product of the gene DR0467. Elimination of the function of this gene increases the sensitivity of the cells to ionizing radiation, and reduced the rate of repair of double strand breaks. The purified DNA polymerase is stimulated by $MnCl_2$, a property of other X family polymerases.

8. Summary

The capacity of *Deinococcus radiodurans* to withstand IR exposure and accurately reconstitute its genome is extraordinary. There are many extant ideas for the molecular basis of this phenotype. However, research on this organism is still in its early stages. Interesting proteins have been identified, and plausible pathways for repair have been proposed. However, it is clear that much remains to be done.

Dedication

This article is dedicated to the memory of Drs. N. W. Timofeeff-Ressovsky and Vladimir Ivanovich Korogodin.

Acknowledgements

Work from the authors' laboratories was supported by a grant, GM067085) from the National Institutes of Health of the United States of America

References

Albuquerque, L., Simoes, C., Nobre, M.F., Pino, N.M., Battista, J.R., Silva, M.T., Rainey, F.A. and da Costa, M.S., 2005, Truepera radiovictrix gen. nov., sp nov., a new radiation resistant species and the proposal of Trueperaceae fam. nov. *FEMS Microbiology Letters* **247**(2): 161–169.

Anderson, A.W., Nordon, H.C., Cain, R.F., Parrish, G. and Duggan, D., 1956, Studies on a radio-resistant micrococcus. I. Isolation, morphology, cultural characteristics, and resistance to gamma radiation. *Food Technology* **10**: 575–578.

Battista, J.R. and Rainey, F.A., 2001, *Phylum BIV.Deinococcus-Thermus Family 1. Deinococcaceae* Brooks and Murray 1981, 356, Rainey, Nobre, Schumann, Stackebrandt and da

Costa 1997, 513. *Bergey's Manual of Systematic Bacteriology*. D.R. Boone and Castenholz, R.W. New York, Springer. **1**: 395–414.

Bernstein, D.A., Eggington, J.M., Killoran, M.P., Misic, A.M., Cox, M.M. and Keck, J.L., 2004, Crystal structure of the *D. radiodurans* single-stranded DNA binding protein suggests a novel mechanism for coping with DNA damage. *Proceedings of the National Academy of Sciences of the United States of America* **101**: 8575–8580.

Bernstein, D.A. and Keck, J.L., 2003, Domain mapping of Escherichia coli RecQ defines the roles of conserved N- and C-terminal regions in the RecQ family. *Nucleic Acids Research* **31**(11): 2778–2785.

Bernstein, D.A., Zittel, M.C. and Keck, J.L., 2003, High-resolution structure of the E.coli RecQ helicase catalytic core. *The EMBO Journal* **22**(19): 4910–4921.

Brendel, V., Brocchieri, L., Sandler, S.J., Clark, A.J. and Karlin, S., 1997, Evolutionary comparisons of RecA-like proteins across all major kingdoms of living organisms. *Journal of Molecular Evolution* **44**(5): 528–541.

Carreto, L., Moore, E., Nobre, M.F., Wait, R., Riley, P.W., Sharp, R.J. and Dacosta, M.S., 1996, Rubrobacter xylanophilus sp nov - a new thermophilic species isolated from a thermally polluted effluent. *International Journal of Systematic Bacteriology* **46**(2): 460–465.

Clark, A.J., 1971, Toward a metabolic interpretation of genetic recombination of E. coli and its phages. *Annual Review of Microbiology* **25**: 437–64.

Clark, A.J. and Sandler, S.J., 1994, Homologous genetic recombination: the pieces begin to fall into place. *Critical Reviews in Microbiology* **20**(2): 125–42.

Courcelle, J. and Hanawalt, P.C., 1999, RecQ and RecJ process blocked replication forks prior to the resumption of replication in UV-irradiated Escherichia coli. *Mol Gen Genet* **262**(3): 543–51.

Cox, M.M., 1997, Recombinational crossroads - eukaryotic enzymes and the limits of bacterial precedents. *Proceedings of the National Academy of Sciences of the United States of America* **94**(22): 11764–11766.

Cox, M.M., 1999, Recombinational DNA repair in bacteria and the RecA protein. *Progress in Nucleic Acids Research and Molecular Biology* **63**: 310–366.

Cox, M.M. and Battista, J.R., 2005, Deinococcus radiodurans - The consummate survivor. *Nature Reviews Microbiology*. **3**(11): 882–892.

Daly, M.J., Gaidamakova, E.K., Matrosova, V.Y., Vasilenko, A., Zhai, M., Venkateswaran, A., Hess, M., Omelchenko, M.V., Kostandarithes, H.M., Makarova, K.S., *et al.* 2004, Accumulation of Mn(II) in, Deinococcus radiodurans facilitates gamma-radiation resistance. *Science* **306**(5698): 1025–1028.

Daly, M.J., Ling, O. and Minton, K.W., 1994, Interplasmidic recombination following irradiation of the radioresistant bacterium Deinococcus radiodurans. *Journal of Bacteriology* **176**(24): 7506–7515.

Daly, M.J. and Minton, K.W., 1995, Interchromosomal recombination in the extremely radioresistant bacterium Deinococcus radiodurans. *Journal of Bacteriology* **177**(19): 5495–5505.

Daly, M.J. and Minton, K.W., 1996, An alternative pathway of recombination of chromosomal fragments precedes recA-dependent recombination in the radioresistant bacterium Deinococcus radiodurans. *Journal of Bacteriology* **178**(15): 4461–4471.

Daly, M.J. and Minton, K.W., 1997, Recombination between a resident plasmid and the chromosome following irradiation of the radioresistant bacterium Deinococcus radiodurans. *Gene* **187**(2): 225–229.

Daly, M.J., Ouyang, L., Fuchs, P. and Minton, K.W., 1994, In vivo damage and recA-dependent repair of plasmid and chromosomal DNA in the radiation-resistant bacterium Deinococcus radiodurans. *Journal of Bacteriology* **176**(12): 3508–3517.

Earl, A.M., Mohundro, M.M., Mian, I.S. and Battista, J.R., 2002, The IrrE protein of Deinococcus radiodurans R1 is a novel regulator of recA expression. *J Bacteriol* **184**(22): 6216–24.

Earl, A.M., Rankin, S.K., Kim, K.-P., Lamendola, O.N. and Battista, J.R., 2002, Genetic evidence that the *uvsE* gene product of *Deinococcus radiodurans* R1 is a UV damage endonuclease. *Journal of Bacteriology* **184**: 1003–1009.

Eggington, J.M., Haruta, N., Wood, E.A. and Cox, M.M., 2004, The single-stranded DNA-binding protein of *Deinococcus radiodurans*. *BMC Microbiology* **4**: 2.

Eltsov, M. and Dubochet, J., 2005, Fine Structure of the Deinococcus radiodurans Nucleoid Revealed by Cryoelectron Microscopy of Vitreous Sections. *Journal of Bacteriology* **187**: 8047–8054.

Ferreira, A.C., Nobre, M.F., Moore, E., Rainey, F.A., Battista, J.R. and da Costa, M.S., 1999, Characterization and radiation resistance of new isolates of Rubrobacter radiotolerans and Rubrobacter xylanophilus. *Extremophiles* **3**(4): 235–238.

Funayama, T., Narumi, I., Kikuchi, M., Kitayama, S., Watanabe, H. and Yamamoto, K., 1999, Identification and disruption analysis of the recN gene in the extremely radioresistant bacterium Deinococcus radiodurans. *Mutation Research* **435**(2): 151–161.

Gerard, E., Jolivet, E., Prieur, D. and Forterre, P., 2001, DNA protection mechanisms are not involved in the radioresistance of the hyper thermophilic archaea Pyrococcus abyssi and P-furiosus. *Molecular Genetics & Genomics: MGG* **266**(1): 72–78.

Gutman, P.D., Carroll, J.D., Masters, C.I. and Minton, K.W., 1994, Sequencing, targeted mutagenesis and expression of a recA gene required for the extreme radioresistance of Deinococcus radiodurans. *Gene* **141**(1): 31–37.

Hansen, M.T., 1978, Multiplicity of genome equivalents in the radiation-resistant bacterium Micrococcus radiodurans. *Journal of Bacteriology* **134**(1): 71–75.

Harris, D.R., Tanaka, M., Saveliev, S.V., Jolivet, E., Earl, A.M., Cox, M.M. and Battista, J.R., 2004, Preserving Genome Integrity: The DdrA Protein of Deinococcus radiodurans R1. *PLoS Biology* **2**(10): e304.

Harsojo, Kitayama, S. and Matsuyama, A., 1981, Genome multiplicity and radiation resistance in Micrococcus radiodurans. *Journal of Biochemistry* **90**(3): 877–80.

Henne, A., Bruggemann, H., Raasch, C., Wiezer, A., Hartsch, T., Liesegang, H., Johann, A., Lienard, T., Gohl, O., Martinez-Arias, R. *et al.*, 2004, The genome sequence of the extreme thermophile Thermus thermophilus. *Nature Biotechnology* **22**(5): 547–553.

Horii, Z. and Clark, A.J., 1973, Genetic analysis of the RecF pathway to genetic recombination in *Escherichia coli* K12: isolation and characterization of mutants. *J. Mol. Biol.* **80**(2): 327–44.

Iyer, L.M., Koonin, E.V. and Aravind, L., 2002, Classification and evolutionary history of the single-strand annealing proteins, RecT, Redβ, ERF and RAD52. *BMC Genomics* **3**(1): 8.

Kim, J.-I. and Cox, M.M., 2002, The RecA proteins of *Deinococcus radiodurans* and *Escherichia coli* promote DNA strand exchange via inverse pathways. *Proceedings of the National Academy of Sciences of the United States of America* **99**: 7917–7921.

Kim, J.I., Sharma, A.K., Abbott, S.N., Wood, E.A., Dwyer, D.W., Jambura, A., Minton, K.W., Inman, R.B., Daly, M.J. and Cox, M.M., 2002, RecA protein from the extremely radioresistant bacterium Deinococcus radiodurans: Expression, purification, and characterization. *Journal of Bacteriology* **184**(6): 1649–1660.

Kitayama, S., Narumi, I., Kikuchi, M. and Watanabe, H., 2000, Mutation in recR gene of Deinococcus radiodurans and possible involvement of its product in the repair of DNA interstrand cross-links. *Mutation Research* **461**(3): 179–187.

Kowalczykowski, S.C., Dixon, D.A., Eggleston, A.K., Lauder, S.D. and Rehrauer, W.M., 1994, Biochemistry of homologous recombination in Escherichia coli. *Microbiological Reviews* **58**(3): 401–65.

Kowalczykowski, S.C. and Eggleston, A.K., 1994, Homologous Pairing and DNA Strand-Exchange Proteins. *Annu Rev. Biochemg* 63: 991–1043.

Krasin, F. and Hutchinson, F., 1981, Repair of DNA double-strand breaks in Escherichia coli cells requires synthesis of proteins that can be induced by UV light. *Proceedings of the National Academy of Sciences of the United States of America* **78**(6): 3450–3.

Kuzminov, A., 1999, Recombinational repair of DNA damage in Escherichia coli and bacteriophage lambda. *Microbiology & Molecular Biology Reviews* **63**(4): 751–813.

Lecointe, F., Shevelev, I.V., Bailone, A., Sommer, S. and Hubscher, U., 2004, Involvement of an X family DNA polymerase in double-stranded break repair in the radioresistant organism Deinococcus radiodurans. *Molecular Microbiology* **53**(6): 1721–1730.

Levin-Zaidman, S., Englander, J., Shimoni, E., Sharma, A.K., Minton, K.W. and Minsky, A., 2003, Ringlike structure of the Deinococcus radiodurans genome: A key to radioresistance? *Science* **299**(5604): 254–256.

Lin, J.Y., Qi, R., Aston, C., Jing, J.P., Anantharaman, T.S., Mishra, B., White, O., Daly, M.J., Minton, K.W., Venter, J.C. *et al.*, 1999, Whole-genome shotgun optical mapping of Deinococcus radiodurans. *Science* **285**(5433): 1558–1562.

Lipton, M.S., Pasa-Tolic, L., Anderson, G.A., Anderson, D.J., Auberry, D.L., Battista, J.R., Daly, M.J., Fredrickson, J., Hixson, K.K., Kostandarithes, H. *et al.*, 2002, Global analysis of the Deinococcus radiodurans proteome by using accurate mass tags. *Proc Natl Acad Sci U S A* **99**(17): 11049–11054.

Liu, Y.Q., Zhou, J.Z., Omelchenko, M.V., Beliaev, A.S., Venkateswaran, A., Stair, J., Wu, L.Y., Thompson, D.K., Xu, D., Rogozin, I.B. *et al.*, 2003, Transcriptome dynamics of Deinococcus radiodurans recovering from ionizing radiation. *Proceedings of the National Academy of Sciences of the United States of America* **100**(7): 4191–4196.

Liu, Z., Macias, M.J., Bottomley, M.J., Stier, G., Linge, J.P., Nilges, M., Bork, P. and Sattler, M., 1999, The three-dimensional structure of the HRDC domain and implications for the Werner and Bloom syndrome proteins. *Structure Fold Des* **7**(12): 1557–66.

Losada, A. and Hirano, T., 2001, Shaping the metaphase chromosome: coordination of cohesion and condensation. *Bioessays* **23**(10): 924–35.

Lusetti, S.L. and Cox, M.M., 2002, The bacterial RecA protein and the recombinational DNA repair of stalled replication forks. *Annual Review of Biochemistry* **71:** 71–100.

Makarova, K.S., Aravind, L., Wolf, Y.I., Tatusov, R.L., Minton, K.W., Koonin, E.V. and Daly, M.J., 2001, Genome of the extremely radiation-resistant bacterium Deinococcus radiodurans viewed from the perspective of comparative genomics. *Microbiology & Molecular Biology Reviews* **65**(1): 44–79.

Mattimore, V. and Battista, J.R., 1996, Radioresistance of *Deinococcus radiodurans*: functions necessary to survive ionizing radiation are also necessary to survive prolonged desiccation. *Journal of Bacteriology* **178**: 633–637.

Minton, K.W., 1996, Repair of ionizing-radiation damage in the radiation resistant bacterium Deinococcus radiodurans. *Mutation Research* **363**(1): 1–7.

Morozov, V., Mushegian, A.R., Koonin, E.V. and Bork, P., 1997, A putative nucleic acid-binding domain in Bloom's and Werner's syndrome helicases. *Trends Biochem Sci* **22**(11): 417–418.

Mustard, J.A. and Little, J.W., 2000, Analysis of Escherichia coli RecA interactions with LexA, lambda CI, and UmuD by site-directed mutagenesis of recA. *Journal of Bacteriology* **182**(6): 1659–1670.

Narumi, I., Satoh, K., Cui, S., Funayama, T., Kitayama, S. and Watanabe, H., 2004, PprA: a novel protein from Deinococcus radiodurans that stimulates DNA ligation. *Molecular Microbiology* **54**(1): 278–285.

Narumi, I., Satoh, K., Kikuchi, M., Funayama, T., Yanagisawa, T., Kobayashi, Y., Watanabe, H. and Yamamoto, K., 2001, The LexA protein from Deinococcus radiodurans is not involved in RecA induction following gamma irradiation. *Journal of Bacteriology* **183(**23): 6951–6956.

Paques, F. and Haber, J.E., 1999, Multiple pathways of recombination induced by double-strand breaks in Saccharomyces cerevisiae. *Microbiology & Molecular Biology Review,Washington, DC)* **63(**2): 349–404.

Passy, S.I., Yu, X.N., Li, Z.F., Radding, C.M. and Egelman, E.H., 1999, Rings and filaments of beta protein from bacteriophage lambda suggest a superfamily of recombination proteins. *Proceedings of the National Academy of Sciences of the United States of America.* **96**(8): 4279–4284.

Pollard, E.C. and Fugate, J.K., Jr., 1978, Relative rates of repair of single-strand breaks and postirradiation DNA degradation in normal and induced cells of Escherichia coli. *Biophysical Journal.* **24**(2): 429–437.

Pollard, E.C. and Kraus, K., 1973, All-or-nothing character of DNA degradation in bacteria after ionizing radiation. *Biophysical Journal.* **13**(4): 332–9.

Pollard, E.C. and Randall, E.P., 1973, Studies on the inducible inhibitor of radiation-induced DNA degradation of Escherichia coli. *Radiation Research.* **55**(2): 265–279.

Rainey, F.A., Nobre, M.F., Schumann, P., Stackebrandt, E. and Dacosta, M.S., 1997, Phylogenetic diversity of the deinococci as determined by 16s ribosomal dna sequence comparison. *International Journal of Systematic Bacteriology* **47**(2): 510–514.

Rainey, F.A., Ray, K., Ferreira, M., Gatz, B.Z., Nobre, M.F., Bagaley, D., Rash, B.A., Park, M.J., Earl, A.M., Shank, N.C. *et al.*, 2005, Extensive diversity of ionizing-radiation-resistant bacteria recovered from Sonoran Desert soil and description of nine new species of the genus Deinococcus obtained from a single soil sample. *Applied & Environmental Microbiology* **71**(9): 5225–5235.

Roca, A.I. and Cox, M.M., 1997, RecA protein: structure, function, and role in recombinational DNA repair. *Progress in Nucleic Acid Research & Molecular Biology* **56**(129): 129–223.

Satoh, K., Narumi, I., Kikuchi, M., Kitayama, S., Yanagisawa, T., Yamamoto, K. and Watanabe, H., 2002, Characterization of RecA424 and RecA670 Proteins from Deinococcus radiodurans. *Journal of Biochemistry,Tokyo)* **131**: 121–129.

Smith, G.R., 2001, Homologous recombination near and far from DNA breaks: alternative roles and contrasting views. *Annual Review of Genetics* **35**: 243–274.

Tanaka, M., Earl, A.M., Howell, H.A., Park, M.J., Eisen, J.A., Peterson, S.N. and Battista, J.R., 2004, Analysis of Deinococcus radiodurans's transcriptional response to ionizing radiation and desiccation reveals novel proteins that contribute to extreme radioresistance. *Genetics* **168**(1): 21–33.

Tang, M.J., Shen, X., Frank, E.G., O'Donnell, M., Woodgate, R. and Goodman, M.F., 1999, UmuD $'_2$C is an error-prone DNA polymerase, Escherichia coli pol V. *Proceedings of the National Academy of Sciences of the United States of America.* **96**(16): 8919–8924.

van den Bosch, M., Lohman, P.H.M. and Pastink, A., 2002, DNA double-strand break repair by homologous recombination. *Biological Chemistry* **383**(6): 873–892.

Weller, G.R., Kysela, B., Roy, R., Tonkin, L.M., Scanlan, E., Della, M., Devine, S.K., Day, J.P., Wilkinson, A., di Fagagna, F.D. *et al.*, 2002, Identification of a DNA nonhomologous end-joining complex in bacteria. *Science* **297**(5587): 1686–1689.

White, O., Eisen, J.A., Heidelberg, J.F., Hickey, E.K., Peterson, J.D., Dodson, R.J., Haft, D.H., Gwinn, M.L., Nelson, W.C., Richardson, D.L. *et al.*, 1999, Genome sequence of the radioresistant bacterium Deinococcus radiodurans R1. *Science* **286**(5444): 1571–1577.

Williams, J.G., Shibata, T. and Radding, C.M., 1981, Escherichia coli recA protein protects single-stranded DNA or gapped duplex DNA from degradation by RecBC DNase. *Journal of Biological Chemistry*. **256**(14): 7573–82.

Zimmerman, J.M. and Battista, J.R., 2005, A ring-like nucleoid is not necessary for radioresistance in the Deinococcaceae. *BMC Microbiology* **5**: 17.

LIST OF AUTHORS

ALEXAKHIN, Rudolf M.
Russian Institute of Agricultural Radiology and Agroecology,
249032, Obninsk, Kaluga region,
Kievskoe str., 109 km,
Russia
e-mail:alexakhin@riar.obninsk.org

AKSYUTIKT, V.
Institute of Genetics and Cytology
National Academy of Sciences,
Minsk, , Akademicheskaya str.27,
Belarus

ANOSHENKO, I.P.
National Academy of Sciences,
Minsk, , Akademicheskaya str.27,
Belarus

AROUTIOUNIAN, Rouben
Department of Genetics and Cytology,
Yerevan State University,
375025, Yerevan, Armenia
E-mail: genetik@ysu.am

BABKOFF, Vassily
Institute of the History of Science & Technology,
1/5 Staropansky, Moscow K-12,
Russia
E-mail: heba@ihst.ru

BAMBLEVSKY, V.P.

Joint Institute for Nuclear Research (JINR),
141980 Dubna, Moscow region,
Russia

BATTISTA, John R.
Department of Biological Sciences,
Louisiana State University,
202 Life Sciences Building,
Baton Rouge, Louisiana, 70803,
USA

BAYKOVA, Tatyana A.
Russian Institute of Agricultural Radiology and Agroecology,
249032, Obninsk,
Russia

CAREY, Charles C.
Division of Basic Sciences,
Fred Hutchinson Cancer Research Center,
1100 Fairview Ave. N.,
Seattle, WA 98109-1024
USA

CHEBOTINA, Margarita
Institute of Plant and Animal Ecology, Russian Academy Science,
Ural Division,
8 Marta 202,Ekaterinburg,
Russia
E-Mail: chebotina@ipae.uran.ru

CIGNA, Arrigo A.
Past President Union Internationale de Radioécologie
Strada Bottino 2, Frazione Tuffo
I-14023 COCCONATO (Asti)
Italy
E-mail: *arrigocigna@tiscali.it*

COX, Michael M.
Department of Biochemistry,
University of Wisconsin-Madison,
433 Babcock Dr, Madison,
Wisconsin, 53706-1544,
USA
E-mail: cox@biochem.wisc.edu

DIKAREV, Vladimir G.
Russian Institute of Agricultural Radiology and Agroecology,
249032, Obninsk,
Russia

DIKAREVA, Nina S.
Russian Institute of Agricultural Radiology and Agroecology,
249032, Obninsk,
Russia

DRAKE, John W.
Laboratory of Molecular Genetics,
National Institute of Environmental Health Sciences,
Research Triangle Park,
North Carolina 27709-2233
USA
E-mail: drake@niehs.nih.gov

DURANTE, Marco
Department of Physics,
University Federico II,
Napoli,
Italy
E-mail:durante@na.infn.it

DZYUBENKO, Olena
Institute Of Hydrobiology,
12 Geroyev Stalingrada Avenue,
04210 Kiev,
Ukraine

EDWARDS, A.A.
Health Protection Agency,
Radiation Protection Division
Chilton, Didcot, OX11 0RQ.
UK

FLORKO, B.V.

Joint Institute for Nuclear Research (JINR),
141980 Dubna, Moscow region,
Russia
E-mail: borvlad@mail.ru

GERAS'KIN*, Stanislav A.
Russian Institute of Agricultural Radiology and Agroecology,
249032, Obninsk,
Russia E-mail: stgeraskin@gmail.com

GLAZKO, Tatiana T.
Institute of Agriecology and Biotechnology
UAAS, Kiev,
Ukraine.
E-mail: glazko@biotech.relc.com

GLAZKO, V.I.
Institute of agriecology and biotechnology uaas,
Metrologicheskaja St. 12,
Kiev, 03143,
Ukraine
E-mail: glazko@biotech.relc.com

GLUSHKOVA, I.V.
National Academy of Sciences,
Minsk, , Akademicheskaya str.27,
Belarus

GUDKOV, Dmitri
Institute of Hydrobiology,
12 Geroyev Stalingrada Avenue,
04210 Kiev,
Ukraine E-mail: digudkov@svitonline.com

GUDKOV, Igor N.
National Agricultural University, Geroev oboroni
15, 03041 Kiev,
Ukraine,
E-mail: ingudkov@i.com.ua

GUEGAMIAN, G. V.
1, Place du Petit Pont,
F- 94140 Alfortville,
France
E-mail: ggnalfort@aol.com

INGE-VECHTOMOV, Sergey G.
Department of Genetics and Breeding, St-Petersburg State University,
University embankment, 199034 St-Petersburg,
Russia
E-mail: inge@mail333.com

IMANAKA, Tetsuji
Research Reactor Institute, Kyoto University,
Kumatori-cho, Osaka 590-0494
Japan
E-mail: imanaka@rri.kyoto-u.ac.jp

IVANOV, Victor
Medical Radiological Research Centre
Russian Academy of Medical Sciences
4 Korolev str., Obninsk, 249036
Russia
E-mail: nrer@obninsk.com

KARAVAEVA, Elena
Institute of Plant and Animal Ecology
Russian Academy of Sciences, Ural Division,
8 Marta 202,Ekaterinburg,
Russia

KHOLMURODOV, Kholmirzo
Joint Institute for Nuclear Research, Laboratory of Radiation Biology,
M. Reg., Dubna 141980;
Russia
E-mail: mirzo@jinr.ru

KIEFER, Jürgen
University of Giessen, Germany
Am Dornacker 4, S-35435 Wettenberg,
Germany
E-mail: juergen.kiefer@strz.uni-giessen.de

KOLTOVAYA, Natalia
Joint Institute for Nuclear Research,
Laboratory of Radiation Biology,
M. Reg., Dubna 141980;
Russia

KOROGODINA Viktoria.L.
Joint Institute for Nuclear Research (JINR),
141980 Dubna, Moscow region,
Russia
E-mail:korogod@jinr.ru

KOVALOVA, Oxana A.
Institute of Agriecology and Biotechnology
UAAS, Kiev,
Ukraine.
E-mail: glazko@biotech.relc.com

KOVARIK, Ales
Academy of Sciences of the Czech Republic,
CZ-61265 Brno,
Czech Republic

KRETOV, Dmitry
Joint Institute for Nuclear Research,
Laboratory of Radiation Biology,
M. Reg., Dubna 141980;
Russia

KRIKUNOVA, Ludmila I.
Medical Radiological Research Center of RAMS,
Korolev str.-4, Obninsk, 249036,
Russia

KUTLAKHMEDOV, Yuriy
Institute of cell biology and genetic engineering NASU,
Kiev,
Ukraine
E-*mail: ecoetic@mail.vtv.kiev.ua*

KUZMENKO, Mikhail
Institute of Hydrobiology,
12 Geroyev Stalingrada Avenue,
04210 Kiev,
Ukraine

LEITCH, Andrew R.
School of Biological and Chemical Sciences,
Queen Mary,
University of London, E1 4NS,
UK
E-mail: a.r.leitch@qmul.ac.uk

LLOYD, D.C.
Health Protection Agency,
Radiation Protection Division
Chilton, Didcot, OX11 0RQ.
UK
E-mail: David.Lloyd@hpa-rp.org.uk

MATIC, Ivan
INSERM U571 Faculté de Médecine,
Université Paris 5
156 rue Vaugirard, 75015 Paris,
France
E-mail: matic@necker.fr

MCMANUS, Kevin
Division of Basic Sciences,
Fred Hutchinson Cancer Research Center,
1100 Fairview Ave. N.,
Seattle, WA 98109-1024
USA

MKRTCHYAN, Liana S.
Medical Radiological Research Center of RAMS,
Korolev str.-4, Obninsk, 249036,
Russia

MOLCHANOVA, Inna
Institute of Plant and Animal Ecology
Russian Academy of Sciences, Ural Division,
8 Marta 202,Ekaterinburg,
Russia
E-mail: molchanova_i_v@mail.ru

MOSSE , I.B.
Institute of Genetics and Cytology
National Academy of Sciences,
Minsk, , Akademicheskaya str.27,
Belarus
E-mail: i.mosse@igc.bas-net.by

MOTHERSILL, Carmel
Medical Physics and Applied Radiation Sciences Department,
McMaster University, Hamilton,
Ontario, Canada, L8S 2K1
E-mail: mothers@mcmaster.ca

NAZAROV, Alexandr
Chernobyl Radioecological Centre,
6 Shkol'naya Street,
07270 Chernobyl,
Ukraine

NICOLIN, OLEG
Institute of Plant and Animal Ecology, Russian Academy Science,
Ural Division,
8 Marta 202,Ekaterinburg,
Russia

ORLOVA, Nina V.
Medical Radiological Research Center of RAMS,
Korolev str.-4, Obninsk, 249036,
Russia

OUDALOVA, Alla A.
Russian Institute of Agricultural Radiology and Agroecology,
249032, Obninsk,
Russia

PCHELOVSKYA, Svetlana
Institute of cell biology and genetic engineering NASU,
Kiev,
Ukraine

POPOV, Igor
Institute for the Lake and River Fisheries (GosNIORKh)
Laboratory for salmon population monitoring
199053 Saint-Petersburg, Makarova n., 26
Russia
E-msail: igorioshapopov@mail.ru

RODINA, Viktor
Division Institute of geography NASU,
Kiev,
Ukraine

RUTHERFORD, Suzannah
Division of Basic Sciences,
Fred Hutchinson Cancer Research Center,
1100 Fairview Ave. N.,
Seattle, WA 98109-1024
USA
E-Mail: srutherf@fhcrc.org

SAENKO, Alexander S.
Medical Radiological Research Center of RAMS,
Korolev str.-4, Obninsk, 249036,

Russia

SEYMOUR, Colin
Medical Physics and Applied Radiation Sciences Department,
McMaster University, Hamilton,
Ontario, Canada, L8S 2K1
E-mail: seymouc@mcmaster.ca

SHEVTSOVA, Natalia
Institute of Hydrobiology,
12 Geroyev Stalingrada Avenue,
04210 Kiev,
Ukraine

SCANNELL, Chris A.
Division of Basic Sciences,
Fred Hutchinson Cancer Research Center,
1100 Fairview Ave. N.,
Seattle, WA 98109-1024
USA

SKALICKA, Kamila
Institute of Biophysics,
Academy of Sciences of the Czech Republic,
CZ-61265 Brno,
Czech Republic

SMIRNOVA, Olga
Research and Technical Center
of Radiation-Chemical Safety and Hygiene,
Shchukinskaya st., 40, Moscow 123182,
Russia

SMIRNOVA, Svetlana G.
Medical Radiological Research Center of RAMS,
Korolev str.-4, Obninsk, 249036,
Russia

SZLUINSKA, M.
Health Protection Agency,
Radiation Protection Division
Chilton, Didcot, OX11 0RQ.
UK

TRAPEZNIKOV, Alexander
Institute of Plant and Animal Ecology
Russian Academy of Sciences, Ural Division,
8 Marta 202,Ekaterinburg,
Russia

TSYB, Anatoly
Medical Radiological Research Centre
Russian Academy of Medical Sciences
4 Korolev str., Obninsk, 249036
Russia
E-mail: nrer@obninsk.com

YOONG LIM, Kar
School of Biological and Chemical Sciences,
Queen Mary,
University of London, E1 4NS,
UK

YONEZAWA, Morio
Research Institute for Advanced Science and Technology,
Osaka Prefecture University,
1-2 Gakuen-cho, Sakai, Osaka 599-8570,
Japan

VASILIEV, Denis V.
Russian Institute of Agricultural Radiology and Agroecology,
249032, Obninsk,
Russia

ZAMULAEVA, Irina A.
Medical Radiological Research Center of RAMS,
Korolev str.-4, Obninsk, 249036,
Russia
E-mail: zamulaeva@mail.ru

Editio emendata huius libri ad finem
perducta atque impressa die
vigesimo primo mensis
martii cum ver incipit
anno Domini duo
millesimo
sexto

Deo gratias